The Springer Series in Applied Machine Learning

Series Editor: Oge Marques

The Springer Series in Applied Machine Learning focuses on monographs, textbooks, edited volumes, and reference books that provide suitable content and educate the reader on how the theoretical approaches, algorithms, and techniques of machine learning can be applied to address real-world problems in a principled way.

The goal is a series of state-of-the-art books for the libraries of both machine learning scientists interested in applying the latest techniques to novel problems as well as practitioners, technologists, and researchers in various fields interested in leveraging the latest advances in machine learning for developing solutions that work for practical problems. The scope spans the breadth of machine learning and AI as it pertains to all application areas — both through books that address techniques specific to one application domain — and books that show the applicability of different types of machine learning methods to a wide array of domains.

Ömer Faruk Ertuğrul • Josep M. Guerrero
Musa Yilmaz
Editors

Shallow Learning vs. Deep Learning

A Practical Guide for Machine Learning Solutions

Springer

Editors
Ömer Faruk Ertuğrul
Department of Electrical and Electronics Engineering
Batman University
Batman, Türkiye

Josep M. Guerrero
Center for Research on Microgrids (CROM)
AAU Energy
Aalborg University
Aalborg East, Denmark

Musa Yilmaz
Center for Environmental Research and Technology
University of California, Riverside
Riverside, CA, USA

ISSN 2520-1298 ISSN 2520-1301 (electronic)
The Springer Series in Applied Machine Learning
ISBN 978-3-031-69498-1 ISBN 978-3-031-69499-8 (eBook)
https://doi.org/10.1007/978-3-031-69499-8

This Springer imprint is published by the registered company Springer Nature Switzerland AG
The registered company address is: Gewerbestrasse 11, 6330 Cham, Switzerland

Preface

In the dynamic field of machine learning, the debate between shallow learning and deep learning is ongoing and complex. Both methodologies have proven their merit across various applications, yet the choice between them is not always straightforward. This book, "Shallow Learning vs Deep Learning: A Practical Guide for Machine Learning Solutions," seeks to illuminate this choice by exploring these methodologies across a spectrum of applications, emphasizing that the selection of a machine learning approach should not be biased by prevailing trends but rather informed by the specific characteristics of the dataset, the operational environment, and the unique requirements of each application.

Our journey begins with a comprehensive survey of machine learning methods, setting the stage for an in-depth exploration of both shallow and deep learning across diverse domains—from engineering and real-world scenarios to social applications and beyond. Each chapter delves into these domains, comparing and contrasting the effectiveness of shallow and deep learning in addressing specific challenges and opportunities.

The goal of this book is not merely to present a collection of methodologies but to provide a framework for thoughtful selection and application of these technologies. It is tailored for researchers, practitioners, and students who are already familiar with the basics of machine learning and are looking to deepen their understanding of when and how to apply different machine learning strategies effectively.

As we navigate through the chapters, we invite readers to engage with the comparative studies, reflect on the detailed analyses, and apply the insights gained to make informed decisions in their respective fields. By the end of this book, you will be better equipped to choose the most appropriate machine learning technique for your specific needs, free from the common biases that often influence these decisions.

We thank all contributors whose expertise and insights have made this comprehensive guide possible. We hope that this book serves as a valuable resource in your continued exploration of machine learning technologies.

Batman, Türkiye — Ömer Faruk Ertuğrul
Aalborg East, Denmark — Josep M. Guerrero
Riverside, CA, USA — Musa Yilmaz

Contents

Machine Learning Methods from Shallow Learning to Deep Learning

Tahir Cetin Akinci, Oguzhan Topsakal, and M. Ilhan Akbas

1 Introduction

The journey of Machine Learning began in the mid-twentieth century, rooted in the quest to enable machines to learn from and make decisions based on data. The initial phase, now known as shallow learning, was characterized by algorithms that relied heavily on hand-engineered features and linear models. Techniques like Linear Regression, Logistic Regression, and Naïve Bayes were among the first to be developed, forming the backbone of early ML applications. These methods were instrumental in simple pattern recognition and decision-making tasks, marking the first steps toward automated learning [1].

As the field progressed, ML expanded to include more complex algorithms, such as Decision Trees, SVM, and kNN. These methods, which still fall under the umbrella of SL, represented a significant leap forward. They enabled machines to handle nonlinear relationships and more complex decision boundaries, thereby broadening the scope of ML applications. During this era, the focus was on enhancing the algorithms' ability to generalize from limited data, a challenge that was central to the success of ML [2, 3].

T. C. Akinci (✉)
Bourns College of Engineering - Center for Environmental Research and Technology (Ce-Cert), University of California Riverside, Riverside, CA, USA
e-mail: tahircetin.akinci@ucr.edu

O. Topsakal
College of Engineering, Computer Science and Engineering, University of South Florida, Tampa, FL, USA
e-mail: topsakal@usf.edu

M. I. Akbas
Department of Electrical Engineering and Computer Science, Embry-Riddle Aeronautical University, Daytona Beach, FL, USA
e-mail: akbasm@erau.edu

Ö. F. Ertuğrul et al. (eds.), *Shallow Learning vs. Deep Learning*, The Springer Series in Applied Machine Learning, https://doi.org/10.1007/978-3-031-69499-8_1

The concept of neural networks, inspired by the workings of the human brain, emerged as a groundbreaking development in ML. Initially, these networks were simple, comprising a few layers of interconnected nodes, and were limited in their capabilities. However, they laid the foundation for what would eventually evolve into DL [4, 5].

The advent of the twenty-first century witnessed a seismic shift in ML with the introduction of DL. This paradigm shift was fueled by the convergence of three critical elements: the availability of massive datasets, substantial improvements in computational power, and the development of advanced algorithms. DL, characterized by deep neural networks with multiple layers, allowed for the processing of data in complex, hierarchical ways. This enabled the extraction of features and patterns at various levels of abstraction, a task that shallow learning methods could not efficiently perform [6].

DL architectures such as ConvNets, RNNs, and Generative Adversarial Networks (GANs) revolutionized fields such as image and speech recognition, NLP, and even creative Artificial Intelligence applications. The development of Transformer-Based Neural Networks and Large Language Models (LLMs) marked another leap in the capabilities of AI, allowing for unprecedented performance in tasks like language translation, content generation, and semantic analysis [7, 8].

This chapter investigates the distinctions and relationships between shallow and deep learning while also placing their evolution within the wider context of ML history. Understanding this progression sheds light on the present dynamics and future directions of the field. The evolution from shallow to DL highlights the rapid and transformative growth of ML.

Throughout this chapter, we examine the intricacies and applications of various ML methods, always mindful of their historical context. This approach will foster a deeper understanding of these technologies' sophistication and pave the way for a more comprehensive exploration in subsequent chapters.

The chapter also sheds light on the shift from traditional to more advanced ML techniques, emphasizing how the field adapts to handle complex data. This narrative not only establishes a foundational understanding for the rest of the book but also bridges the gap between theoretical concepts and real-world applications. It underscores the synergy between SL and DL, extending the discussion to cover cutting-edge trends and potential future developments. Consequently, readers will gain a well-rounded view of ML's dynamic evolution within the realm of AI.

2 Relationships Between Artificial Intelligence and Its Concepts

AI is a broad field focused on the imitation of human-like intelligence by computers; ML is a specialized branch within AI, encompassing algorithms that enable computers to analyze and learn from data [9, 10]. Within ML, DL is emerging as a

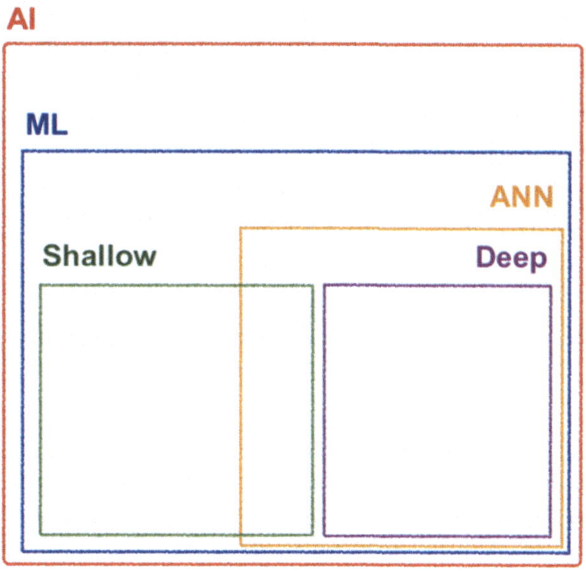

Fig. 1 AI and its sub-learning methodologies [11, 12]

distinct subfield characterized primarily by the use of neural networks, which are the basis for DL algorithms. These networks are distinguished by the number or depth of layers of nodes; DL algorithms, unlike standard neural networks, often require more than three layers to function effectively. Furthermore, ML includes SL, which offers simple but effective models, and uses various learning methods such as supervised and unsupervised learning [11, 12].

The complex relationships and differences between these components of AI, including the specialized structures o4f multilayer ANN, SL, and DL are presented in Fig. 1, AI and its sub-learning methodologies.

2.1 The Difference Between Artificial Intelligence and Machine Learning

ML is a subset of AI as shown in Fig. 1. While AI is a broader concept, ML represents a specialized subfield focused on performing specific tasks and learning from data. Rule-based chatbots and the computational models for NLP are a couple of examples for AI without ML. Hence, main differences between AI and ML can be described as follows:

- AI aims to endow computer systems with the ability to mimic human intelligence. ML, on the other hand, targets enabling computer systems to learn a specific task through data analysis and pattern recognition.

- AI typically employs complex algorithms and extensive datasets. It encompasses a broader and more general learning spectrum. ML includes algorithms capable of learning from datasets, usually with a focus on specific tasks.
- While AI provides a more general and versatile approach, ML focuses on specific tasks, learning from data to improve its performance.

In summary, AI offers a broader and more versatile approach, while ML is more specific and focused on particular tasks [13–16].

2.2 *The Difference Between Deep Learning and Machine Learning*

DL and ML are interrelated concepts. DL algorithms represent a specialized subset within the broader spectrum of ML algorithms. The unique characteristics of DL can be distinguished by its neural network architecture, reduced necessity for human intervention, and greater data requirements [15, 16]. The differences between traditional ML and DL can be summarized as follows:

- Structural Complexity: Traditional machine learning algorithms usually use less complex structures such as decision trees or linear regression. DL models, on the other hand, are built on artificial neural networks (ANNs) with multiple, complex levels that resemble the connected nodes (similar to neurons) that make up the human brain.
- Human Interaction: Due to their ability to self-learn and adapt, DL models generally require much less human interaction compared to ML models.
- Feature Extraction: This is an automated method in deep learning, where the algorithm automatically learns from its mistakes and makes changes without the need for special software and programmers for feature setup and extraction.
- Data Requirements: DL models require much larger datasets than standard ML algorithms. This is due to their complex, multi-layered structure, as mentioned in references [17–19], in order to minimize fluctuations and derive accurate interpretations.

2.3 *The Difference Between Deep Learning and Neural Networks*

Neural networks and DL are closely related concepts, but they are not identical. An ANN is typically considered a DL algorithm when it has more than three layers, including the input and output layers. Additionally, most deep neural networks are of the feedforward type, where data flows in one direction from input to output. The fundamental differences between neural networks and DL can be outlined as follows:

- DL networks have multiple layers, making DL more complex than a standard neural network.
- Due to their complexities, DL networks require significant training time, whereas standard neural networks require much less training [20–22].

2.4 The Difference Between AI and Neural Networks

AI broadly encompasses the development of machines capable of emulating human cognitive functions. AI, fundamentally based on computer systems possessing human-like intelligence, represents the broadest set encompassing the aim for computers to perform tasks, learn, respond to their environments, and solve complex problems, exhibiting intelligent behavior [27–29]. AI, along with its subfields, is continuously evolving. ANNs, on the other hand, are composed of interconnected artificial neurons or nodes, drawing inspiration from the biological neural networks found in the human brain. The primary role of an ANN is to classify and categorize data by identifying similarities.

One of the significant strengths of ANNs is their adaptability to evolving output patterns. They can be trained using either supervised or unsupervised learning methods, which reduces the need for continuous manual adjustments based on the input they receive. The distinctions between these concepts can be summarized as follows:

- AI represents the broader concept of crafting machines that exhibit intelligent behavior.
- ANNs form a subset of ML that leverages extensive datasets and intricate algorithms to train models.
- ANNs involve algorithms that process and analyze data in a manner akin to human logical reasoning, as seen in data science research and trial-and-error methodologies, as referenced in [23–26].

ANNs can predict data distributions in complex systems, support new designs, and work with various probability functions to identify multiple possible scenarios and optimal strategies. ANNs can solve complex equations that are even beyond the capability of traditional numerical methods [27]. Despite the challenges, ANNs empower the execution of efficient, intelligent, and highly autonomous experimental design and data collection. Operating under human supervision, ANN systems are adept at evaluating outcomes, which endows them with the capability to make real-time decisions [28].

3 Definition and Basic Concept of Machine Learning

The ability of computers to perform complex tasks requiring human intelligence forms the foundation of AI. AI encompasses systems that mimic cognitive functions associated with the human mind. Traditional AI is programmed using "if-then" rules, often based on a knowledge base created by expert knowledge [30–32].

ML emphasizes the capability of computers to autonomously learn from data and experiences. As a subset of AI, it involves crafting algorithms that can construct models using sample data, commonly referred to as "training data." These models are adept at making predictions or decisions without the need for explicit programming, enabling them to adapt and improve from their experiences. Hence, ML uses training datasets instead of rigid rules, unlike traditional programming algorithms that use strict rules and instructions to perform a task [38]. In scenarios where traditional algorithms struggle to extract knowledge from data, ML methods provide effective solutions across various applications [33, 34].

ML enables computers to perform specific tasks by learning from data [36, 37]. Its integration with various disciplines such as statistics, computer science, and information theory has made ML a highly dynamic and influential field. ML is related to various disciplines, including mathematical optimization and data mining [35]. Data mining is oriented toward exploratory data analysis through unsupervised learning. Computational statistics, a subset of ML, focuses on making predictions.

ML engages in a learning process using training data, fundamentally using data-centric learning principles and the ability to generalize, unlike traditional programming where tasks are clearly defined. Their effectiveness on new datasets is attributed to their capacity to generalize, as referred to in [39, 40]. ML is broadly categorized into several types:

- Supervised Learning: This model is trained with a dataset where each instance is associated with the corresponding output label. The goal of the model is to learn the mapping from input to output during the training phase. This provides the ability to accurately predict the output of unseen data.
- Unsupervised Learning: Here the model is trained on data without any labels. So, it learns without knowing what to predict. However, it is expected to find the relationships in the dataset on its own.
- Semi-supervised Learning: This approach lies at the intersection of supervised and unsupervised learning. Models are trained with a small amount of labeled data while being supported with a larger amount of unlabeled data. This learning method is used when obtaining a fully labeled dataset is time-consuming.
- Reinforcement Learning (RL) is a method in ML where an agent is trained to make decisions by performing certain actions in a given environment, aiming to achieve a predefined goal. It uses a system of rewards and punishments in response to the actions it performs. The overall goal is to optimize the total accumulated reward. This learning method is similar to the process by which humans learn from the consequences of their actions [11, 30, 41–43].

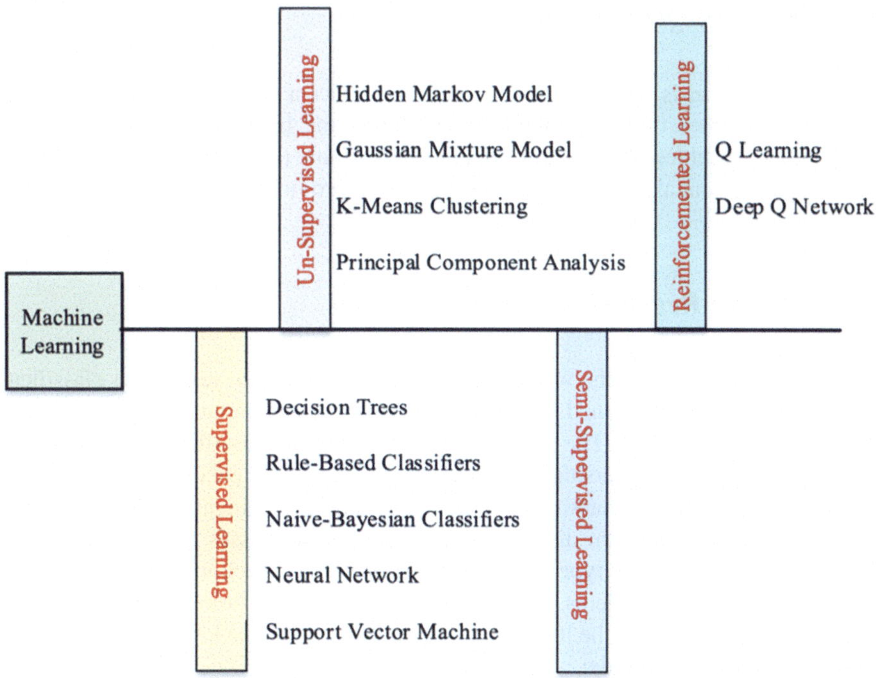

Fig. 2 Classification of ML algorithms [30, 42–44]

Figure 2 presents the classification of the ML system and the well-known algorithms in the subcategories of ML. In this chapter, our focus is supervised learning.

4 Shallow Learning

In the field of ML, Shallow Learning comprises models with fewer layers and simpler structures, especially when compared to DL. These types of learning models are typically used to predict outputs directly from input data and are not suited for automatically learning complex or abstract features in an AI model. While SL models require less data and computational power in practice, they do not perform as highly as DL models in solving complex problems [11, 45, 46].

While we explore Shallow Learning in the following subsections, we focus on the main supervised learning algorithms used in Shallow Learning.

4.1 Linear Regression

Linear Regression is a basic method used to predict a continuous output. Linear regression operates on the principle of modeling the linear relationship between independent variables (inputs) and a dependent variable (output) [11, 47]. An example of regression analysis with a sample dataset is provided in Fig. 3.

4.2 Logistic Regression

Logistic Regression models, which fall under the category of regression classification, are algorithms designed to ascertain whether given input data belongs to a specific category or class. The primary objective of these classification models is to determine the class membership of particular input data. This encompasses two types of classification: binary classification and multi-class classification. In multi-class classification, the algorithm discerns which of multiple classes the output belongs to. Logistic Regression, a type of classification model, is predominantly utilized in binary classification problems. Its nomenclature is derived from a specialized activation function known as the logistic or sigmoid function. The key characteristics of Logistic Regression include:

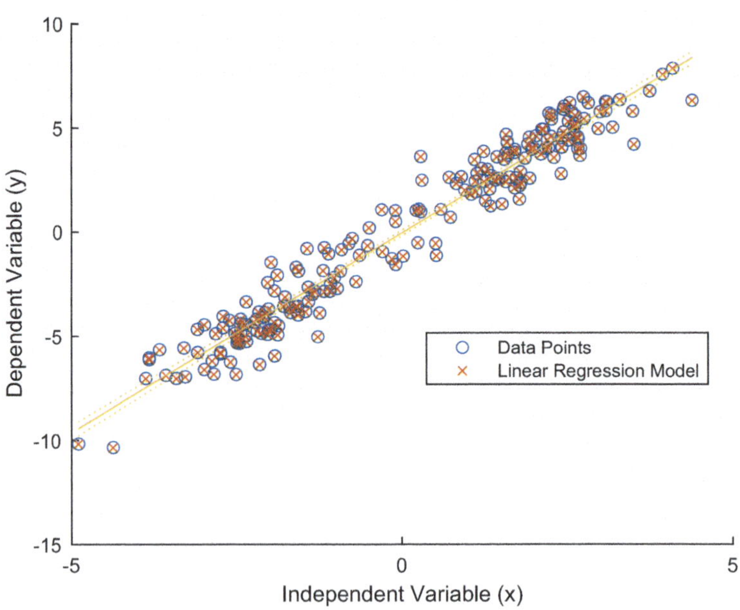

Fig. 3 Linear Regression model

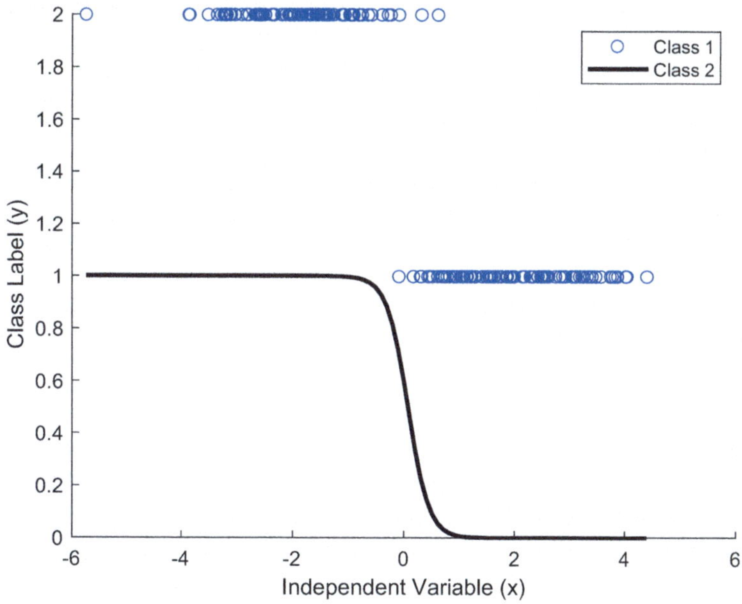

Fig. 4 Logistic Regression model

- Predicting the likelihood of belonging to a particular categorical output using input data.
- In binary classification problems, the probability is often interpreted as "0" or "1".
- Logistic Regression is used to model the log-odds ratio.
- The sigmoid function compresses the output to a probability value between 0 and 1.
- Logistic Regression learns the parameters on the weights of input features and a bias term.
- These parameters are derived from samples in the training data, typically learned based on maximum likelihood estimation.
- The fundamental mathematical expression of Logistic Regression is represented by the sigmoid (logistic) function as in Eq. 1. This function takes a number (log-odds) and converts it into a probability value between 0 and 1.

$$g(x) = \frac{1}{(1 + \mathrm{e}^\wedge(-\mathrm{x}))} \tag{1}$$

- Once learned on training data, the Logistic Regression model can be used to predict probabilities for new input data [11, 48–51].

An example of dataset analysis for the Logistic Regression model is shown in Fig. 4, and the model parameters are presented in Table 1. In Table 1, *tStat* and

Table 1 Logistic Regression model parameters

	Estimate	*SE*	*tStat*	*pValue*
(intercept)	0.38683	0.69126	0.5596	0.57575
x1	−5.3016	1.6026	−3.3082	0.00093907

pValue are not significant for the Intercept but are significant for $x1$. *Estimate* and *SE* play a decisive role in both cases.

4.3 Decision Trees

The Decision Tree algorithm is a member of the supervised learning algorithm family, distinguished by its versatility in addressing both regression and classification challenges, unlike many other algorithms confined to a single type of problem. It operates by formulating predictions through the application of straightforward decision rules, which are inferred from the dataset's features [52].

The fundamental objective of employing the Decision Tree algorithm is to construct a training model capable of predicting the class or value of the target variable. This prediction is facilitated by the algorithm's ability to learn simple decision rules from the training data. Decision Trees are categorized based on the nature of the target variable into two types: the Categorical Variable Decision Tree and the Continuous Variable Decision Tree, as detailed in references [53, 54]. An example of a decision tree analysis output with sample data is presented in Fig. 5.

4.4 Support Vector Machines

Support Vector Machines are models with strong and flexible structures, effectively used in both classification and regression tasks. These models aim to identify a hyperplane that optimally separates data points.

The SVM approach allows for the analysis of the geometric properties and patterns of data, facilitating the evaluation of these data in terms of both size and direction. When examining the positioning of data relative to the dividing line in the sample dataset presented in Fig. 6, in cases where the data are exactly on the dividing line, the x data can be expressed by an equation formed by a constant b value and the w point vector. This constant value shifts the w vector from its starting point in the direction away from the starting point of the w point. Accordingly, if the data are on the side of class 1 (x+), they will take a value greater than 0, while data on the side of class 2 ($x-$) will take a value less than 0. These situations are considered as the parameters of the model or values that the model needs to calculate from the data. The dividing line is positioned to provide the largest gap between the two groups. This gap between the two groups is calculated based on three data points on the boundary of the group. These three data points are the fundamental elements that

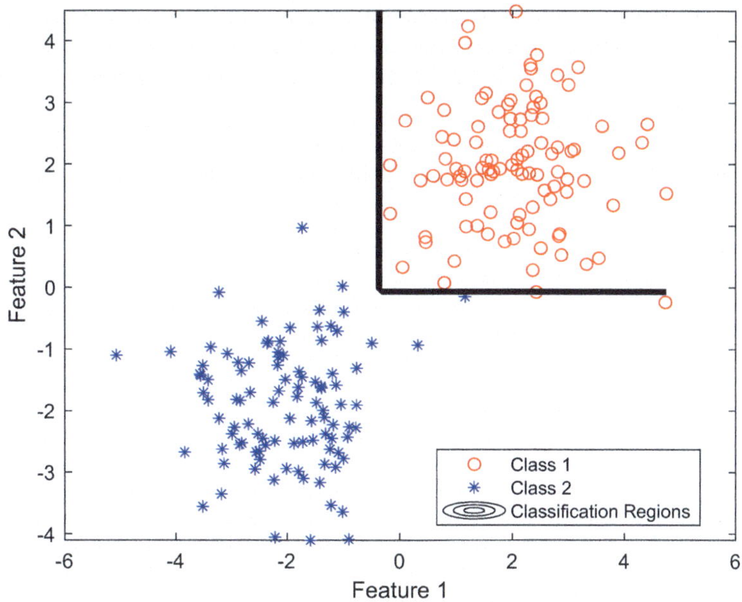

Fig. 5 Decision Tree classification

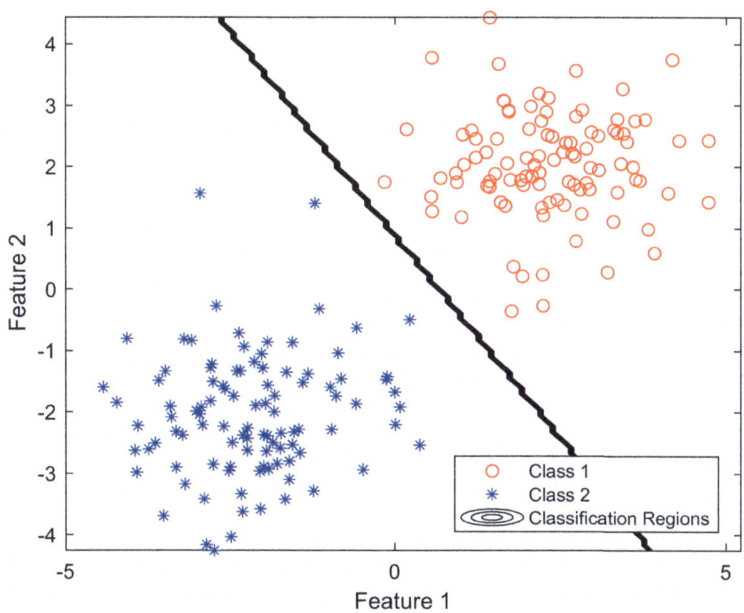

Fig. 6 Support Vector Machine classification

determine the position and direction of the dividing line. The name of this method, "support vector machine," derives from the fact that it views data as vectors, and the dividing line is formed with the support of these data vectors [11, 55–57]. Lagrange Multipliers (Eq. 2) are often used in solving such problems. Compared to DL methods, SVM requires less computational intensity and achieves successful results, especially in datasets containing a reasonable level of noise.

$$\text{Minimize } \frac{1}{2}\|\vec{w}\|^2$$
$$y\left(\vec{w}.\vec{x}+b=\geq 1\right.$$

$$L=\frac{1}{2}\|\vec{w}\|^2-\sum_i \lambda_i\left[y_i\left(\vec{w}.\vec{x}+b\right)-1\right] \tag{2}$$

4.5 K-Nearest Neighbors

The kNN algorithm works under the assumption that data entities with similar input characteristics are likely to produce comparable outputs. Adhering to the concept that similar entities are close to each other, this algorithm applies a proximity-based method to determine the similarities between data points. This similarity is quantified through a distance metric between data points. In other words, a smaller distance metric indicates that data points are more likely to belong to the same class, while a larger distance indicates dissimilarity, implying that data points are probably not members of the same class.

The distance between these points is calculated using the Euclidean Distance, the most fundamental technique for finding the shortest distance, as shown in Eq. 3, where p and q are data points in an n-dimensional plane. Other distance calculation methods like Manhattan Distance, Minkowski Distance, and Chebyshev Distance also perform calculations in a similar manner. These equations are quite sensitive to large variations.

$$(p,q)+d(q,p)=\sqrt{(q_1-p_1)^2+(q_2-p_2)^2+\ldots+(q_n-p_n)^2} \tag{3}$$

Prior to employing kNN, it is imperative to ensure that the dataset is thoroughly scaled, given that kNN classifies data based on proximity measures. kNN is versatile, applicable for both classification and regression tasks, and is characterized as a nonparametric learning algorithm, relying on the structure of the data rather than predefined parameters. It makes no assumptions about the underlying distribution and does not attempt to predict it. However, it is easily affected by outliers or

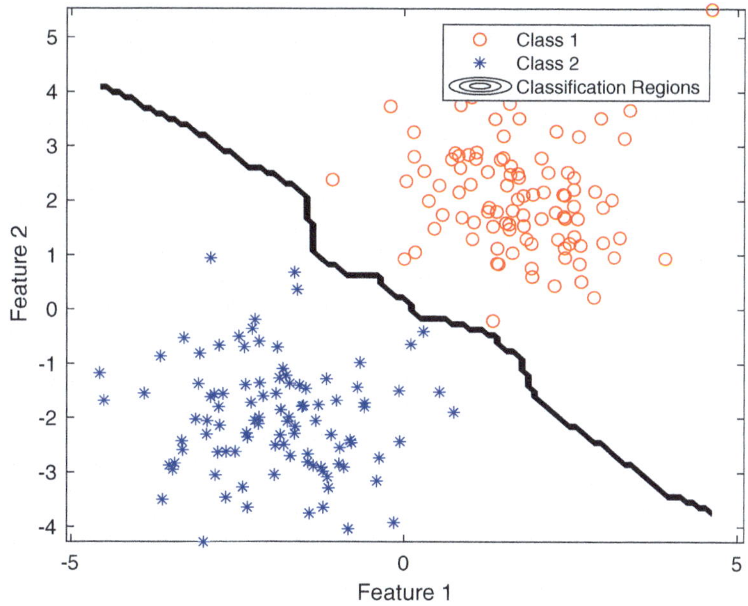

Fig. 7 *k*-Nearest Neighbors classification

imbalanced datasets and does not perform well with large datasets [58–60]. An example of kNN classification with a sample dataset is provided in Fig. 7.

The kNN algorithm is a unique type of ML model that is characterized by its method of "instance-based" and "lazy learning," meaning it doesn't build a generalized model but instead stores instances of training data. kNN operates on the principle of analyzing the closest "*k*" data points to predict the classification or value of a new data point. Unlike other algorithms, kNN is nonparametric with no explicit training phase, which makes it computationally intensive during the prediction stage due to the real-time calculation of distances between data points [11, 58]. It requires careful preprocessing like feature scaling to ensure accuracy, and its performance is highly dependent on the choice of '*k*' and the distance metric used. kNN's adaptability to multi-class problems is an advantage, although it contrasts with other algorithms that involve significant training to create predictive models, such as SVMs, NN, Decision Trees, or ensemble methods like Random Forests and Gradient Boosting Machines [59, 60].

4.6 Naïve Bayes

The Naive Bayes is a basic probability-based classification method. It relies on the assumption of independence among features and is widely used particularly in text classification.

The Naive Bayes classifier belongs to the simple probabilistic classifier family and is used in supervised learning algorithms based on the conditional probability rule. The Naive Bayes classifier is commonly applied in analyses such as spam detection, face recognition, and character recognition. An example of a dataset created for the Naive Bayes classifier is shown in Fig. 8, as represented by Eq. (4). Here, *P*(*A*) denotes the a priori probability of the occurrence of event *A* independent of event *B*. *P*(*A*|*B*) represents the conditional probability of event A given the prior occurrence of event B. *P*(*B*) is the prior probability of the predictor event *B*, independent of any other events. P(*B*|*A*) is articulated as the likelihood, which quantifies the probability of the predictor event B given the occurrence of event *A*.

$$P(A/B) = \frac{P(B/A).P(B)}{P(A)} \tag{4}$$

Shallow learning models are utilized in problems with less complex data structures and are often preferred due to their quick training times, low computational

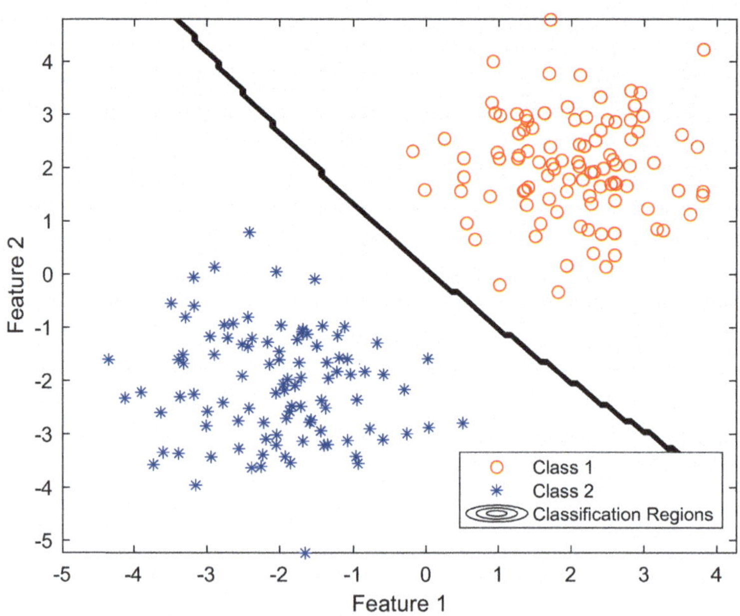

Fig. 8 Naïve Bayes classification

requirements, and meaningful results. Compared to DL models with multilayered structures, shallow learning models find more application in learning less complex and abstract features, especially when contrasted with fields like image and sound processing used in DL models [61–63].

5 Deep Learning

Representation and complex reasoning are at the core of fundamental debates in AI and cognitive science. Paradigms inspired by logic emphasize symbols and their rule-based manipulations; each symbol possesses the property of being the same as or different from other symbol instances and acts as a variable within inference rules. In contrast, approaches inspired by neural networks facilitate more intuitive and effortless reasoning using large activity vectors, weighted matrices, and nonlinear scalar functions. For instance, prior to the development of neural language models, statistical modeling of language typically relied on N-grams, treating each word as an atomic unit and lacking the ability to generalize between semantically related words. However, neural language models, representing each word with a vector of real-valued features, ensure that semantically related words are proximate in vector space. The future of AI will be shaped by the development of systems that combine representational learning with complex reasoning. This necessitates new paradigms that go beyond DL and simple reasoning techniques in areas like speech and handwriting recognition, involving the manipulation of symbolic expressions and operations on large vectors.

DL plays a significant role in AI and ML, particularly in applications requiring learning from large and complex datasets. In this paradigm, computer models analyze and learn from given sample data to perform various tasks. DL is fundamentally a sub-branch of ANN. These networks are designed to successfully perform complex tasks through connections between units, guided by weights determined during the learning process. Neural networks are typically organized in layers; initial layers detect low-level features, while subsequent layers combine these features to form more complex models [65, 66]. Larger number of layers in DL increases the model's capacity to learn deeper and more complex relationships [64, 65].

DL offers successful outcomes in areas where traditional methods fail or are insufficient. In fields like image and sound processing, NLP, and complex pattern recognition, DL models provide data analysis and interpretation capabilities previously unattainable. This highlights the critical role of DL in advancing AI, especially in data-intensive and complex applications [67, 68].

5.1 Brief Overview of How Deep Learning Networks Function

Supervised learning, particularly in the form of DL, represents a significant advancement in ML. The process begins with a large dataset of labeled images, such as houses, cars, people, and pets. During training, the machine is shown an image and produces a score vector for each category. The goal is for the desired category to have the highest score, which may not occur initially. An objective function measures the error between the output scores and the desired model, guiding the machine to adjust its internal parameters, or weights, to minimize this error. These weights act as "knobs" defining the machine's input-output function. In DL systems, there can be millions of these weights, trained with millions of labeled examples [11, 67].

The learning algorithm calculates a gradient vector for each weight, indicating how the error would change with a slight adjustment of the weight. The weights are then modified in the direction opposite to the gradient vector. The objective function, averaged over all training examples, resembles a hilly landscape in the high-dimensional space of weight values, with the negative gradient vector pointing toward the steepest descent, leading to a minimum where the average output error is low [9, 68].

Since the 1960s, it has been known that linear classifiers can only divide input spaces into simple regions. However, problems like image and speech recognition require sensitivity to certain small changes and insensitivity to irrelevant changes. Shallow classifiers need a good feature extractor to solve this dilemma. DL avoids the need for hand-designed feature extractors by automatically learning good features through a general-purpose learning procedure, which is a major advantage of this approach [69].

DL architectures consist of multilayered structures of simple modules that perform mostly nonlinear input-output operations. Each module increases both selectivity and invariance, allowing the system to perform complex functions insensitive to detailed and background changes [9, 11, 67].

Forward propagation calculates outputs for specific inputs, while backpropagation, a form of supervised learning, requires known correct outputs for learning. In backpropagation, the network's output is compared with the correct output, and weights are updated accordingly. This process is repeated until the error rate is minimized, enhancing the network's performance and learning model [68].

Backpropagation involves calculating the derivative of the target with respect to the input of a module, backward from the gradient with respect to the output. This application of the chain rule allows gradients to propagate through all modules, from the output layer back to the input layer. Once these gradients are calculated, the gradient with respect to the weights of each module can be computed, forming the foundation of the DL architecture and enhancing its capacity to process complex data structures [69, 70].

5.2 Convolutional Neural Networks

ConvNets are advanced AI systems capable of processing multidimensional data arrays. These networks are designed to process various data formats, including 2D arrays like pixel intensities in color images, as well as 1D signals and arrays, 3D videos, or volumetric images. Fundamental principles of ConvNets include local connections, shared weights, pooling, and the use of multilayered structures [71].

The ConvNets architecture is defined as a multistage combination of convolutional and pooling sublayers. In the convolutional sublayers, node elements are arranged into feature matrices, each interfacing with proximal segments in the preceding matrices through an ensemble of filters. This proximal aggregation is modulated through a nonlinear mechanism, in particular, ReLU. Uniformity in the filter ensembles is maintained across a single feature matrix, while diversity across matrices within the same layer is achieved through different filter ensembles.

This architecture is based on the premise that local groups of values in sequence data, such as images, are highly correlated and form easily detectable distinctive local motifs [9, 72].

The role of convolution layers is to detect local connections of features from the previous layer, while pooling layers aim to combine semantically similar features into a single feature. This ensures reliable detection of the motif, as the relative positions of features forming a motif can vary slightly. Pooling units compute the maximum of a local patch of units in a feature map, thereby reducing the representation's dimensionality and creating invariance to small shifts and distortions [11].

The underlying idea of ConvNets is that natural signals are compositional hierarchies. For instance, in images, local combinations of edges form motifs, which in turn combine into parts, and parts form objects. Similar hierarchies are found in other domains, such as speech and text. The pooling process allows representations to change very little when the position and appearance of elements in the previous layer change, making the systems more stable and reliable [9, 73].

Since the early 2000s, ConvNets have achieved great success in detecting, segmenting, and recognizing objects in images. They have been effective in areas such as traffic sign recognition, biological image segmentation, and the detection of faces, text, pedestrians, and human bodies in natural images [74]. ConvNets have also spread to technological applications like autonomous mobile robots and self-driving cars [75]. In these applications, ConvNet-based methods play a significant role. Their use has also increased in areas like natural language understanding and speech recognition. Particularly in face recognition, significant practical successes have been achieved.

ConvNets, when applied to large-scale image datasets, have achieved significantly lower error rates compared to competing methods. New applications like the creation of image captions have been realized by combining ConvNets with RNNs. These systems are trained to convert high-level representations of images into captions, focusing on different parts of the image during word production. ConvNets have become a dominant method in various recognition and detection tasks. In some applications, results close to human performance have been achieved [69, 72, 73].

Advanced ConvNet architectures contain numerous ReLU layers, hundreds of millions of weights, and billions of connections between units. The training of these large networks has been reduced from processes that previously took weeks to just a few hours, thanks to hardware and software advancements. These developments have solidified the significant position of ConvNets in the fields of AI and ML [76, 77].

5.3 Recurrent Neural Networks

RNNs are dynamic and powerful systems specifically designed for processing sequential data. Effective in tasks involving sequential inputs like speech and language, RNNs handle each input sequence separately, storing information about past elements of the sequence in a "state vector" within their hidden units. This feature allows RNNs to understand the context of past data and predict future outputs. The backpropagation algorithm plays a critical role in the training process of RNNs; it calculates the derivative of the network's total error, adjusting the network's parameters to form the correct sequence of outputs. However, training RNNs can be challenging, as the gradients propagated backward in time can grow or shrink, leading to gradient explosion or vanishing [11, 70].

Advances in the architectures and training methods of RNNs have made them highly successful at predicting the next character in a text or the next word in a sentence. This development supports the idea that sentence meaning and everyday reasoning can occur through more intuitive processes involving multiple simultaneous analogies, rather than through the manipulation of symbolic expressions using inference rules [69, 70].

RNNs can also translate the meaning of visual data into text. In this process, the pixels of an image are converted into an activity vector in the last hidden layer of a deep ConvNet, which is processed by a decoder in RNNs used for machine translation and neural language modeling. When unfolded over time, RNNs function as very deep feed-forward networks where all layers share the same weights. The effectiveness of RNNs in processing sequential data and complex tasks is an important capability in AI and ML [76].

5.4 Memory-Based Enhanced Networks

Memory-based networks are a discipline in AI and algorithmic learning, with the ability to store long-term information and advanced logical inference capabilities. These systems use specialized hidden units that can retain inputs over a long period of time, mimicking natural memory functions. In this field, Long Short-Term

Memory (LSTM) networks have a specialized component called a memory cell. This component functions as a cumulative or gate-controlled leaky neuron that maintains its state and collects external signals. This self-connection capability is regulated by another unit that determines when the contents of the memory should be reset.

LSTM networks, with their multilayer structure at each time step, are used in a wide range of speech recognition applications that convert acoustic signals into written text and outperform traditional RNNs. They also serve as encoding and decoding networks in the field of machine translation [77].

The application scope of memory-based networks extends beyond simple memorization to tasks that require logical reasoning and symbolic operations. Memory networks can monitor environmental states in interactive text-based game-like environments and are very good at answering questions that require coordination [40, 41, 78, 79].

5.5 Transformer-Based Neural Networks

Transformer-based neural networks, introduced in the paper "Attention Is All You Need" by Vaswani et al. in 2017 [81], have brought a revolution in the field of ML, especially in NLP. These networks are built on the foundation of attention mechanisms, which allow the model to focus on different parts of the input sequence when predicting an output. This is a significant shift from previous models like RNNs and LSTMs that processed data sequentially and faced limitations due to long-term dependency and computational inefficiency.

The architecture of transformers comprises two main components: the encoder and the decoder. The encoder processes the input data, such as a sentence in language translation, and the decoder then generates the output, like the translated sentence. Both these components consist of layers that include attention mechanisms and fully connected networks.

One of the transformative aspects of transformers is their ability to process all elements of the input data simultaneously, in parallel. This contrasts with the sequential processing of RNNs and allows for faster training and greater efficiency with large datasets.

A key feature of transformers is self-attention, enabling each position in the input sequence to attend to all positions in the previous layer of the encoder. This ability significantly enhances the model's understanding of context and the relationships between words in a sentence [80].

Transformers have been widely adopted for various NLP tasks like language translation, text summarization, sentiment analysis, and question-answering. They serve as the foundation for many state-of-the-art language models, including Google's BERT and OpenAI's GPT series [81].

The success of transformer-based neural networks lies in their capability to handle long-range dependencies in text and their efficiency in training on extensive datasets. This has not only led to improved performance in traditional NLP tasks but also opened up new avenues in machine understanding and interaction with human language.

5.6 Generative Adversarial Networks

GANs, a subset of AI algorithms, are used in unsupervised ML. Developed by Ian J. Goodfellow and colleagues in 2014, GANs consist of two different neural networks: Generative and Discriminative [82, 83].

The Generator network is tasked with generating new data samples, aiming to produce data that is indistinguishable from real data. Initially, the generated data may not be of high quality, but its accuracy typically increases gradually. The Discriminator network evaluates the incoming data by estimating the probability that the data is real or synthetic data. Its training is focused on distinguishing between real data and data generated by the producer.

In the training phase of GANs, the two networks operate in an alternating manner. The Generator strives to fabricate data increasingly akin to real data, while the Discriminator endeavors to discern genuine data from fabricated data. This iterative process persists until the Generator can produce data that the Discriminator is unable to categorize as synthetic with high reliability.

GANs are applied in a broad spectrum of domains. They are capable of generating images so lifelike that they become virtually indistinguishable from actual photographs. Additionally, they can modify and augment images, thereby enhancing image resolution. Owing to their proficiency in generating new data that closely resembles the input data, GANs find extensive applications in image processing, artistic creation, scientific research, and various other fields [82, 83].

5.7 Large Language Model Networks

LLM Networks in DL are advanced AI models specifically designed for tasks involving NLP. LLMs are distinguished by their enormous size, often containing billions or even trillions of parameters. They are trained on extensive datasets featuring a wide variety of text sources, enabling them to understand, interpret, generate, and interact with language in contextually relevant and nuanced ways.

Built using deep neural networks, particularly transformer models, LLMs are adept at handling sequential data, allowing them to consider the context of words within a sentence. This capability makes them highly effective for a range of applications including translation, summarization, question answering, and creative writing. A significant portion of their training involves self-supervised learning, where the model learns to predict parts of the text from other parts, helping it understand language structure and semantics without explicit labeling.

LLMs have found applications in various domains such as chatbots, writing assistants, content creation tools, and search engine enhancements. However, they come with their own set of challenges and ethical considerations. The computational resources required for training and operating these models are substantial, and there are concerns about biases in the training data, potential misuse, and the impact they might have on human labor in certain sectors [82, 83].

As a result, while LLMs represent a significant advancement in AI's ability to interact with human language, they also bring forth important challenges that need addressing as their use becomes increasingly widespread [84].

5.8 Deep Learning and Its Future

DL plays an important role in AI and ML disciplines, especially in processes that require knowledge extraction from large and complex datasets. In this methodology, models learn from existing examples to successfully complete various tasks. As is well known, human and animal learning is largely unsupervised, and unsupervised learning is expected to see even greater value in the future. Inspired by human visual processing, DL enables the development of systems that integrate deep and reinforcement learning. This offers significant innovations in areas such as NLP and complex thought processes. This contributes to the emergence of new and innovative methodologies beyond the current application areas of DL and is expected to have the potential to be actively used in broader application areas in the future [84, 85].

6 Comparison of Machine Learning, Shallow Learning, and Deep Learning

Shallow Learning, with its diverse algorithms, plays a significant role in scenarios with medium-scale datasets and clearly defined features. Methods such as SVM and Random Forests have shown remarkable success in such scenarios. Shallow Learning methods are known for their capacity to effectively classify datasets based on distinct features. These algorithms tend to provide stable and reliable results in scenarios with limited datasets.

DL is an approach developed to process large and complex datasets. Particularly in situations requiring complex pattern recognition and hierarchical feature extraction, DL models have a transformative effect. Models like Con Nets and RNNs exhibit unique abilities in areas such as image processing and sequence modeling. These models achieve high success by using interlayer interactions to understand the internal structure of data and solve complexities [11, 86].

Conventional ML methodologies exhibited constraints in their capacity to process unrefined natural data. The development of pattern recognition or ML frameworks necessitated the formulation of feature extraction mechanisms, which converted raw data into a fitting internal representation or feature vector. In contrast, DL approaches enhance the paradigm of representation learning. This advancement permits the direct input of unprocessed data into a machine, enabling it to autonomously ascertain the requisite representations for tasks such as detection or classification [87, 88].

In traditional ML systems, feature extraction is usually done manually. This process requires expert knowledge and can be time-consuming. DL models automate this process. During the learning phase, the model learns from raw data and identifies the necessary features on its own. This offers a significant advantage, particularly for large and complex datasets.

7 Relationships and Differences

DL and Shallow Learning are important methodologies in the ML subset. Both approaches are based on the principle of data-driven learning, where algorithms are trained based on input data to make predictions or decisions.

DL, characterized by the use of multilayer neural networks, is very good at extracting detailed features directly from data. This ability to extract features enables DL models to succeed in complex tasks such as image and speech recognition or NLP, where data is typically unstructured and abundant. However, the depth of these models requires significant computational resources and typically sophisticated hardware such as GPUs for efficient processing. The complexity of DL models has the major drawback of predictability, as the multilayered structure can obscure the decision-making process.

Shallow Learning, on the other hand, consists of simpler, usually linear models that do not have the multi-layered architecture of DL. Techniques such as decision trees, Linear Regression or Support Vector Machines, with their simpler structure, often contain more interpretable features. Shallow Learning is more effective when the relationship between input features and output is less complex or when data availability is limited. In contrast to DL, Shallow Learning often relies on manual feature extraction and selection and requires a more hands-on approach to model design.

The choice between Deep and Shallow Learning depends on the specific requirements of the task at hand. DL is capable of handling large-scale and complex datasets, while SL is effective for more defined, less data-intensive problems [11, 86].

8 Challenges and Future Directions and Ethical Considerations

ML, Shallow Learning, and DL provide data-driven solutions and create transformations across various industries. However, the application of these technologies brings significant challenges and ethical issues. ML, especially DL, requires large datasets and high computational power, while Shallow Learning needs less data and computational resources. Data quality and quantity directly affect model performance. Model transparency and interpretability, particularly in DL models, complicate understanding decision-making processes. This increases the need for ethical and reliable AI development.

Biases in training data can make algorithms less fair and unbiased, leading to incorrect and unjust results in sensitive applications. Security and privacy are major concerns, especially when personal data is involved. Ensuring the security of ML systems and protecting them against malicious attacks are central focuses of research in this field.

Future directions aim to enhance efficiency and generalization capabilities, strengthen model robustness, and reduce energy consumption. Ethical considerations are important in terms of data privacy, transparency, accountability, and societal impacts. The socioeconomic effects of these technologies, their impact on the workforce, and sustainability issues are critical areas that need to be addressed to ensure responsible and beneficial uses [89].

9 Conclusion

This chapter discusses the fundamental differences between traditional Shallow Learning and more complex DL, as well as an overview of ML methods. It presents a broad perspective on the nuanced strengths, limitations, and application areas of both paradigms. Additionally, by focusing on synthetic datasets supported with examples, particularly for shallow learning topics, it contributes to a better understanding of the subject for readers. The examination of Shallow Learning techniques, along with DL architectures such as Conv Nets and RNNs, establishes a solid foundation for subsequent chapters.

This chapter provides a solid foundation for those wishing to start learning about ML methods and their application. It is noted that Shallow Learning methods are particularly successful with small datasets, while DL models are capable of processing large and complex datasets and complex patterns. Additionally, the adaptability of Shallow Learning methods and the feature hierarchy abstraction capabilities of DL models offer various approaches for ML.

References

1. Harvard University. (2017). *The history of artificial intelligence*. https://sitn.hms.harvard.edu/flash/2017/history-artificial-intelligence/
2. Roser, M. (2022). *The brief history of artificial intelligence: The world has changed fast—What might be next? Our world in data*. https://ourworldindata.org/brief-history-of-ai
3. LeCun, Y., Bengio, Y., & Hinton, G. (2015). Deep learning. *Nature, 521*, 436–444. https://doi.org/10.1038/nature14539. This survey summarizes key elements of deep learning and its development in speech recognition, computer vision, and natural language processing.
4. Raschka, S., Patterson, J., & Nolet, C. (2020). Machine learning in Python: Main developments and technology trends in data science, machine learning, and artificial intelligence. *Information, 11*(4), 193. https://doi.org/10.3390/info11040193
5. Wang, H., Fu, T., Du, Y., et al. (2023). Scientific discovery in the age of artificial intelligence. *Nature, 620*, 47–60. https://doi.org/10.1038/s41586-023-06221-2
6. Chen, P., Wu, L., & Wang, L. (2023). AI fairness in data management and analytics: A review on challenges, methodologies, and applications. *Applied Sciences, 13*(18), 10258. https://doi.org/10.3390/app131810258
7. Zhu, S., Yu, T., Xu, T., Chen, H., Dustdar, S., Gigan, S., et al. (2023). Intelligent computing: The latest advances, challenges, and future. *Intelligent Computing, 2*, 0006. https://doi.org/10.34133/icomputing.0006
8. Zador, A. M. (2019). A critique of pure learning and what artificial neural networks can learn from animal brains. *Nature Communications, 10*, 3770. https://doi.org/10.1038/s41467-019-11786-6
9. Taye, M. M. (2023). Understanding of machine learning with deep learning: Architectures, workflow, applications and future directions. *Computers, 12*, 91. https://doi.org/10.3390/computers12050091
10. Simplilearn. (n.d.). *Recursive neural network in deep learning*. https://www.simplilearn.com/recursive-neural-network-in-deep-learning-article. Accessed Sept 2023.
11. Şen, Z. (2023). *Shallow and deep learning principles: Scientific, philosophical, and logical perspectives* (pp. 575–638). Springer.
12. T. R. N, & Gupta, R. (2020). *A survey on machine learning approaches and its techniques*. In IEEE International Students' Conference on Electrical, Electronics and Computer Science (SCEECS) (pp 1–6). https://doi.org/10.1109/SCEECS48394.2020.190.
13. Tiwari, T., Tiwari, T., & Tiwari, S. (2018). How artificial intelligence, machine learning and deep learning are radically different? *International Journal of Advanced Research in Computer Science and Software Engineering, 8*(2), 1.
14. Zapier. (n.d.) *Machine learning vs. AI: What's the difference?* https://zapier.com/blog/machine-learning-vs-ai/. Accessed Sept 2023.
15. Früh, A., & Haux, D. (2022). *Foundations of artificial intelligence and machine learning*.
16. Srivastava, G., Sapra, D., Sinha, A., Anup, M., & Sinwar, D. (2024). Artificial intelligence and IoT-assisted sustainable manufacturing for industry 4.0. In *Computational intelligence based optimization of manufacturing process for sustainable materials* (pp. 15–34). CRC Press.
17. Xie, X. (2024). *A review of methods and applications of sentiment analysis based on deep learning models*. In Internet finance and digital economy: Advances in digital economy and data analysis technology The 2nd International Conference on Internet Finance and Digital Economy, Kuala Lumpur, Malaysia, 19–21 August 2022 (pp. 993–1009).
18. Brown, A., Gupta, M., & Abdelsalam, M. (2024). Automated machine learning for deep learning based malware detection. *Computers & Security, 137*, 103582.
19. Pichler, M., & Hartig, F. (2023). Machine learning and deep learning – A review for ecologists. *Methods in Ecology and Evolution, 14*(4), 994–1016.
20. Taherdoost, H. (2023). Deep learning and neural networks: Decision-making implications. *Symmetry, 15*(9), 1723.

21. Triwiyanto, T., Caesarendra, W., Ahmed, A. A., & Abdullayev, V. H. (2023). How deep learning and neural networks can improve prosthetics and exoskeletons: A review of state-of-the-art methods and challenges. *Journal of Electronics, Electromedical Engineering, and Medical Informatics, 5*(4).
22. Talaei Khoei, T., Ould Slimane, H., & Kaabouch, N. (2023). Deep learning: Systematic review, models, challenges, and research directions. *Neural Computing and Applications, 35*(31), 23103–23124.
23. Makwana, J. J., Tiwari, M. K., & Deora, B. S. (2023). Development and comparison of artificial intelligence models for estimating daily reference evapotranspiration from limited input variables. *Smart Agricultural Technology, 3*, 100115.
24. Krichen, M. (2023). Convolutional neural networks: A survey. *Computers, 12*(8), 151.
25. Wang, H., Fu, T., Du, Y., Gao, W., Huang, K., Liu, Z., et al. (2023). Scientific discovery in the age of artificial intelligence. *Nature, 620*(7972), 47–60.
26. Gawlikowski, J., Tassi, C. R. N., Ali, M., Lee, J., Humt, M., Feng, J., et al. (2023). A survey of uncertainty in deep neural networks. *Artificial Intelligence Review, 56*(Suppl 1), 1513–1589.
27. Moein, M. M., Saradar, A., Rahmati, K., Mousavinejad, S. H. G., Bristow, J., Aramali, V., & Karakouzian, M. (2023). Predictive models for concrete properties using machine learning and deep learning approaches: A review. *Journal of Building Engineering, 63*, 105444.
28. Surianarayanan, C., Lawrence, J. J., Chelliah, P. R., Prakash, E., & Hewage, C. (2023). A survey on optimization techniques for edge artificial intelligence (AI). *Sensors, 23*(3), 1279.
29. Soori, M., Arezoo, B., & Dastres, R. (2023). Artificial intelligence, machine learning and deep learning in advanced robotics: A review. *Cognitive Robotics.*
30. Aguiar-Pérez, J. M., Pérez-Juárez, M. A., Alonso-Felipe, M., Del-Pozo-Velázquez, J., Rozada-Raneros, S., & Barrio-Conde, M. (2023). Understanding machine learning concepts. In *Encyclopedia of data science and machine learning* (pp. 1007–1022). IGI Global.
31. Mendonça, M. O., Netto, S. L., Diniz, P. S., & Theodoridis, S. (2024). Machine learning: Review and trends. In *Signal processing and machine learning theory* (pp. 869–959). Springer.
32. Xu, W., Gao, Z., & Ge, L. (2024). New research paradigms and agenda of human factors science in the intelligence era. *Acta Psychologica Sinica, 56*(3), 363.
33. Devi, M. K., Dhar, M. M., Unnamalai, K., & Shobana, S. (2024). A human-centered approach to artificial intelligence in the supply chain. In *Data-driven technologies and artificial intelligence in supply chain* (pp. 1–30). CRC Press.
34. Harel, D., Yerushalmi, R., Marron, A., & Elyasaf, A. (2024). Categorizing methods for integrating machine learning with executable specifications. *Science China Information Sciences, 67*(1), 1–15.
35. Pan, W., Xiang, Y., Gong, W., & Shen, H. (2024). Risk evaluation of elevators based on fuzzy theory and machine learning algorithms. *Mathematics, 12*(1), 113.
36. Kunze, H., La Torre, D., Riccoboni, A., Galán, M. R., & (Eds.). (2023). *Engineering mathematics and artificial intelligence: Foundations, methods, and applications*. CRC Press.
37. Risser-Maroix, O., & Chamand, B. (2023). *What can we learn by predicting accuracy?* In Proceedings of the IEEE/CVF winter conference on applications of computer vision (pp. 2390–2399).
38. Piccolo, R., Scuotto, V., Papa, A., Pironti, M., & Del Giudice, M. (2023). AI and innovation design for new product and service development in digital ecosystems. In *Handbook of research on artificial intelligence, innovation, and entrepreneurship* (pp. 88–101). Edward Elgar Publishing.
39. Wang, J. T., & Jia, R. (2023, April). Data Banzhaf: A robust data valuation framework for machine learning. In *International conference on artificial intelligence and statistics* (pp. 6388–6421). PMLR.
40. Topsakal, O. Learning machine learning via memorable stories, Amazon (pp 21–87).
41. Duarte, J. M., & Berton, L. (2023). A review of semi-supervised learning for text classification. *Artificial Intelligence Review*, 1–69.

42. Taha, K. (2023). Semi-supervised and un-supervised clustering: A review and experimental evaluation. *Information Systems*, 102178.
43. Xiong, Q., Yuan, C., He, B., Xiong, H., & Kong, Q. (2023). GTRF: A general deep learning framework for tuples recognition towards supervised, semi-supervised, and unsupervised paradigms. *Engineering Applications of Artificial Intelligence, 124*, 106500.
44. Morales, E. F., & Escalante, H. J. (2022). A brief introduction to supervised, unsupervised, and reinforcement learning. In *Biosignal processing and classification using computational learning and intelligence* (pp. 111–129). Academic Press.
45. Alzubaidi, L., Zhang, J., Humaidi, A. J., et al. (2021). Review of deep learning: Concepts, CNN architectures, challenges, applications, future directions. *Journal of Big Data, 8*, 53. https://doi.org/10.1186/s40537-021-00444-8
46. DeLancey, E. R., Simms, J. F., Mahdianpari, M., Brisco, B., Mahoney, C., & Kariyeva, J. (2019). Comparing deep learning and shallow learning for large-scale wetland classification in Alberta, Canada. *Remote Sensing, 12*(1), 2.
47. James, G., Witten, D., Hastie, T., Tibshirani, R., & Taylor, J. (2023). Linear regression. In *An introduction to statistical learning: With applications in Python* (pp. 69–134). Springer.
48. Loh, W. Y. (2023). Logistic regression tree analysis. In *Springer handbook of engineering statistics* (pp. 593–604). Springer.
49. Varady, N. H., Pareek, A., Eckhardt, C. M., Williams, R. J., III, Madjarova, S. J., Ollivier, M., et al. (2023). Multivariable regression: Understanding one of medicine's most fundamental statistical tools. *Knee Surgery, Sports Traumatology, Arthroscopy, 31*(1), 7–11.
50. Pampel, F. C. (2020). *Logistic regression: A primer* (Vol. 132).
51. Wang, H. (2020, November). Logistic regression for massive data with rare events. In *International conference on machine learning* (pp. 9829–9836). PMLR.
52. Mienye, I. D., Sun, Y., & Wang, Z. (2019). Prediction performance of improved decision tree-based algorithms: A review. *Procedia Manufacturing, 35*, 698–703.
53. Somvanshi, M., Chavan, P., Tambade, S., & Shinde, S. V. (2016, August). *A review of machine learning techniques using decision tree and support vector machine*. In 2016 international conference on computing communication control and automation (ICCUBEA) (pp. 1–7). IEEE.
54. Charbuty, B., & Abdulazeez, A. (2021). Classification based on the decision tree algorithm for machine learning. *Journal of Applied Science and Technology Trends, 2*(01), 20–28.
55. Li, H. (2024). Support vector machine. In *Machine learning methods*. Springer. https://doi.org/10.1007/978-981-99-3917-6_7
56. https://towardsdatascience.com/support-vector-machine-introduction-to-machine-learning-algorithms-934a444fca47. Accessed Oct 2023.
57. https://www.analyticsvidhya.com/blog/2021/10/support-vector-machinessvm-a-complete-guide-for-beginners/. Accessed Oct 2023.
58. Kramer, O., & Kramer, O. (2013). K-nearest neighbors. In *Dimensionality reduction with unsupervised nearest neighbors* (pp. 13–23). Springer.
59. Zhang, Z. (2016). Introduction to machine learning: K-nearest neighbors. *Annals of Translational medicine, 4*(11).
60. https://www.analyticsvidhya.com/blog/2018/03/introduction-k-neighbours-algorithm-clustering/. Accessed Sept 2023.
61. https://www.analyticsvidhya.com/blog/2017/09/naive-bayes-explained/
62. https://www.simplilearn.com/tutorials/machine-learning-tutorial/naive-bayes-classifier. Accessed Sept 2023.
63. Schonlau, M. (2023). The naive Bayes classifier. In *Applied statistical learning: With case studies in stata* (pp. 143–160). Springer.
64. Topsakal, O., (2023). *Learning deep learning via memorable stories, Amazon* (pp 12–98).
65. Sharifani, K., & Amini, M. (2023). Machine learning and deep learning: A review of methods and applications. *World Information Technology and Engineering Journal, 10*(07), 3897–3904.
66. Yao, K., & Zheng, Y. (2023). Fundamentals of machine learning. In *Nanophotonics and machine learning: Concepts, fundamentals, and applications* (pp. 77–112). Springer.

67. Kufel, J., Bargieł-Łączek, K., Kocot, S., Koźlik, M., Bartnikowska, W., Janik, M., et al. (2023). What is machine learning, artificial neural networks and deep learning? – Examples of practical applications in medicine. *Diagnostics, 13*(15), 2582.
68. LeCun, Y., Bengio, Y., & Hinton, G. (2015). Deep learning. *Nature, 521*, 436–444. https://doi.org/10.1038/nature14539
69. https://www.analyticsvidhya.com/blog/2022/11/analyzing-and-comparing-deep-learning-models/. Accessed Sept 2023.
70. Wani, M. A., Bhat, F. A., Afzal, S., & Khan, A. I. (2020). *Advances in deep learning*. Springer.
71. https://medium.com/@oscarleo/deep-learning-vs-neural-network-3ea20b5ec19a. Accessed Sept 2023.
72. Dewangan, O. (2023). Study and innovative approach to deep learning algorithms and architecture. In *Exploring future opportunities of brain-inspired artificial intelligence* (pp. 28–45). IGI Global.
73. Zhang, H., Sun, A., Jing, W., & Zhou, J. T. (2023). Temporal sentence grounding in videos: A survey and future directions. *IEEE Transactions on Pattern Analysis and Machine Intelligence*.
74. https://www.flatworldsolutions.com/data-science/articles/7-applications-of-convolutional-neural-networks.php. Accessed Sept 2023.
75. Medrano-Berumen, C., & Akbas, M. I. (2020). Validation of decision making in artificial intelligence based autonomous vehicles. *Journal of Information and Telecommunication*, 1–21.
76. Yang, J., & Li, J. (2017, December). *Application of deep convolution neural network.* In 2017 14th International Computer Conference on Wavelet Active Media Technology and Information Processing (ICCWAMTIP) (pp 229–232). IEEE.
77. Heaton, J. (2020). Applications of deep neural networks. arXiv preprint arXiv:2009.05673.
78. Sun, J., Zhang, X., & Wang, J. (2023). Lightweight bidirectional long short-term memory based on automated model pruning with application to bearing remaining useful life prediction. *Engineering Applications of Artificial Intelligence, 118*, 105662.
79. Vaswani, A., Shazeer, N., Parmar, N., Uszkoreit, J., Jones, L., Gomez, A. N., ... & Polosukhin, I. (2017). Attention is all you need. In *Advances in neural information processing systems* (p. 30).
80. Sattarzadeh, A. R., Pathirana, P. N., Kutadinata, R., & Huynh, V. T. (2023). Extracting long-term spatiotemporal characteristics of traffic flow using attention-based convolutional transformer. *IET Intelligent Transport Systems*.
81. Tunstall, L., Von Werra, L., & Wolf, T. (2022). *Natural language processing with transformers*. O'Reilly Media.
82. Goodfellow, I., Pouget-Abadie, J., Mirza, M., Xu, B., Warde-Farley, D., Ozair, S., et al. (2020). Generative adversarial networks. *Communications of the ACM, 63*(11), 139–144.
83. Eizaguirre Peral, I. (2022). *Design, implementation and validation of ML models based on adversarial network for generating synthetic biometric behaviour* (Master's thesis, Universitat Politècnica de Catalunya).
84. Topsakal, O., & Akinci, T. C. (2023). *Creating large language model applications utilizing langchain: A primer on developing LLM apps fast*. In Proceedings of the international conference on applied engineering and natural sciences, Konya, Turkey, pp 10–12.
85. Patil, D. D., Dhotre, D. R., Gawande, G. S., Mate, D. S., Shelke, M. V., & Bhoye, T. S. (2024). Transformative trends in generative AI: Harnessing large language models for natural language understanding and generation. *International Journal of Intelligent Systems and Applications in Engineering, 12*(4s), 309–319.
86. Xu, Y., Zhou, Y., Sekula, P., & Ding, L. (2021). Machine learning in construction: From shallow to deep learning. *Developments in the Built Environment, 6*, 100045.
87. Suganeshwari, G., & Divya, D. (2024). 7 deep learning in big. In *Big data computing: Advances in technologies, methodologies, and applications* (p. 132). CRC Press.

88. Hazra, A., Choudhary, P., & Sheetal Singh, M. (2021). Recent advances in deep learning techniques and its applications: An overview. In *Advances in biomedical engineering and technology: Select proceedings of ICBEST 2018* (pp. 103–122). https://doi.org/10.1007/978-981-15-6329-4_10
89. Niu, Y., Fan, Y., & Ju, X. (2024). Critical review on data-driven approaches for learning from accidents: Comparative analysis and future research. *Safety Science, 171*, 106381.

Shallow Learning vs. Deep Learning in Engineering Applications

Fereshteh Jafari, Kamran Moradi, and Qobad Shafiee

1 Introduction

Artificial intelligence (AI) stands as an ever-evolving field that intersects with engineering, pushing the boundaries of technology and innovation. AI involves the development of algorithms and systems that enable machines to exhibit human-like intelligence, learning, and decision-making capabilities. It encompasses various subfields, including ML, natural language processing, computer vision, robotics, and more. ML, a subset of AI, focuses on enabling systems to learn and improve from experience without being explicitly programmed. It involves algorithms that allow computers to identify patterns, extract insights, and make predictions from data. This technology aids in various applications, from recognizing speech and images to driving autonomous vehicles and optimizing industrial processes. Shallow learning, often referred to as traditional ML, comprises models that lack complex architectures and typically work with structured data. These models, such as logistic regression, decision trees, and support vector machines (SVMs), are well-suited for solving relatively simpler problems and are effective when the data characteristics are apparent. In contrast, deep learning, a powerful branch of ML, involves complex neural networks (NNs) inspired by the human brain. These networks comprise multiple layers capable of automatically learning intricate patterns from unstructured data, such as images, audio, and text. Convolutional neural networks (CNNs) and recurrent neural networks (RNNs) are among the sophisticated architectures used in deep learning, offering superior performance in various applications, especially when handling vast amounts of complex data. The beauty of these technologies lies in their ability to transform industries by offering data-driven

F. Jafari · K. Moradi · Q. Shafiee (✉)
Smart/Mirco Grids Research Center, Department of Electrical Engineering,
University of Kurdistan, Sanandaj, Iran
e-mail: f.jafari@uok.ac.ir; k.moradi@uok.ac.ir; q.shafiee@uok.ac.ir

Ö. F. Ertuğrul et al. (eds.), *Shallow Learning vs. Deep Learning*, The Springer Series
in Applied Machine Learning, https://doi.org/10.1007/978-3-031-69499-8_2

insights, automation of processes, and the ability to solve intricate problems. With continuous advancements and integration into various sectors, the future of AI, ML, and deep learning presents an exciting journey of innovation and discovery for engineers and researchers alike. In continuing, different applications of shallow learning and deep learning in most engineering fields have been discussed.

2 Application of Machine Learning in Engineering Applications

2.1 Application of Shallow Learning in Engineering

Shallow learning methods have significant applications within engineering, offering practical solutions for various challenges. These methods, built upon traditional ML techniques, serve as foundational tools for addressing engineering problems. Common shallow learning approaches include:

Linear Regression: This method is essential for modeling linear relationships between variables.

Decision Trees (DT): This is often used for fault diagnosis and system health monitoring, providing a clear visualization of decision-making processes. They are relatively easy to understand and interpret.

SVM: These are commonly applied for anomaly detection and classification. SVMs are powerful in separating complex datasets by finding an optimal hyperplane.

K-Nearest Neighbors (KNN): Particularly useful for pattern recognition and classification, KNN algorithms assign classifications based on the majority class among its k-Nearest Neighbors.

Although shallow learning methods provide several advantages as illustrated in Fig. 1, they may struggle when confronted with large datasets or those that possess highly intricate relationships. They often lack the ability to automatically learn abstract features from raw data, which limits their performance with unstructured data like images, audio, or text. These methods might also face issues when dealing with high-dimensional data, where deep learning models excel.

In the engineering domain, where data complexity and sizes can vary significantly across different applications, the choice between shallow learning and deep learning approaches often depends on the specific problem, the nature of the data, and the required accuracy of the solution. While shallow learning remains a fundamental part of engineering workflows, the advancements in deep learning have expanded the horizons, enabling the field to tackle more complex problems and deliver more sophisticated solutions.

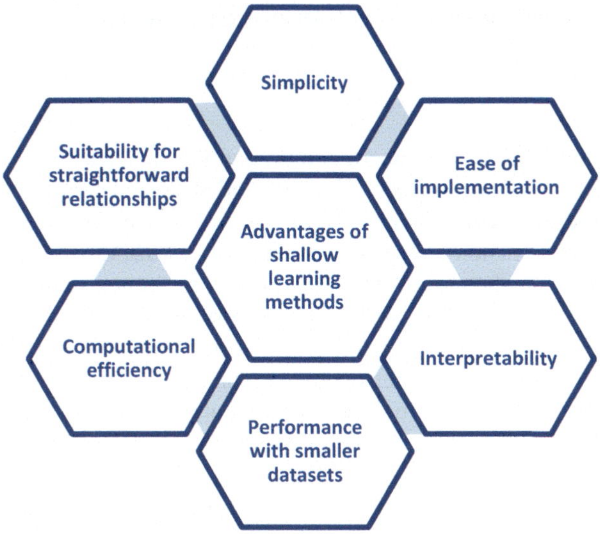

Fig. 1 Advantages of shallow learning methods

2.2 Application of Deep Learning in Engineering

Deep learning has rapidly emerged as a transformative force in various engineering domains, revolutionizing traditional practices and augmenting capabilities across industries. In engineering, its applications span multiple disciplines, showcasing remarkable potential in engineering areas such as mechanical, chemical, biomedical, materials, civil, computer, aerospace, automotive, marine, manufacturing, system, architectural, petroleum, nuclear, robotics, agricultural, and electrical engineering.

However, the effective deployment of deep learning in engineering is not without challenges. It requires robust data collection, labeled datasets, computational resources, and expertise in model architecture and optimization. Furthermore, ensuring the reliability, interpretability, and safety of deep learning-driven systems remains an ongoing concern. The integration of deep learning in engineering heralds a new era of innovation and efficiency, offering solutions to complex problems across diverse engineering disciplines. As technology continues to evolve, its seamless integration with engineering practices promises groundbreaking advancements and transformative solutions to societal challenges. In this chapter, we explore the utilization of ML methodologies, including shallow and deep learning, across various engineering domains.

3 Machine Learning in Mechanical Engineering

3.1 Mechatronics

In the realm of mechanical engineering, a groundbreaking transformation is underway through the integration of deep learning and shallow learning methodologies, heralding a new era across diverse applications. Picture this: a visionary smart walker, employing a fusion of hybrid motion models (HMM) and ML, remarkably easing the burden for the elderly [1]. Then, envision the real-time marvel of NNs solving inverse kinematics for robot manipulators, revolutionizing dynamic tasks [2]. Meanwhile, the estimation of wind speed for determining the feasibility of wind energy harnessing is becoming remarkably accurate, courtesy of extreme learning machines [3]. Not stopping there, a pioneering control system harnesses the output of a single sensor's time series, simplifying the complexities within printing systems [4].

The advancements continue. Battery manufacturing quality achieves new heights with a RUBoost ensemble, empowering sensitivity analysis [5], while fault diagnosis in multichannel motor-rotor systems undergoes a remarkable evolution with the multi-manifold deep extreme learning machine algorithm [6]. Further enhancing this landscape, the output tracking of soft magnetic actuators experiences unprecedented refinement through iterative machine learning (IML) coupled with complex-valued Gaussian process regression [7]. However, that is not all—enter meta-learning with adaptive learning rates, a champion in addressing few-shot fault diagnosis [8], adapting astoundingly to limited fault data, a feat previously deemed improbable.

These applications span diverse fields, showcasing ML's versatility—from enhancing mobility aids for the elderly and optimizing wind energy feasibility to improving fault diagnosis in complex systems like robots and actuators. ML-driven solutions optimize efficiency, reduce complexity, and enable accurate predictions, enhancing systems' performance in mechanical engineering contexts.

3.2 Microelectromechanical

Deep learning techniques have made a profound impact on the rapid development of microelectromechanical systems (MEMS), notably accelerometers crucial in health monitoring. Referenced in [9], this method combines artificial neural networks (ANNs) with optimized 3D printing for a swift workflow to swiftly and affordably produce custom MEMS devices, particularly accelerometers, catering to IoT demands. Bidirectional ANNs play a dual role: predicting device performance from design parameters and reverse-engineering design from specified performance needs.

3.3 Biomechanical

In biomechanical engineering, ML techniques exhibit diverse and transformative applications. A supervised multiclass classifier is developed for activity and gait phase recognition, specifically for knee-controlled powered motion assistive devices [10], a robust pre-impact fall detection model, boasting high classification accuracy, aiming to reduce fall-related injuries [11], and rupture risk in abdominal aortic aneurysms (AAAs) predicted using NN modeling, outperforming standard ML algorithms in predicting patient-specific AAA wall stress [12]. Moreover, an automatic human activity recognition system for exoskeletons improves adaptability by recognizing various human activities accurately [13]. Dynamic-mode decomposition, an unsupervised ML method, accelerates simulations of personalized organ-scale biomechanistic cancer models, offering promising predictions for treatment response and tumor growth in patients [14].

These applications range from crafting assistive devices like motion assistive devices and exoskeletons to enhancing healthcare by predicting fall risks, estimating wall stress in conditions like abdominal aortic aneurysms, enabling exoskeleton adaptability, and simulating personalized cancer models for treatment prediction—all facilitated through the utilization of ML techniques.

3.4 Propulsion

One striking instance involves the convolutional-deep neural network (C-DNN) method utilized to forecast the probabilistic low cycle fatigue (LCF) life of turbine blisks in gas turbine engines [15]. This method amalgamates CNN and DNN, effectively modeling LCF life prediction and proving its efficacy in regression modeling, promising substantial advancements in aero-engine health monitoring. Another vital application lies in AI, particularly ANN, employed to design cooling passages for air-cooled turbine vanes. Leveraging AI in aero-engine design, specifically for intricate air-cooled turbine vanes, involves training ANNs with vast simulation data to accurately predict cooling effectiveness. This development facilitates real-time adjustments in film hole arrangements, significantly impacting temperature distribution along the vane surface [16].

Furthermore, employing ANN to predict the Nusselt number and friction factor for hydrocarbon fuel under supercritical pressure within horizontal circular tubes demonstrates superior accuracy compared to traditional correlations [17].

Employing an intelligent numerical solver based on a multi-layer perceptron (MLP) feed-forward back-propagation ANN to study the heat transfer behavior of hybrid nanofluids in solar-thermal systems highlights the nanofluids' enhanced heat transmission capabilities. This showcases the impact of various physical factors, such as pressure and temperature, on improving thermal energy in solar-thermal systems [18]. These varied applications underscore the pivotal role of ML in predicting critical parameters, optimizing designs, and advancing our comprehension of complex phenomena in mechanical engineering.

4 Machine Learning in Chemical Engineering

In the domain of chemical engineering, ML stands as a critical asset, optimizing processes, boosting efficiency, and ensuring safety across various applications. For instance, an ML model adept in equivariance predicts energies and forces in hydrogen combustion, revolutionizing potential energy surface creation by adopting a "negative design" approach, reducing reliance on chemical intuition, and expediting convergence [19]. ML's impact extends to atom probe tomography, enhancing visualization of atomic arrangements in materials like BCC Fe-Al alloys, providing insights into chemical short-range order and potentially informing the design of high-performance materials [20]. In catalysis, ML aids in discovering efficient catalysts for electrochemical reactions such as CO_2 reduction, leveraging high-throughput data to predict catalyst behavior and synthesize new materials, facilitating reactions like ammonia decomposition [21, 22].

Computational chemistry benefits as ML predicts zeolite elastic responses based on structural features, deepening understanding of stability trends in porous materials [23]. Moreover, ML automates electrochemical data interpretation, streamlining cyclic voltammetry analysis and enhancing EIS model recognition [24, 25]. In pharmaceuticals, it expedites protein redesign by swiftly screening numerous novel protein sequences [26]. Lastly, in understanding uncertainties in chemical property predictions, ML segregates and quantifies uncertainties, offering guidelines to enhance model performance across diverse contexts [27]. These varied applications underscore ML's pivotal role in optimizing chemical engineering processes, advancing material understanding, and fostering discoveries across multifaceted domains in the field.

5 Machine Learning in Biomedical Engineering

5.1 Biomechanics

ML plays a pivotal role in biomedical engineering, particularly in gait analysis and musculoskeletal modeling, addressing clinical hurdles and expediting assessments. For instance, one study [28] focuses on predicting joint moments in cerebral palsy (CP) patients using one-dimensional CNNs, showing potential for future clinical use despite current prediction limitations. Also, ML is utilized to estimate joint contact and muscle forces from gait lab data, employing RNNs to accurately predict patient mechanics for swift clinical and research assessments [29]. Moreover, ML enhances computational efficiency in biomechanical simulations, where ML models predict biomechanical model deformations at a significantly faster pace than finite element models [30].

In the realm of autism spectrum disorder (ASD), ML classifiers proficiently identify ASD-related postural control patterns [31], offering the potential for efficient ASD diagnosis through computer-aided ML. Furthermore, ML techniques with traditional statistical tools analyze healthy and pathological movement data in human movement biomechanics [32]. The aim is to optimize biomechanical research efficiency and establish best practices for training and evaluating ML models.

5.2 Drug Delivery

In drug formulation science, ML accelerates development, reducing reliance on costly wet-lab experiments, and aims for self-driving laboratories, delineating the current landscape and future prospects of ML in drug formulation [33]. In ocular formulation development, ML techniques introduce computational pharmaceutics, employing in silico modeling to enhance drug delivery insights and streamline formulation design for ocular applications, signaling a shift in this domain [34]. Nanomedicine design, traditionally trial-and-error, now integrates data science to predict material properties and pharmacologic parameters. ML algorithms and "big data" approaches in "nanoinformatics" promise advancements by leveraging extensive datasets [35]. AI's pivotal role in combating antimicrobial resistance via innovative drug delivery strategies is highlighted, encompassing drug development, treatment optimization, delivery system design, and outcome prediction, offering substantial impact in infectious disease treatment [36]. In pharmaceutical pursuits, the evolution of AI and ML technologies optimizes analytical methods and dosage forms, transitioning from linear to nonlinear models. These methodologies potentially unravel intricate data relationships, particularly in clinical trials and infectious disease contexts [37].

5.3 Imaging

Notably, in tumor characterization from radiology images, supervised and unsupervised ML algorithms improve cancer staging, prognosis, and personalized treatment planning in precision medicine. Deep learning integration, especially in lung and pancreatic tumor diagnosis, has remarkably boosted performance, achieving state-of-the-art results [38]. Furthermore, within Standard Electromechanical Wave Imaging, ML models such as logistic regression, SVM, and Random Forest automate the zero-crossing selection, vital for isochrone generation. This automated method significantly reduces user variability and time, ensuring accuracy in identifying accessory pathways in patients and localizing pacing electrodes—an efficient alternative to manual selection [39]. Additionally, a study exploring advanced ML methods for automated classification of clustered microcalcifications in mammograms, crucial for computer-aided diagnosis systems in breast cancer diagnosis, highlights the superiority of kernel-based ML techniques like SVM and KFD over NN-based approaches. This significantly enhances the accuracy of breast cancer diagnosis using mammograms [40]. These applications underscore ML's potential in streamlining diagnosis, elevating accuracy, and minimizing manual variability across various biomedical engineering domains.

5.4 Nanotechnology

The fusion of nanotechnology and ML in biomedical engineering has led to groundbreaking progress in the healthcare realms. This amalgamation has revolutionized diagnostics, therapeutics, and biomedical research by enabling precise design and optimization of materials, drug delivery systems, and sensors, catering to tailored biomedical solutions [41]. In nanoscale resistive memory devices, variability is harnessed to create architectures like RRAM-ELM (resistive random-access memory—extreme learning machine), leveraging this variability for efficient, low-power ML tasks [41]. Furthermore, in exploring nanosensors within the bloodstream, ML models detect and pinpoint abnormalities, offering a potential method for effective diagnosis and localization within the intricate network of blood vessels [42].

Data processing architectures witness innovation with solutions like PRINS (processing-in-storage system) using resistive content addressable memory (ReCAM), meeting high-performance demands of large-scale datasets with notable speedups and enhanced power efficiency compared to traditional architectures [43]. Additionally, a proposed distributed in-memory computing architecture based on domain-wall nanowires addresses memory capacity and bandwidth challenges in data-focused applications. This architecture, designed for logic-in-memory operations, showcases significant power reductions and improved energy efficiency, especially in NN-based image resolution enhancement algorithms [44]. These applications demonstrate the transformative potential of combining ML with nanotechnology across diverse biomedical engineering endeavors, paving the way for precise diagnostics, efficient processing, and innovative healthcare solutions.

6 Machine Learning in Materials Engineering

In materials engineering, the integration of ML methodologies has sparked a transformative shift in material design, characterization, and processing, unlocking unprecedented innovation potential. ML algorithms and predictive models have revolutionized materials discovery, optimizing fabrication processes and enabling the development of sustainable, innovative materials across industries [45].

Numerous papers exemplify the synergy between ML and materials engineering. One study enhances the understanding and design of high-performance alloys, addressing existing model limitations for precipitation-hardened alloys, resulting in a novel alloy with enhanced yield strength [45]. Another accelerates high-entropy alloy design, predicting alloys with high strength and strengthening phase fractions [46]. Additionally, ML optimizes lead-free solder alloys for improved mechanical properties [47]. Understanding grain boundary damage in magnesium alloys for enhanced ductility is achieved through ML predicting damage initiation based on specific grain boundary features [48]. In selective laser melting, ML coupled with statistical analysis illuminates pore features' impact on fatigue life in Inconel 718, revealing microstructure-mechanical property relationships [49]. Moreover, a composition-structure-property model optimizes alloy design, yielding a novel alloy with favorable microstructure and mechanical properties [50].

Furthermore, ML techniques in sensor construction offer intelligent detection methodologies, such as a fluorescence sensor for sulfide determination and ML predicting cadmium availability in soil, impacting environmental monitoring and food safety assessment [51, 52]. Another explores ML's predictive capability for pollutant migration from microplastics, aiming to streamline laboratory analyses and enhance environmental and economic efficiency [53]. These diverse applications highlight ML's revolutionizing role in materials engineering, providing predictive insights, accelerating discoveries, and enabling precise material design with profound industry implications.

7 Machine Learning in Civil Engineering

In the domain of civil engineering, the integration of ML stands as a transformative force, reshaping traditional approaches and driving innovation across various domains. With its ability to analyze vast datasets, predict complex outcomes, and optimize decision-making processes, ML has emerged as a powerful tool for revolutionizing how civil engineers plan, design, construct, and manage infrastructure projects. This fusion of advanced computational techniques with the foundational principles of civil engineering is ushering in a new era, enabling the industry to tackle intricate challenges more effectively. From enhancing structural integrity, optimizing traffic flow, and improving resource allocation to enabling predictive maintenance and optimizing construction processes, ML is propelling the field of civil engineering toward greater efficiency, sustainability, and resilience in the face of evolving global demands. The intersection of ML and civil engineering promises a future where infrastructural development is not just efficient but also smarter, safer, and more adaptive to the needs of societies and the environment.

ML methods in civil engineering showcase versatile applications and significant contributions across specialized fields. In structural engineering, ML frameworks accurately predict post-earthquake building damage and aid seismic risk assessments, improving transportation network resilience and retrofit planning for bridges

[54, 55]. ML models estimate shear strength in Reinforced Concrete walls, detect anomalies in structural health monitoring, and address design gaps in ultra high performance concrete beams [56–58].

In hydrologic engineering, ML models effectively estimate evaporation from small reservoirs, predict river flows, and improve rainfall forecasting, offering valuable water management strategies and realistic rainfall predictions [59–61].

ML solutions in hydraulic engineering rectify inaccuracies in sediment transport models, identify flow patterns in water pipelines, aid flood simulation, and classify alluvial stream bedforms, providing practical implications for river management and fluid dynamics applications [62–65].

In bridge engineering, ML techniques excel in predicting shear strength and designing precast concrete joints and beams, enhancing bridge construction accuracy, and improving structural health monitoring with hybrid approaches [66–68]. ML integrated with simulations aids in classifying protection beam performance against overheight trucks, contributing to guidelines for collision protective measures [69]. These diverse ML applications consistently offer accuracy, efficiency, and valuable insights, revolutionizing infrastructure design, assessment, and risk management in civil engineering.

8 Machine Learning in Computer Engineering

Computer engineering focuses on the design, development, and optimization of computer systems and hardware components. It involves the integration of hardware and software to create efficient and effective computing systems. Data mining, on the other hand, is a subfield of computer science that involves discovering patterns, correlations, and insights within large datasets. It utilizes various techniques and algorithms to extract valuable information from data.

In computer engineering, both shallow learning and deep learning techniques find applications in various domains such as:

Computer vision: Deep learning models, especially CNNs, excel in image classification, object detection, segmentation, and image generation tasks.

Natural language processing (NLP): RNNs and transformers (like BERT and GPT) are widely used for tasks like language translation, text summarization, sentiment analysis, and language generation.

Recommendation systems: Both shallow and deep learning techniques are utilized in recommendation systems for personalized content, products, or services.

Anomaly detection: ML, including shallow and deep learning models, is applied to detect anomalies in network traffic, cybersecurity, or system failures.

Healthcare: In healthcare, both shallow and deep learning methods are employed for tasks like disease diagnosis from medical images, patient outcome prediction, and drug discovery.

Both shallow and deep learning approaches have their strengths and limitations, and their choice in application often depends on the nature of the data, the complexity of the problem, the availability of labeled data, computational resources, and the desired level of interpretability.

Specifically, ML addresses critical challenges in distinct domains. One study enhances intrusion detection systems with a novel ML classifier for heterogeneous traffic, strengthening anomaly detection and network traffic monitoring to mitigate security system vulnerabilities [70]. Another study focuses on improving soft tissue tumor (STT) diagnosis using SVM and Decision Tree algorithms, aiming to enhance diagnostic accuracy and automated decision-making in medical diagnostics [71]. In bioinformatics, ML, particularly in genome signal processing, relies on encoding schemes to convert genomic data into suitable numerical forms for computational analyses, emphasizing encoding scheme importance for ML model performance in genomic applications [72]. These diverse applications highlight ML's tailored solutions for specific challenges in computer engineering, offering effective solutions across domains like security, healthcare, and bioinformatics, utilizing both shallow and deep learning methodologies.

9 Machine Learning in Aerospace Engineering

Aerospace Engineering is a multifaceted domain dedicated to the design, development, and upkeep of aircraft, spacecraft, satellites, and associated systems. It spans numerous disciplines, including aerodynamics, propulsion, materials science, avionics, and structural analysis. This field's primary goal is to innovate and refine technologies that enhance flight performance, safety, and efficiency, facilitating advancements in space exploration and aviation.

In Aerospace Engineering, both deep learning and shallow learning technologies play integral roles. Deep learning methods are employed in flight control systems for autonomous navigation, obstacle detection, and decision-making, bolstering aircraft safety. They are also instrumental in aircraft health monitoring, analyzing sensor data to predict faults and failures, and ensuring proactive maintenance. Shallow learning algorithms, on the other hand, aid in aircraft systems diagnostics, aerodynamics, structural analysis, and mission planning. These techniques contribute to optimizing flight paths, predicting airflow patterns, and improving materials to enhance aircraft performance and safety. Both deep learning and shallow learning continually propel aerospace technology forward, enabling more intelligent and efficient aerospace systems. In global navigation satellite systems (GNSS), ML models aim to enhance performance in challenging signal environments, showcasing potential improvements over traditional algorithms [73]. ML plays a pivotal role in detecting ionospheric scintillation events with high accuracy and early identification, potentially automating ionospheric condition monitoring [74]. Furthermore, ML techniques forecast telemetry parameters for low Earth orbit satellites, aiding in risk reduction and preventing satellite failures [75].

Studies highlight ML's effectiveness in predicting ionospheric scintillation and automating chirp jamming signal detection in GNSS, offering quicker identification of jamming attacks [76, 77]. Reinforcement learning optimizes satellite observations, while ML architectures like CNNs facilitate real-time diagnostics in rotating detonation engines, emphasizing diverse ML applications in engine research [78, 79]. ML aids in modeling relationships between design parameters and objective functions in supersonic combustion engines, providing crucial insights into engine performance [80]. Moreover, tailored interpretable learning algorithms for aircraft systems improve ML adoption, enhancing performance and computational efficiency for real-time aerospace applications [81]. These applications demonstrate ML's broad impact in aerospace engineering, from navigation systems and satellite telemetry to engine diagnostics and aircraft systems, providing efficiency, accuracy, and automation across various domains.

10 Machine Learning in Automotive Engineering

Automotive Engineering involves the design, development, manufacturing, and maintenance of vehicles, encompassing a wide range of disciplines such as mechanical engineering, electrical engineering, materials science, and computer science. Its core focus is on creating safe, efficient, and reliable vehicles, including cars, trucks, motorcycles, and their subsystems.

Deep learning and shallow learning find applications in various domains within Automotive Engineering:

Autonomous vehicles: Deep learning is pivotal in enabling autonomous driving systems. It is used for object detection, classification, and segmentation in computer vision tasks, allowing vehicles to recognize pedestrians, other vehicles, traffic signs, and lane markings.

Natural language processing (NLP): It plays a role in voice recognition systems used for in-car assistants and human-vehicle interaction interfaces.

Predictive maintenance: Deep learning is applied to predict and detect potential failures in-vehicle components by analyzing sensor data, enabling timely maintenance and reducing breakdowns.

Enhanced navigation systems: Deep learning techniques are used to improve route optimization, traffic prediction, and mapping systems, providing real-time information to drivers.

Driver assistance systems: Shallow learning methods like SVM and DT are utilized in various driver assistance systems for tasks such as lane departure warning, adaptive cruise control, and collision avoidance.

Fuel efficiency optimization: Shallow learning techniques help in optimizing engine performance, predicting fuel consumption, and enhancing energy efficiency through data analysis.

Pattern recognition: Shallow learning algorithms are employed in recognizing patterns from various sensor inputs, aiding in fault diagnosis, engine health monitoring, and emission control systems.

Both deep learning and shallow learning play vital roles in advancing automotive technology and enhancing vehicle safety, performance, and user experience. They contribute to the ongoing evolution of the automotive industry toward more intelligent and autonomous vehicles. Some research in this field is mentioned below.

In Automotive Engineering, ML revolutionizes transportation and vehicle systems. ML methods detect railway wagon wheel defects, preventing infrastructure damage and reducing noise. NNs and SVMs analyze sensor data, significantly improving defect classification over traditional methods and predicting flat spots, shelling, and non-roundness [82]. ML aids real-time driver distraction detection, enhancing driver-centered assistance systems. Semi-supervised ML methods like Laplacian SVM and semi-supervised extreme learning machine leverage unlabeled data to enhance detection performance, offering cost-efficient model development and improving driver safety [83]. ML applications in intelligent transport systems (ITS) address authentication and security concerns. Graph-based ML techniques combined with lightweight cryptography enhance security, detecting cyber threats in ITS environments and ensuring secure communication between smart vehicles and roadside units [84].

ML optimizes traffic signal timings during incidents, employing genetic algorithms (GA) and regression models for efficient traffic management. This approach minimizes disruptions in urban traffic networks, providing traffic control centers with practical solutions to handle incidents efficiently [85]. ML techniques in inter-vehicular communications (IVC) enhance link reliability under realistic channel models. Studying relay selection and power allocation in dual-hop IVC systems, ML algorithms significantly improve system performance and reliability [86]. These applications showcase ML's impact on automotive engineering, enhancing safety, security, and traffic flow, demonstrating its versatility and potential for advancement in the automotive industry.

11 Machine Learning in Marine Engineering

ML applications in Marine Engineering span diverse areas like wave forecasting, autonomous underwater vehicle (AUV) operations, ship research, environmental monitoring, and resource development [87]. ML's capacity to process extensive oceanic data enables predictive models for phenomena like wave heights, storm surges, and ship parameters, revolutionizing solutions in marine engineering.

Wave forecasting benefits significantly from ML, particularly NNs' superior performance in predicting wave parameters, overcoming the limitations of traditional methods. ML's role in AUV operations integrates NNs for adaptive trajectory tracking, enhancing teleoperation systems, and remote vehicle control [87], elevating the sophistication of AUV and remotely operated vehicle (ROV) systems.

ML's impact extends to ship research, improving tasks like parameter estimation, docking, maneuvering, trajectory prediction, and fuel consumption estimation [88–94], ensuring maritime safety and operational efficiency.

In marine environmental monitoring, ML contributes to identifying and mitigating issues like red tides, storm surges, and oil spills [95]. Techniques like CNN effectively analyze ocean images for spill detection and offer predictive capabilities for environmental changes.

Deep learning in ML advances image processing and remote sensing, especially in coastal wetland classification and sea surface target detection [96–100]. Innovations in decision fusion algorithms enhance marine oil spill detection, tackling unclear boundaries and limited sample data [101]. Novel frameworks automate identification processes, like turbulent ship wake pattern identification within echo sounder data [102].

These applications underscore ML's transformative role in advancing marine engineering, from accurate wave forecasting to enhancing AUV operations, revolutionizing ship research, environmental monitoring, and resource development, offering immense scientific progress and deeper insights into complex marine phenomena.

12 Machine Learning in Manufacturing Engineering

In manufacturing engineering, the influx of big data offers vast potential for quality improvement, cost estimation, and process optimization [103–105]. However, managing this abundance poses challenges, prompting the integration of ML techniques. Some of the applications of AI and ML in manufacturing engineering are shown in Fig. 2. ML empowers decision-making by deriving patterns from manufacturing data, offering insights that challenge traditional checkpoints and enhance processes [106, 107]. SVMs find significant use in monitoring tasks like tool condition, fault diagnosis, and image recognition for product assessment [107, 108]. ML spans computer vision-based inspection, fault detection, and process optimization, integrating cost-effective sensors for efficient part inspection and continuous monitoring [107]. ML fused with prescriptive analytics fosters faster mass production and minimized waste [109]. Studies showcase ML's impact on fault detection in industrial forging, using CNN and DT with vibration test datasets [110]. Evaluations consistently demonstrate ANN dominance in optimizing manufacturing processes, guiding future research in ML integration [111]. These applications highlight ML's pivotal role in transforming manufacturing, optimizing processes, enhancing efficiency, and enabling sustainable production practices. The advantages and disadvantages of deep and shallow learning techniques in fault detection of manufacturing applications are provided in Table 1.

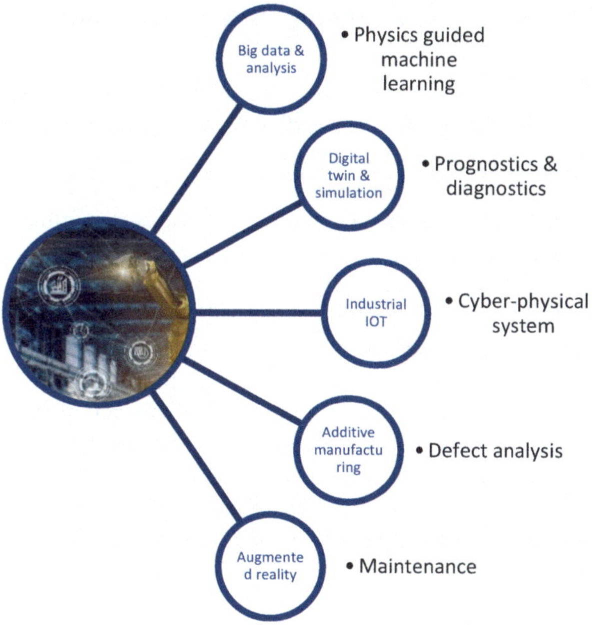

Fig. 2 The multifaceted usage of AI and ML in the manufacturing domain

Table 1 Pros and cons of deep and shallow learning techniques in fault detection in manufacturing applications [112]

Technique		Advantage	Disadvantage
Shallow learning methods	Bayesian Networks	Easily comprehensible and suitable for handling uncertainty, this method adeptly represents multiple causal hierarchies. It demonstrates bidirectional reasoning, catering to both predictive and diagnostic aspects	The tree structure complexity presents initialization challenges, especially in its construction phase
	Hidden Markov Model	Excels in modeling processes involving unobservable states	The training process often demands substantial computational resources
	SVM	Proficiency in modeling linear and nonlinear relationships, boasting faster computation than ANNs. It exhibits strong generalization, even with constrained training data	The model's kernel function parameter selection poses challenges, making it difficult to incorporate domain knowledge and comprehend the learned function
Deep learning method	ANNs	Proficiently tackles complex nonlinear problems with high precision, requiring simpler initialization compared to Bayesian Networks	Interpretability is limited, and it struggles with input uncertainty, alongside being computationally demanding, resulting in slow convergence and susceptibility to overfitting

13 Machine Learning in System Engineering

System engineering is a multidisciplinary approach that applies engineering principles to design, analyze, and manage complex systems over their life cycles. It involves a systematic, holistic view of large-scale systems, considering not only their components but also their interactions and the environment in which they operate. Through the integration of various disciplines and methodologies, system engineering aims to create efficient, reliable, and sustainable systems that meet specified requirements. This field employs rigorous processes, such as requirements analysis, system modeling, and validation, to ensure that the resulting systems satisfy stakeholder needs while considering technical, economic, environmental, and societal aspects. System engineers play a pivotal role in identifying trade-offs, managing risks, and optimizing the performance of systems across diverse domains, ranging from aerospace and defense to healthcare, transportation, and beyond.

In system engineering, ML techniques like shallow learning have been used for tasks like fault detection. Deep learning has brought remarkable advancements, enabling complex pattern recognition and hierarchical feature extraction within systems. These innovations are reshaping System Engineering, enabling smarter and more adaptable systems across diverse domains.

Differences in the structure of deep and shallow learning are also shown in Fig. 3. Deep networks differ from shallow networks primarily based on the number of hidden layers; shallow networks consist of a single hidden layer that transforms inputs through weighted sums, whereas deep networks include at least four or more layers with various types of layers. Shallow networks, often used for regression, employ a diverse range of models. For instance, transformations via a basis function followed by weighted adjustments represent different training methods within the same underlying structure [113]. The node transformation in shallow networks typically involves a local linear weighting of basis functions, leading to local weighted linear regression.

Shallow networks combine feature extraction and feature weighting. Their success often relies on effective feature extraction or utilizing pre-existing relevant features as inputs to a relatively simple network [114]. However, they can struggle

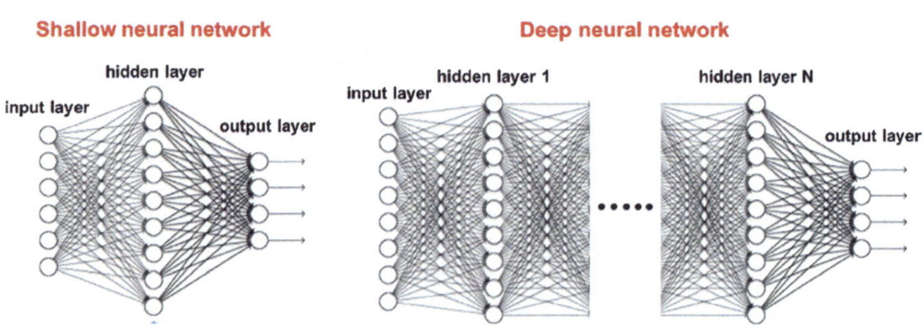

Fig. 3 The structure of shallow NN (left) and deep NN (right) [117]

with domain-specific feature extraction and require a larger number of nodes in fewer layers compared to deeper networks, potentially impacting generalization. In contrast, deep networks organize features hierarchically, enabling feature reuse and potentially requiring fewer parameters for representation compared to shallow networks [115]. The optimal depth varies for each function.

The limitation of shallow networks in capturing non-local features and their architectural inefficiency drive the preference for deeper networks. Although deeper networks can mimic the structure of biological NNs, their complexity and non-convex error functions make them challenging to train using standard optimization techniques [116].

In system engineering, the integration of ML, especially deep learning, has tackled long-standing challenges in training deep NNs [117, 118]. Advances from multilayer architectures to innovations like deep belief networks (DBN) have rekindled interest in deep learning, overcoming issues related to error propagation, saddle points, and suboptimal feature combinations [117–120]. DBN's initial drawbacks were resolved by unsupervised pre-learning via restricted Boltzmann machines, leading to successes in various applications like image classification and strategic games [119]. Recent studies have delved into fault diagnosis in complex systems using deep learning methods [121–124]. However, these approaches often overlook a critical factor: time delays in fault occurrences. Addressing this, tailored solutions considering both feature extraction and time-delayed fault predictions are necessary [122, 123]. In [124], researchers introduce an innovative DL-centered approach. This method concurrently deals with feature extraction and time-delayed fault occurrences through sliding window processing and a CNN long short-term memory (LSTM) model. This represents a significant advancement in fault diagnosis techniques for complex systems, addressing the gap in handling time-delayed fault occurrences left by existing methods.

14 Machine Learning in Architectural Engineering

In the field of architectural engineering, the traditional design process, which previously relied on manual hand drawing, has evolved to incorporate computer-aided techniques like computer-aided design, computer-aided architectural design, and computational design. The emergence of AI has further propelled this transformation by enabling computers to exhibit human-like thinking and problem-solving capabilities. ML, a subset of AI, plays a crucial role in utilizing data to model and address problems within computer-aided systems. By collecting and analyzing data, ML algorithms can make informed decisions based on past experiences, leading to more accurate solutions. However, it is essential to appropriately select and include an adequate number of relevant features from the dataset to ensure effective problem-solving using ML algorithms. ML finds extensive applications across various professional domains, including science, medicine, automotive, engineering, sales and marketing, construction, and architecture [125].

14.1 Transformative Shift in Architectural Practices

The architectural field is experiencing a transformative shift from traditional representations to a data-rich paradigm. Designers now integrate extensive external and internal data into their models, including urban, climate, and 3D scanning data, as well as data generated through scripting and sensors. However, these new practices pose challenges to current architectural models, necessitating larger, broader, and more intricate frameworks to accommodate multidisciplinary information, multiple design phases, and new design scales [125].

14.2 ML in Architectural Fabrication

Architects and engineers are seeking novel models and methods to handle data-intensive processes and extract meaningful information for design synthesis. In the case of Lace Wall (Fig. 4), ML is applied to expand design-integrated simulations into areas marked by numerous potential design solutions and high computational demands [125].

The study's prototype, "A Bridge Too Far" (Fig. 5), explores the integration of ML into the fabrication process to enhance design and manufacturing adaptation [125].

Fig. 4 Lace Wall demonstrator during the complex modeling exhibition Copenhagen in September–December 2016 [125]

Fig. 5 A Bridge Too Far exhibited in Copenhagen at the complex modeling exhibition in September–December 2016 [125]

The use of ML in architectural engineering focuses on refining early-stage design choices to optimize environmental performance throughout a building's lifecycle [126]. Conventionally, reliance on default values or Monte Carlo analysis has yielded unreliable results. To counter this, a new sampling and data-learning method has been proposed to evaluate environmental performance in the initial design phases, offering a dependable alternative without relying on default settings or probability distributions of design parameters. This method integrates parametric design technology with ML algorithms, enabling designers to explore diverse design constraints and uncertainties while uncovering connections between early design features and environmental uncertainty. By employing this approach, designers can assess multiple potential early designs, understand the influence of design decisions on environmental performance, and analyze the importance of early parameters in affecting performance uncertainty. Although this method provides valuable insights into uncertainty for early design decisions, further integration with techniques like bootstrapping procedures could comprehensively consider additional sources of uncertainty, augmenting its decision-making capacity in architectural design [126].

15 Machine Learning in Petroleum Engineering

Deep and shallow learnings have found extensive application within petroleum engineering, revolutionizing traditional approaches in various domains. In reservoir engineering, techniques such as NNs and SVMs enable the prediction of reservoir properties, facilitating more accurate estimations of porosity, permeability, and reservoir fluid behavior. These models aid in reservoir characterization and optimization of recovery strategies. In drilling operations, ML algorithms analyze vast datasets to enhance drilling efficiency, predict equipment failures, and optimize drilling parameters for improved safety and cost-effectiveness. Additionally, in production engineering, ML plays a pivotal role in predictive maintenance, where algorithms forecast equipment failures, detect anomalies in production processes, and optimize production rates. Integrating ML models in these areas of petroleum engineering enables data-driven decision-making, enhances operational efficiency, and leads to more effective resource management.

Petroleum engineering utilizes ML across various domains, enhancing exploration, drilling, reservoir, and production operations in the oil and gas industry. ML methods, particularly ANN, play a significant role in reducing risks and improving success rates in exploration. Innovations in drilling equipment, such as automated rigs and intelligent drill pipes, enhance drilling quality and cost-efficiency. Reservoir engineering benefits from ML algorithms like ANNs, genetic algorithms, and response surface models, integrating seismic, well logs, and production data for precise pressure analysis, reservoir history matching, and production forecasting. ML also addresses drilling challenges like stick-slip vibrations and borehole instability, aiding in tasks such as predicting bit wear and assessing drilling data quality. In production engineering, ML transforms workflows by enabling rapid analysis of extensive datasets, identifying production patterns, predicting closure pressure, and facilitating real-time anomaly detection in electric submersible pump-operated wells. These applications streamline decision-making, enhance operational efficiency, and optimize control strategies to maximize oil production while refining reservoir property maps iteratively [127–134]. A summary of shallow learning and deep learning applications in oil and gas production is provided in Table 2.

15.1 Challenges and Opportunities

ML shows great potential in addressing three major challenges in petroleum engineering: reducing computational costs by creating surrogate models, analyzing problems requiring human intervention, and addressing complex problems that were previously impractical. However, implementing ML in this field comes with numerous hurdles. The petroleum industry deals with heterogeneous environments, expensive decisions like drilling wells, and sporadic data availability, contrary to the ideal settings for ML applications. Overcoming these challenges demands thoughtful design and close collaboration with domain experts. Data issues pose significant barriers, considering the high volume, variety, veracity, and velocity of data influx from downhole and surface sensors. Acquiring accurate and sufficient high-quality data for training ML models remains a prevalent challenge, leading to uncertainties and noise in the training data, impacting model accuracy and generalizability. Additionally, the data's unavailability, confidentiality concerns, and the complexity of geoscientific data hinder the rapid adoption of ML in this domain [136].

Table 2 AI application in oil and gas production [135]

Group	Method	Inputs	Outputs
Deep learning	ANN	GDP growth rate, footage drilled, wells drilled, annual depletion, gas prices, and other resources are all factors to consider	Gas Production
	MLP NN	Regular flowing time; distributed temperature sensing; distributed acoustic sensing	Gas production
	ANN + adaptive network-based fuzzy inference system	Calliper; porosity; gamma-ray; density; neutron; three separate resistivities; gamma-ray; density; neutron	Water saturation
Shallow learning	Backpropagation	Diagenesis; deep; GR log; neutron log; density log; sonic log; deep resistivity log	Porosity; permeability
	Backpropagation	Number of open injection wells, newly opened production wells, and old wells with efficient treatment; remaining geological reserves; total number of production wells; monthly injection production ratio; kernel function; number of open injection wells, newly opened production wells, and old wells with efficient treatment	Monthly oil and liquid-producing capacity
	Backpropagation	Temperature, heat, superficial gas velocity, and superficial liquid velocity are all factors to consider	Liquid holdup

Moreover, the high dimensionality of data in geoscience often leads to misleading correlations and impractical clustering. ML's performance can suffer due to initial data ambiguity, necessitating various preprocessing techniques to identify essential features and enable the model to learn a system's behavior. Missing data and system instability can introduce biases into ML models, making it challenging to extract useful knowledge. Additionally, data diversity from multiple sources raises concerns about privacy, security, and ethical considerations. To address these challenges, hybrid modeling approaches that combine ML with physics-based methods and the adoption of transfer learning techniques using pre-trained models could offer promising solutions in geoscience [136, 137].

16 Machine Learning in Nuclear Engineering

Nuclear engineering encompasses a wide spectrum of disciplines critical to harnessing nuclear energy for various applications, from power generation to medical diagnostics and treatment, as well as industrial applications. In this field, ML holds significant promise and is being increasingly integrated into nuclear engineering processes. ML techniques are employed in tasks such as reactor design optimization, safety analysis, anomaly detection in nuclear power plants, and radiation detection and monitoring. These applications of ML in nuclear engineering aim to enhance efficiency, safety, and reliability by leveraging vast amounts of data generated from nuclear systems. ML algorithms aid in predictive maintenance, fault detection, and optimization of reactor operations, contributing to improved performance and cost-effectiveness while ensuring safety protocols are maintained. The fusion of ML with nuclear engineering enables the field to adapt to complex operational challenges, paving the way for more innovative and efficient utilization of nuclear technologies.

ML plays a crucial role in nuclear engineering, particularly in understanding radiation effects on materials and optimizing scintillator performance for radiation detection systems. Modeling radiation-induced changes in materials due to high-energy radiation, like neutrons and ions, is complex, making ML applications valuable. When combined with physical models, ML can predict irradiation effects based on fundamental materials, properties, and conditions, reducing reliance on costly reactor experiments. Despite data availability challenges, ML models show promise in predicting void formation, thermal conductivity, and other post-irradiation material behaviors, highlighting the need for diverse datasets in nuclear engineering applications [138–140].

Additionally, fault detection and diagnosis (FDD) techniques in the nuclear power industry are evolving. Data-driven approaches, especially when integrated with ML algorithms like NNs, SVMs, and ensemble learning, enhance safety and operational efficiency in nuclear power plants (NPPs). However, there's a need for comprehensive research on the current state and future trends of data-driven ML methods in FDD for NPPs to bridge existing gaps and fully utilize ML's potential in this crucial area [141–144]. The latest applications of deep learning and shallow learning for the FDD of the NPP component and condition monitoring were reported in [145].

17 Machine Learning in Robotics

Robotics, a multidisciplinary field intersecting mechanical engineering, computer science, and AI, focuses on designing, constructing, and operating robots to perform various tasks autonomously or semi-autonomously. ML has significantly transformed robotics by enabling robots to adapt, learn from data, and make decisions in complex and uncertain environments. In robotics, ML techniques, such as reinforcement learning, deep learning, and supervised learning, have been instrumental in enhancing robot perception, manipulation, navigation, and decision-making capabilities. These methods allow robots to learn from experience, recognize patterns, and improve their performance over time, enabling them to handle diverse tasks efficiently. For instance, in object recognition, ML facilitates the identification of objects in varying environments, while in motion planning, it assists robots in navigating cluttered spaces. Moreover, ML empowers robots to learn from human demonstrations, enabling safer human-robot interactions and more intuitive control interfaces. The integration of ML in robotics continues to push the boundaries of what robots can achieve, making them more adaptable, intelligent, and versatile in real-world applications.

AI robots can be categorized based on their intelligence types, ownership, or the tasks they execute. Generally, they fall into three main intelligence groups: artificial narrow intelligence (ANI), artificial general intelligence (AGI), and artificial superintelligence (ASI) [146]. Among these, ASI represents the highest level of AI capability, surpassing even the best human minds across various domains like scientific creativity, social skills, and overall knowledge. While higher-level AI robots possess increased interactivity with users, ASI has not yet been integrated into the tourism and hospitality industry. To assess the impact of robot types on consumer acceptance of current technology, it's crucial to narrow down the types of AI robots used [147].

ML integration in robotics extends across diverse medical applications, notably in surgical procedures. ML algorithms play a vital role in surgical candidate selection and treatment strategies, utilizing medical imaging data to differentiate conditions like fat-poor angiomyolipomas and renal cell carcinomas accurately. These models assist in informed decision-making for patients and physicians regarding treatment plans, post-surgical outcomes prediction, and risk assessment in surgeries like radical cystectomy. In robotic-assisted surgeries (RAS), ML enhances capabilities by automating repetitive tasks like camera positioning and tissue retraction, granting surgeons improved precision and visualization amid diverse anatomical variations among patients [148–152].

Autonomous surgeries benefit significantly from ML algorithms for tasks spanning camera positioning, suturing, tissue dissection, and knot-tying. Leveraging instrument kinematics, laparoscopic video, and surgeon eye-tracking data allows autonomous camera positioning, enhancing procedural visual flow and reducing visual occlusion in the operative field. ML algorithms address accuracy challenges.

In traditional laparoscopic surgeries by adjusting the laparoscopic camera's orientation in real time based on surgical video feedback. Additionally, ML aids in automating intricate surgical tasks like tissue manipulation and trajectory planning through reinforcement learning and computer vision models, although live surgical setting validation is essential for these innovations [153–158]. Figure 6 illustrates the classification of sophisticated robots utilized in the tourism and hospitality sectors, incorporating ML applications.

Beyond medicine, ML in robotics extends to the hospitality and tourism industry. Robots, serving roles like communication, chef, delivery, entertainment, housekeeping, guide, and security, employ ML for functions such as voice and face recognition, object detection, and self-mobility. ML significantly enhances the usability and consumer acceptance of these robots, influencing their effectiveness in human interaction and impacting consumer perceptions and willingness to adopt these technologies [159].

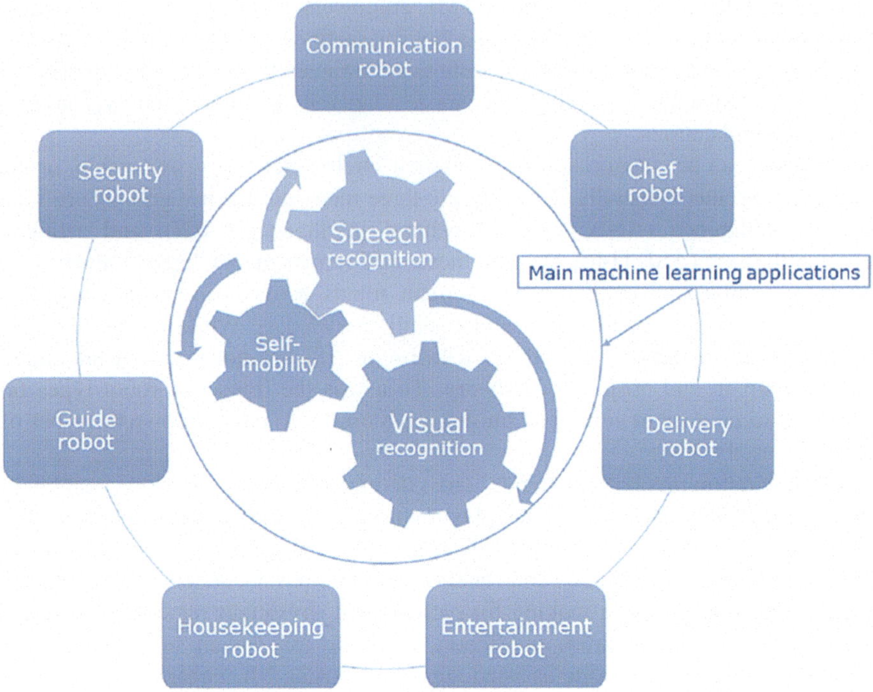

Fig. 6 Typology of advanced robots in tourism and hospitality with ML applications [147]

17.1 Deep Learning in Advanced Robotics

DL plays an essential role in solving complex problems in robotics that traditional ML techniques struggle with, especially in areas like image and speech recognition. The integration of DL, ML, and AI has enabled the development of advanced robotic systems capable of tackling tasks previously deemed challenging. Several applications showcase the synergy between AI, ML, and DL in robotics. For instance, object detection and recognition, facilitated by training NNs on labeled data, empower robots to accurately identify and classify objects in their surroundings. Predictive maintenance, using AI and ML, allows robots to anticipate potential component failures through sensor data analysis, enabling proactive repairs. Gesture and speech recognition, exemplified by robots like Pepper, enhance human-robot interaction in contexts such as customer service or healthcare. In robotic surgery, advanced algorithms leveraging AI, ML, and DL assist human surgeons during intricate procedures, improving precision and outcomes. The medical field benefits from DL techniques in analyzing medical images for disease identification. Military robotics utilize AI and ML algorithms for reconnaissance, surveillance, and decision-making in operations. Agriculture sees the development of autonomous robots using AI and ML for tasks like planting, harvesting, and crop management.

Service robotics, including cleaning and food delivery, leverage AI and ML for human interaction and understanding needs. Autonomous driving relies on AI and ML to navigate roads and adapt to diverse driving conditions.

Robotics in manufacturing employs AI and ML for automation, optimizing tasks like assembly and welding. Each application underscores the significance of AI, ML, and DL in shaping the capabilities of advanced robotic systems. Additionally, specific performance data for these methods in advanced robotics is discussed, covering areas such as object recognition, motion planning, control, and localization, showcasing the effectiveness of techniques like CNN, reinforcement learning (RL), deep deterministic policy gradient, proximal policy optimization, SVMs, and Random Forests [160].

17.2 Advantages and Challenges of AI in Robotics Applications

AI, ML, and DL have ushered in significant advancements in robotics. Their application offers several benefits in advanced robotics, including automation, freeing human resources from repetitive tasks to focus on more complex ones. These technologies elevate accuracy and precision in robotic systems, reducing errors and enhancing overall performance. Adaptability is a key feature, enabling AI-powered robots to flexibly handle diverse environments and tasks, enhancing their utility across various industries [161, 162]. Predictive maintenance, driven by ML, allows robots to anticipate maintenance needs, thereby reducing downtime and costs [163]. Moreover, AI and ML algorithms enable robots to analyze vast datasets for improved decision-making and appropriate actions [164]. Optimization through these technologies not only boosts efficiency but also curtails waste, resulting in cost savings and heightened productivity. Emphasizing better decision-making and enhanced adaptability, AI, ML, and DL empower robots to excel across a wider array of tasks and environments, thus bolstering overall performance and safety [165, 166]. Furthermore, their implementation significantly reduces labor and maintenance costs [167]. This integration of AI and ML in robotics holds the promise of revolutionizing the field, unlocking new levels of performance, efficiency, and safety.

The integration of AI, ML, and DL in robotics presents numerous advantages, yet it comes with considerable hurdles. Foremost among these challenges is the demand for substantial amounts of high-quality data to train these algorithms, which can be expensive and time intensive. Obtaining clean and unbiased data is particularly complex in robotics due to the difficulty in data collection, compounded by noise and uncertainty in real-world robotic applications [161]. Moreover, the real-time processing requirement in robotics demands substantial computational power, often limited by the energy and computing constraints of robots. Adaptability in dynamic environments adds another layer of complexity, as AI/ML/DL models need to dynamically learn from experience, posing a significant challenge. Safety remains a critical concern, requiring robust algorithms to prevent accidents and enable robots to detect and respond to potential hazards. Additionally, ethical considerations surrounding job displacement, biases in AI systems, and potential misuse in military or surveillance applications pose societal challenges. In summary, while these technologies offer immense potential, addressing these challenges necessitates the development of resilient algorithms, sophisticated hardware, and ethical frameworks to ensure their safe and effective deployment in robotics [160].

18 Machine Learning in Agricultural Engineering

In contemporary agricultural engineering, the integration of ML techniques, including both DL and shallow learning methods, has ushered in a paradigm shift, revolutionizing various facets of agricultural practices and systems. ML applications in this domain encompass a wide spectrum, from crop management to resource optimization, pest control, yield prediction, and precision agriculture.

Shallow learning algorithms, like SVMs and DT, have been extensively employed in agricultural engineering for diverse tasks. These methods excel in classification problems, aiding in the identification and classification of crops, diseases, and pests. For instance, SVM models have been essential in crop disease detection through image analysis, enabling early identification of infected plants based on leaf patterns or color variations. DT has facilitated decision-making in crop management by predicting yield based on soil types, climate data, and agricultural practices. Additionally, these algorithms have been instrumental in weed detection and classification, crucial for targeted herbicide application, minimizing environmental impact and optimizing crop health [168, 169].

Conversely, deep learning techniques, such as CNN and RNNs, have unlocked new dimensions in agricultural engineering due to their ability to process complex and voluminous data. CNNs have significantly enhanced image-based analysis in agriculture, enabling precise identification of diseases, nutrient deficiencies, or invasive species in crops by learning intricate patterns from high-resolution images. They have also been pivotal in plant phenotyping, accurately assessing plant growth, health, and stress factors. RNNs, with their sequential data processing capabilities, have been harnessed in climate modeling and prediction, aiding farmers in making informed decisions related to planting, irrigation, and harvesting schedules based on long-term weather patterns. Furthermore, the fusion of these ML methodologies has driven the development of smart agricultural systems, advancing precision agriculture. Integration of sensors, drones, and IoT devices coupled with ML algorithms allows real-time monitoring of crop conditions, soil moisture levels, and environmental parameters. This amalgamation facilitates the creation of predictive models for crop yield estimation, optimizing resource usage and fostering sustainable agricultural practices. Despite these advancements, challenges persist. Acquiring labeled data for training ML models, ensuring robustness in diverse environmental conditions, and deploying complex algorithms on resource-constrained devices in rural areas are among the prominent hurdles. Moreover, ethical considerations related to data privacy, equitable access to technology, and the societal impact of automation in agriculture demand thoughtful deliberation and responsible implementation.

ML methods, especially deep learning, and NN models have significantly transformed plant disease diagnosis and management in agricultural engineering. For instance, in [169], CNN models, trained on a vast database of over 87,000 plant images, achieved an exceptional accuracy rate of 99.53% in identifying specific plant diseases or healthy plants. Such accuracy positions these models as valuable tools for disease management's early warning systems, potentially applicable in real cultivation scenarios. Other studies, like [170], delve into deep learning's application in agricultural disease diagnosis and management, aiming for a comprehensive assessment of tool development. These studies address critical aspects such as dataset requirements, imaging sensors, deep learning methods, disease severity estimation, and ongoing research directions. The primary objective is to bridge research gaps, enhance diagnostic tools using deep learning, and offer guidance to farmers for more effective plant disease management through technological advancements.

19 Machine Learning in Electrical Engineering

In electrical engineering, the integration of ML marks a substantial departure from conventional approaches, bringing advancements across numerous aspects of power systems. By handling extensive data, discerning intricate patterns, and providing predictive insights, ML reshapes practices in electrical engineering. Its applications range widely, from enhancing grid reliability and efficiency to enabling predictive maintenance and driving smart grid technologies. For instance, research like [171] concentrates on predicting voltage stability margin in power grids. They utilize an online probabilistic extreme learning machine (ELM) algorithm that adapts to dynamic conditions, especially beneficial for grids with high renewable energy generation. Another study, [172], employs synchrophasor measurements for real-time event identification in power systems using ML models, demonstrating promising results in detecting line trips and generation loss events. Additionally, the research highlighted in [173] showcases how ANN, ML, and DL contribute to accurate forecasts of renewable energy efficiency in microgrid forecasting, aiming to fortify network stability and reliability. These techniques empower engineers to make informed decisions for resilient and sustainable microgrid ecosystems.

Moreover, ML addresses critical challenges such as DC arc fault detection in power systems [174], control strategies in multi-active-bridge converters [175], and modeling electrochemical mechanisms in lithium-ion batteries [176]. These applications range from ensemble ML for fault detection to using feedforward NNs for power flow control without precise circuit models. ML also offers solutions for nonminimum phase systems and robotic tasks like human-robot interactions or handling delicate objects [177, 178]. Additionally, studies like [179] focus on variability-aware machine-learning approaches to predict electrical parameters in flash memories, highlighting the effectiveness of machine-learning algorithms in capturing variability effects across structural conditions. Overall, ML in electrical engineering spans diverse areas, propelling progress and offering innovative solutions to pressing challenges. Its integration fosters enhanced efficiency, predictive capabilities, and adaptive control, promising transformative impacts within the field.

In control engineering, ML serves as a substitute for traditional approaches, replacing the need for physical modeling of dynamic behavior and subject-specific control design methods. Using deep and shallow learning methods in modeling is the branch that is chosen to be described more deeply in this chapter [180].

Modern industrial processes often operate under different modes, which brings challenges to model predictive control (MPC). Recently, most MPC-related methods would establish prediction models independently for different modes, which results in their control effect highly relying on switching strategies. In order to deal with the multimode control problem, a DL-based MPC method, LSTM-MPC, is applied by researchers in this matter [181].

It seems like there is a snippet related to intrusion detection in industrial control systems, particularly mentioning the challenge of detecting abnormal activities within control layers. [182] introduces the concept of a measurement intrusion detection system (MIDS) as a solution to identify anomalies in the system, even if attackers attempt to mimic normal behavior to deceive network-based intrusion detection systems. MIDS serves as an additional layer of security within industrial control systems, specifically aiming to detect and flag any unusual or malicious activities occurring within the control layer. This technology represents an advancement in control engineering, offering a more robust and resilient approach to identifying and addressing potential attacks or abnormalities in industrial settings [182].

The applicability of genetic algorithms (GAs) in control engineering, highlighting their effectiveness in addressing challenges posed by nonlinear systems and uncertainties, is another application of ML methods. GAs offer a powerful tool for control engineers, as they do not necessitate stringent prerequisites beyond a fitness measure, making them advantageous for handling complex problems that traditional methods struggle with. The flexibility of GAs in accommodating various problem types without imposing limitations underscores their potential to overcome control challenges where conventional techniques fall short. However, the resource-intensive nature of GAs suggests that their application might demand significant computational resources. Overall, the recommendation is for control engineers to consider utilizing GAs when confronted with control problems that conventional methods struggle to address effectively, given the willingness to accommodate the computational demands of GAs [183].

20 Shallow Learning vs. Deep Learning in Modeling of PV Panels: An Example

The pressing need for sustainable energy production has underscored the significance of renewable sources like photovoltaic (PV) technology in future energy frameworks. Ensuring optimal utilization of solar energy, especially achieving the maximum power point in PV panels, is crucial given the substantial investment and low efficiency. Incorporating ML to model solar cells/modules has become pivotal to ensure their overall performance and yield optimal outputs under varying environmental conditions [184]. Predicting the overall power of PV panels relies on their modeling, typically using mathematical equations and technical specifications. While these models provide reasonable power predictions under realistic conditions, greater accuracy is attained through real-world measurements and testing [185]. However, performance prediction extends beyond power output, encompassing a wider scope of PV panel behavior. It serves as a primary application of ML in solar cell setups, evaluating efficiency and considering factors like voltage, current, and various key points on the I-V (current-voltage) characteristic curve.

The chosen PV module utilized in this investigation, sourced from the Oregon/Eugene dataset [186], is the xSi12922 single-crystalline silicon PV module provided by NREL. Table 3 details the specifications of this particular panel.

The I-V curve of a solar panel delineates the correlation between the panel's current and voltage outputs under varying levels of illumination and diverse load conditions. This curve signifies an increase in the panel's output current as the output voltage rises until it attains the maximum power point (MPP). At the MPP, the panel optimally delivers the highest power output for a specific irradiance level and load [187]. Figure 7 graphically portrays the I-V curve alongside the corresponding output power for a PV module.

20.1 PV Modeling with Polynomial Regression

To model and estimate critical parameters like open-circuit voltage, short-circuit current, maximum power point voltage, and maximum power point current within the current-voltage curves of PV panels across diverse temperature and irradiance settings, a second-degree polynomial regression (PR) is employed as a shallow learning method for currents (Isc, I_{MPP}). For voltages (Voc, V_{MPP}), a third-degree PR is utilized. The general formulation of the regression equations is depicted as follows:

Table 3 NREL xSi12922 panel specification

Date	Irr (W/m^2)	Temp (°C)	Voc (V)	Isc (A)	Vmp (V)	Imp (A)	Pmp (W)
1/31/2014	1000	24.9	22.06	5.127	17.58	4.724	83.04

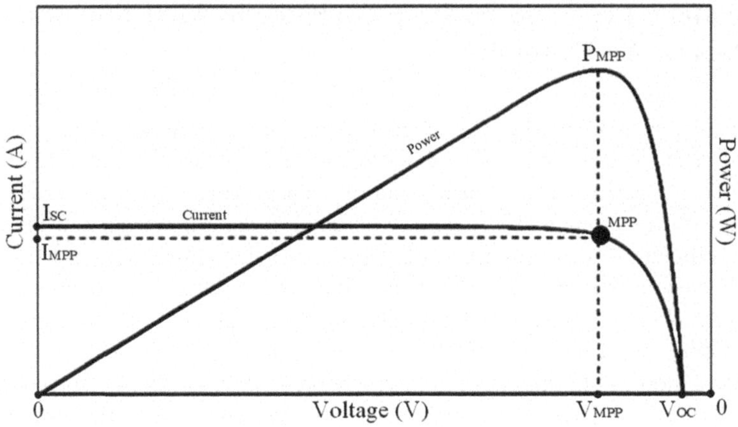

Fig. 7 Characteristic curves: I-V and P-V curves of PV modules

$$I = d_0 + d_1 J + d_2 T + d_3 JT + d_4 J^2 + d_5 T^2 \quad (1)$$

$$V = d_0 + d_1 J + d_2 T + d_3 JT + d_4 J^2 + d_5 T^2 + d_6 J^3 + d_7 T^3 + d_8 T^2 J + d_9 J^2 T \quad (2)$$

where *J*, *T*, and *R* represent solar radiation, panel temperature, and load, respectively. However, d_0, d_1, ..., and d_{11} are coefficients of the polynomial terms, and *V* and *I* represent the voltage and the current, respectively.

In an empirical examination using authentic data sourced from NREL, an array of irradiance and temperature values was evaluated spanning from December 2012 to January 2014 in Oregon/Eugene, set at a tilt of 44°. The derived outcomes from the PR model were juxtaposed with actual data points, showcased in Figs. 8, 9, 10 and 11.

For gauging the model's precision, the root mean square error (RMSE) (3) was computed between the forecasted and observed values across distinct parameters: 0.0762 for short-circuit current, 0.2201 for open-circuit voltage, 0.0736 for current at the maximum power point, 0.2787 for voltage at the maximum power point, and 1.4382 for the maximum power point. These RMSE values were derived from 8637 data points spanning various temperatures and irradiance levels using the PR model (PRM).

$$\text{RMSE} = \sqrt{\frac{1}{n}\sum_{i=1}^{n} \left(X_i - X_{pi}\right)^2} \quad (3)$$

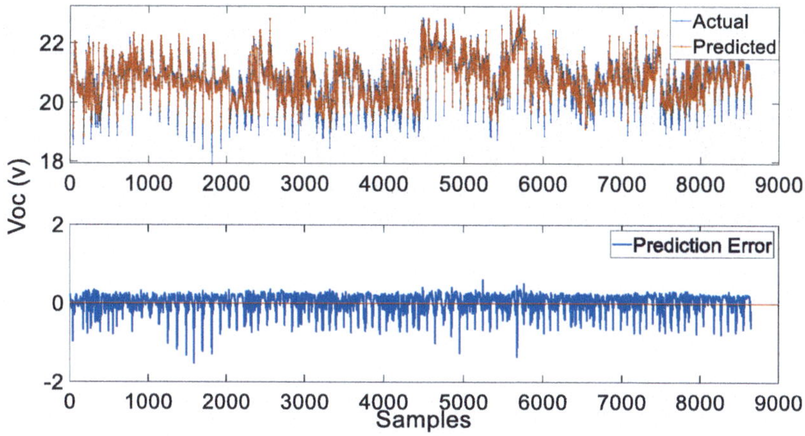

Fig. 8 Real and predicted open-circuit voltage by the PRM with test dataset by NREL

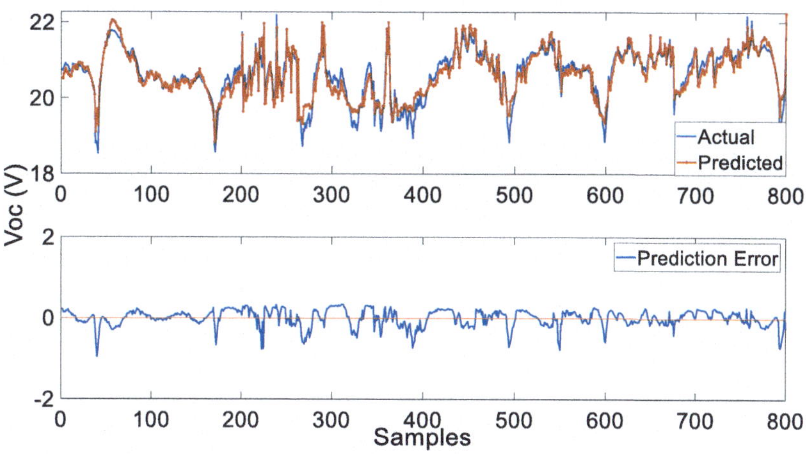

Fig. 9 A magnified version of Fig. 8

20.2 PV Modeling with NNs

The structure and functioning of ANNs involve three essential layers: input, hidden, and output, with each neuron having weighted input connections, an activation function, and an output connection [188]. Neurons (Fig. 12a) in the input layer receive inputs that are multiplied by weights and then summed, undergoing an activation function to produce an output. This process continues through hidden layers until reaching the output layer, making ANNs effective for both linear and nonlinear problems [189]. A multilayer perceptron (Fig. 12b), a common type of ANN, includes input, hidden, and output layers, with careful consideration needed to avoid overfitting [190]. The NN's complexity, determined by hidden layers and

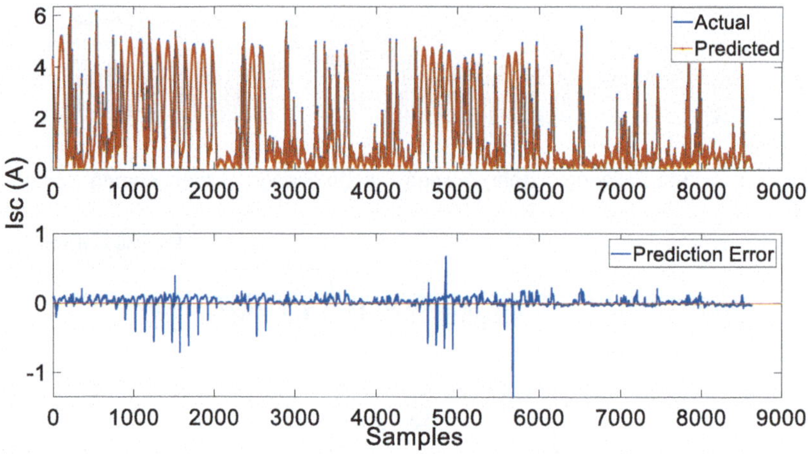

Fig. 10 Real and predicted short-circuit current by the PRM with test dataset by NREL

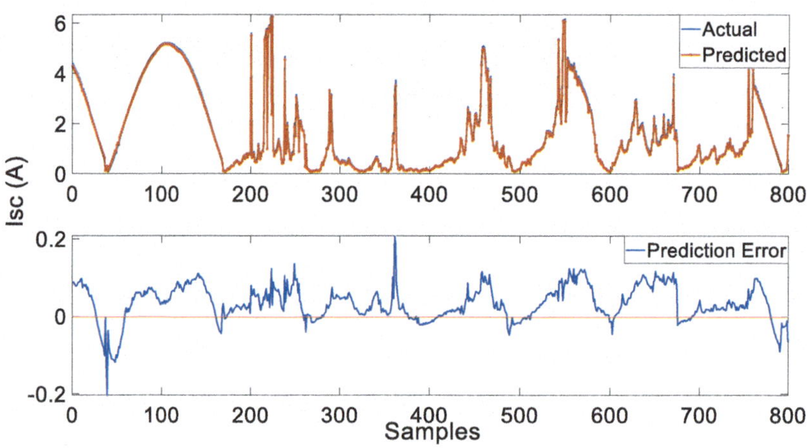

Fig. 11 A magnified version of Fig. 10

weight dimensions, is crucial for performance, and researchers employ various methods to optimize topology [188].

For modeling PVs and obtaining voltage and current, a MATLAB/Simulink-simulated dataset is used, allocating 70% for training, 10% for validation, and 20% for testing. In this model, two hidden layers are considered for the current model and five for the voltage network.

A standard mapping function for input x and output y with d input and m hidden layer is defined as:

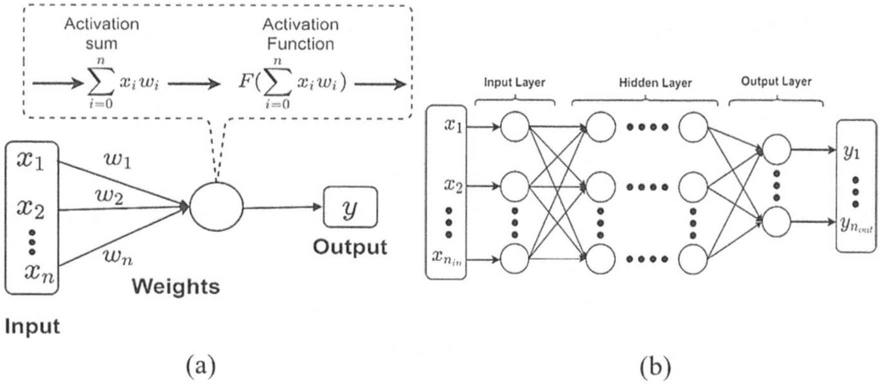

Fig. 12 (**a**) The structure of a neuron, (**b**) the structure of a multilayer perceptron NN [189]

$$y = y(x; w) = a_0 + \sum_{j=1}^{m} (w_j f) \sum_{i=1}^{d} w_{ij} x_i \tag{4}$$

Gradient descent is an iterative optimization algorithm widely employed in ML to find the best result (usually the minimum of a curve). In essence, the gradient corresponds to the slope or descent of a curve, and descent signifies a reduction or decrease. Since gradient descent is an iterative algorithm, it must be run multiple times to achieve the optimal result. The quality of the iterations in gradient descent can serve as a metric to help the system optimally fit the curve to the data. This algorithm also features a parameter known as the learning rate [191].

In ML, due to the large volume of data, it is not feasible to transfer all data to the system at once. Therefore, to tackle this issue, data is divided into smaller portions and gradually fed into the system. The weights of NNs are updated at the end of each stage to synchronize with input data. When the entire dataset is transferred forward and backward through a NN, it is called an epoch. Since an epoch is often too large to be handled at once, it is further divided into smaller subsets called batches. Transmitting the entire dataset through a NN alone is insufficient, and it needs to be done multiple times. However, due to the utilization of a limited dataset for learning optimization, gradient descent is employed. Therefore, updating the weights with just one epoch is not enough [191]. In this method, the specified number of epochs for training the current and voltage NN is 2000 and 200, respectively, and the structure is shown in Fig. 13.

The "Levenberg-Marquardt Backpropagation Algorithm" is a popular optimization method often used in training ANNs, especially for solving nonlinear regression problems and curve fitting. In this research, this algorithm has also been utilized for training the NN. The Levenberg-Marquardt algorithm is an extension of the traditional backpropagation algorithm, commonly used for NN training. While this method uses gradient descent to update the network weights, the

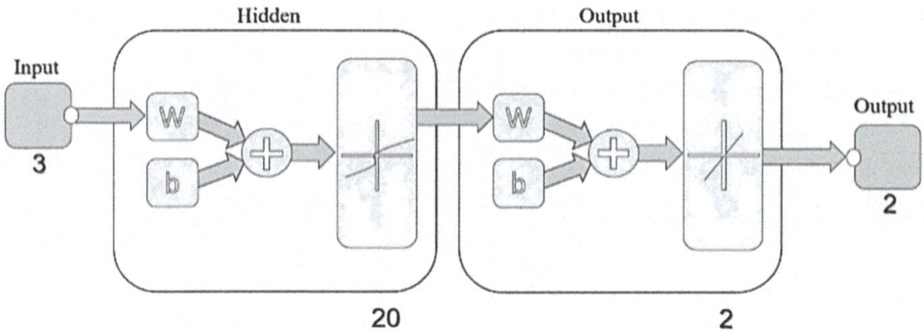

Fig. 13 NN structure for the panel current and voltage modeling

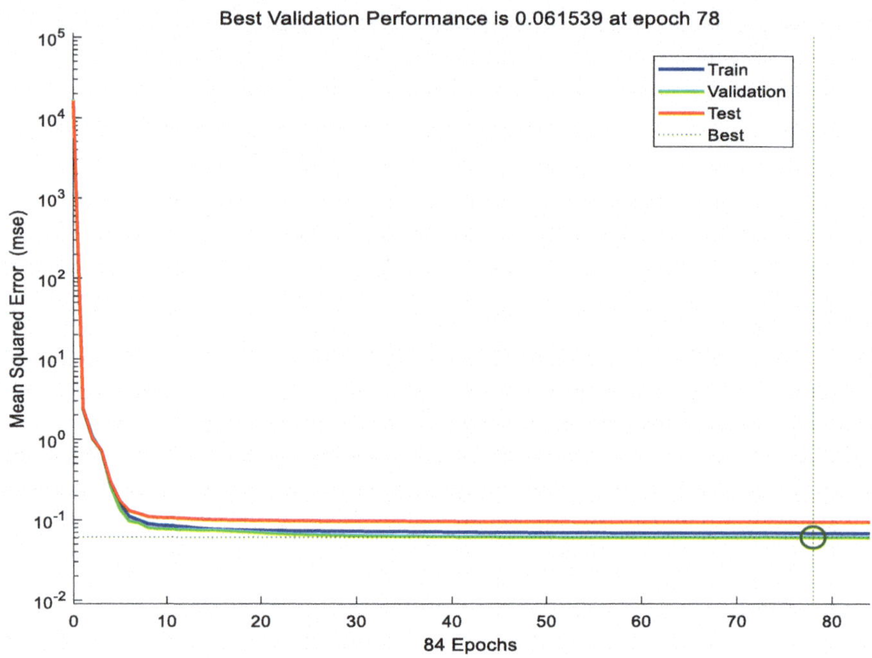

Fig. 14 Performance validation plot for the NN model with the best performance of 0.061539 in 78 epochs for key-point modeling of PV panel

Levenberg-Marquardt algorithm combines the advantages of both gradient descent and Gauss-Newton methods to enhance convergence speed and stability.

As the NN progresses through the training epochs, the mean squared error (MSE) is calculated at each iteration and plotted on the performance chart, as shown in Fig. 14. In this figure, the green circular point indicates where the model exhibits the highest level of generalization while avoiding overfitting the training data. As depicted in this figure, the best validation performance with an MSE of approximately 0.061539 was achieved at epoch 78.

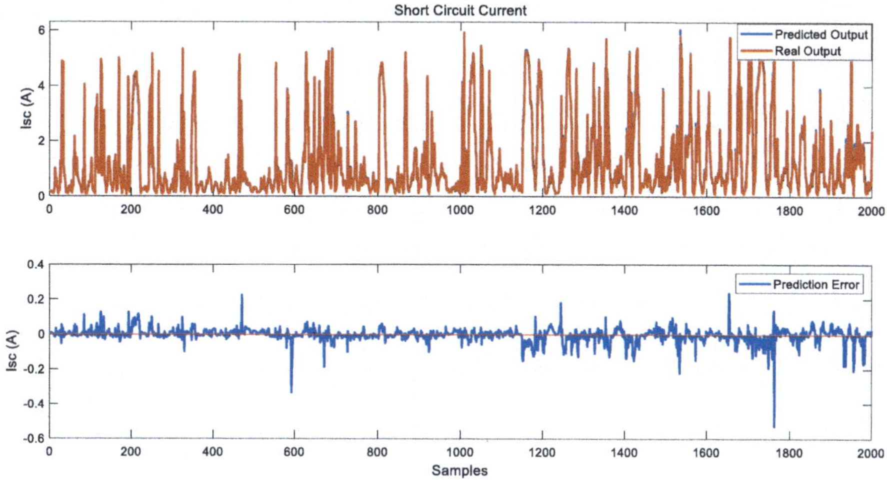

Fig. 15 Actual and predicted short circuit current value (top figure) and prediction error (bottom figure) by NN model

The model's performance evaluation indicates exceptional results across different datasets. The R-squared value for the training dataset is 0.9998, highlighting a robust fit and effective capture of fundamental patterns. The R-squared value for the validation dataset is 0.99982, demonstrating strong generalization capabilities on unfamiliar data. This underscores the model's accuracy in making predictions on new data, affirming its reliability. Furthermore, the R-squared value for the testing dataset, an independent set not utilized in training or validation, is 0.99973. This result confirms the model's proficiency in performing well on entirely unseen data, emphasizing its avoidance of overfitting to the training or validation datasets [192].

After modeling the key points of the panel, the results are obtained from the model and real values with the amount of error in Figs. 15, 16, 17, 18, 19 and 20.

Table 4 provides a comparison between the error of modeling of Isc and Voc with both PR and NN. Due to the range of errors for each parameter, it is observed that the error with NN modeling is less than the PRM.

21 Conclusion

In this chapter, the application of ML in various engineering domains has ushered in transformative advancements, offering solutions to intricate problems and paving the way for unprecedented innovation. Shallow learning methods, characterized by their simplicity, ease of implementation, and interpretability, have demonstrated efficacy in scenarios where datasets are limited, computational resources are constrained, and relationships between variables are relatively straightforward. These methods provide a valuable toolkit for engineers seeking practical solutions with comprehensible

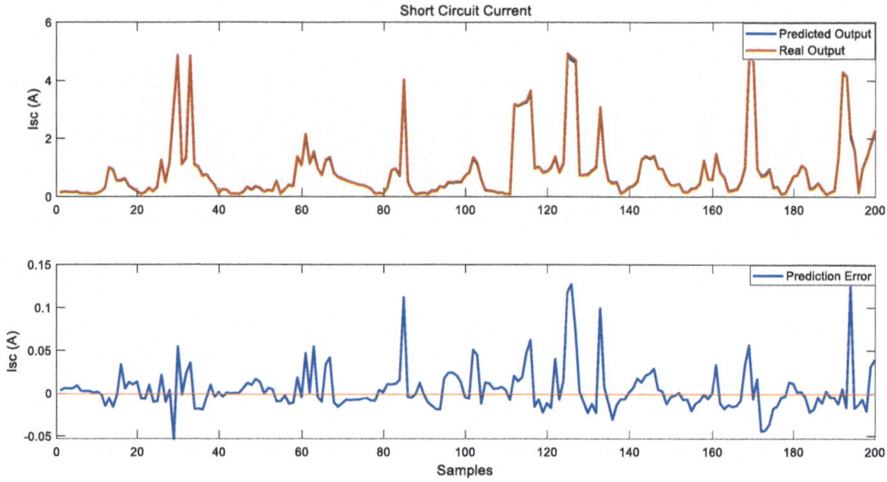

Fig. 16 Magnified diagram Fig. 15

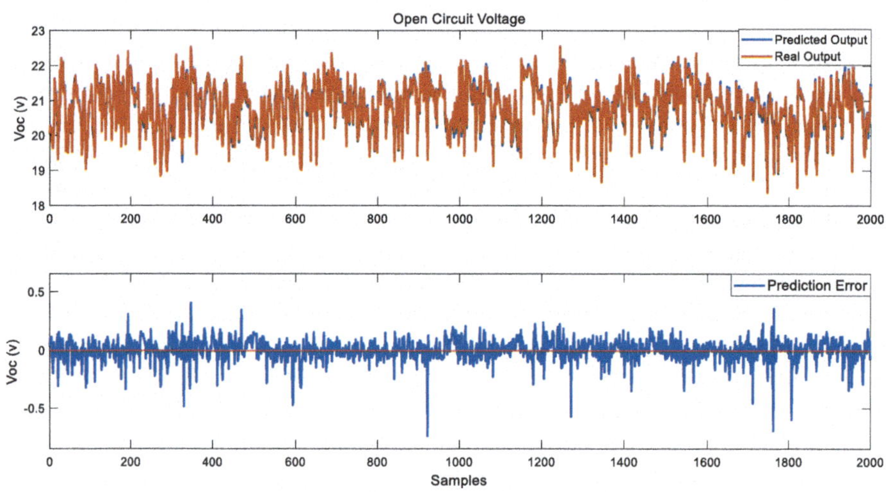

Fig. 17 Actual and predicted open-circuit voltage value (top figure) and prediction error (bottom figure) by NN model

insights. Their ability to perform well with smaller datasets makes them particularly suitable for applications in domains where data collection may be challenging or expensive.

On the other hand, the advent of deep learning has significantly broadened the horizons of ML applications, enabling the extraction of intricate patterns and representations from vast and complex datasets. Deep learning excels in scenarios where the underlying relationships between variables are highly intricate, offering unparalleled accuracy and predictive capabilities. However, it comes with the cost of

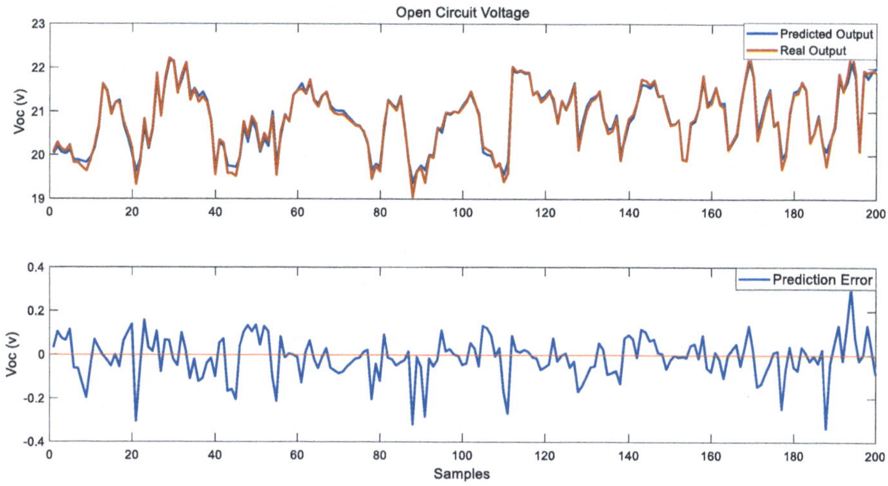

Fig. 18 Magnified diagram Fig. 17

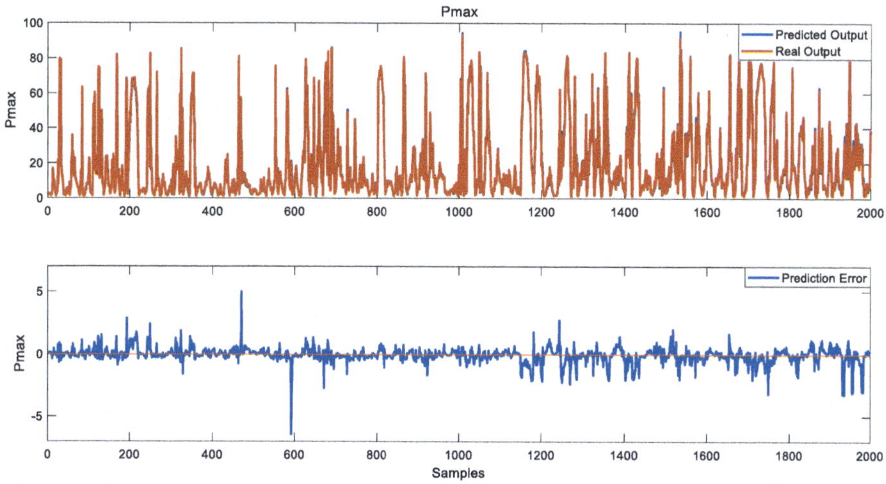

Fig. 19 Actual and predicted maximum power value (top figure) and prediction error (bottom figure) by NN model

increased computational demands, a need for substantial labeled data, and a level of complexity that may limit interpretability. As engineering applications continue to evolve, the choice between shallow and deep learning methodologies hinges on the specific requirements, constraints, and nature of the problem at hand. In navigating the landscape of engineering applications, practitioners must weigh the advantages and disadvantages of each approach to harness the full potential of machine learning for innovative solutions.

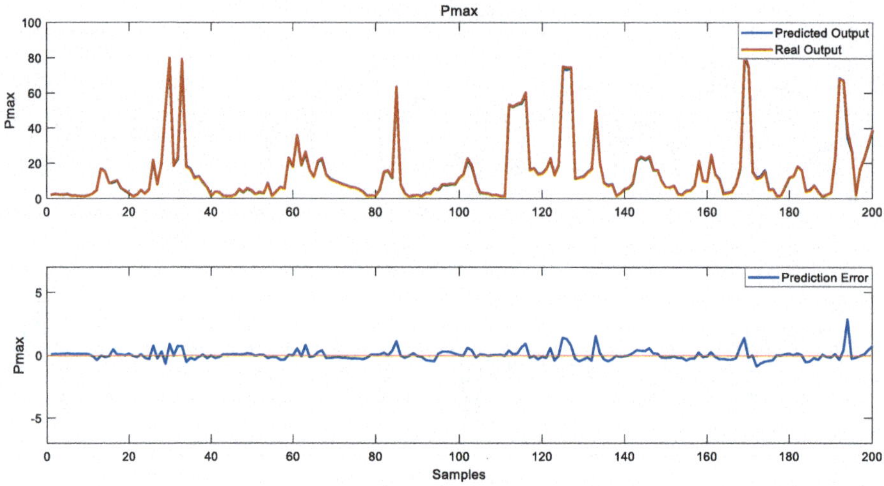

Fig. 20 Magnified diagram Fig. 19

Table 4 PRM vs. NN modeling of Isc and Voc of PV panels

Parameters	Approximate error	
	PRM	NN Model
Isc	From 0.75 to −1.5	From 0.2 to −1.5
Voc	From 0.5 to −1.5	From 0.5 to −0.75

References

1. Cheng, J., & Shen, T. (2023). A smart walker based on a hybrid motion model and ML method. *Mechatronics, 96*, 103069.
2. Vu, M. N., Beck, F., Schwegel, M., Hartl-Nesic, C., Nguyen, A., & Kugi, A. (2023). ML-based framework for optimally solving the analytical inverse kinematics for redundant manipulators. *Mechatronics, 91*, 102970.
3. Nikolić, V., Motamedi, S., Shamshirband, S., Petković, D., Ch, S., & Arif, M. (2016). Extreme learning machine approach for sensorless wind speed estimation. *Mechatronics, 34*, 78–83.
4. Murrell, N., Bradley, R., Bajaj, N., Whitney, J. G., & Chiu, G. T. C. (2018). A method for sensor reduction in a supervised ML classification system. *IEEE/ASME Transactions on Mechatronics, 24*(1), 197–206.
5. Liu, K., Hu, X., Meng, J., Guerrero, J. M., & Teodorescu, R. (2021). RUBoost-based ensemble ML for electrode quality classification in Li-ion battery manufacturing. *IEEE/ASME Transactions on Mechatronics, 27*(5), 2474–2483.
6. Zhao, X., Jia, M., Ding, P., Yang, C., She, D., & Liu, Z. (2020). Intelligent fault diagnosis of multichannel motor–rotor system based on multimanifold deep extreme learning machine. *IEEE/ASME Transactions on Mechatronics, 25*(5), 2177–2187.
7. Banka, N., & Devasia, S. (2018). Application of iterative ML for output tracking with magnetic soft actuators. *IEEE/ASME Transactions on Mechatronics, 23*(5), 2186–2195.
8. Chang, L., & Lin, Y. H. (2022). Meta-learning with adaptive learning rates for few-shot fault diagnosis. *IEEE/ASME Transactions on Mechatronics, 27*(6), 5948–5958.

9. Liu, G., Wang, C., Jia, Z., Wang, K., Ma, W., & Li, Z. (2021). A rapid design and fabrication method for a capacitive accelerometer based on ML and 3D printing techniques. *IEEE Sensors Journal, 21*(16), 17695–17702.
10. Bauman, V. V., & Brandon, S. C. (2022). Gait phase detection in walking and stairs using ML. *Journal of Biomechanical Engineering, 144*(12), 121007.
11. Kim, T. H., Choi, A., Heo, H. M., Kim, K., Lee, K., & Mun, J. H. (2019). ML-based pre-impact fall detection model to discriminate various types of fall. *Journal of Biomechanical Engineering, 141*(8), 081010.
12. Rengarajan, B., Patnaik, S. S., & Finol, E. A. (2021). A predictive analysis of wall stress in abdominal aortic aneurysms using a neural network model. *Journal of Biomechanical Engineering, 143*(12), 121004.
13. Vakacherla, S. S., Kantharaju, P., Mevada, M., & Kim, M. (2023). Single accelerometer to recognize human activities using neural networks. *Journal of Biomechanical Engineering, 145*(6), 061005.
14. Viguerie, A., Grave, M., Barros, G. F., Lorenzo, G., Reali, A., & Coutinho, A. L. (2022). Data-driven simulation of Fisher–Kolmogorov tumor growth models using dynamic mode decomposition. *Journal of Biomechanical Engineering, 144*(12), 121001.
15. Fei, C. W., Han, Y. J., Wen, J. R., Li, C., Han, L., & Choy, Y. S. (2023). Deep learning-based modeling method for probabilistic LCF life prediction of turbine blisk. *Propulsion and Power Research, 13*, 12–25.
16. Li, D., Qiu, L., Tao, K., & Zhu, J. (2020). Artificial intelligence aided design of film cooling scheme on turbine guide vane. *Propulsion and Power Research, 9*(4), 344–354.
17. Tao, K., Zhu, J., Cheng, Z., & Li, D. (2022). Artificial neural network analysis of the Nusselt number and friction factor of hydrocarbon fuel under supercritical pressure. *Propulsion and Power Research, 11*(3), 325–336.
18. Jakeer, S., Rupa, M. L., Reddy, S. R. R., & Rashad, A. M. (2023). Artificial neural network model of non-Darcy MHD Sutterby hybrid nanofluid flow over a curved permeable surface: Solar energy applications. *Propulsion and Power Research, 12*(3), 410–427.
19. Guan, X., Heindel, J. P., Ko, T., Yang, C., & Head-Gordon, T. (2023). Using ML to go beyond potential energy surface benchmarking for chemical reactivity. *Nature Computational Science*, 1–10.
20. Li, Y., Wei, Y., Wang, Z., Liu, X., Colnaghi, T., Han, L., et al. (2023). Quantitative three-dimensional imaging of chemical short-range order via ML enhanced atom probe tomography. *Nature Communications, 14*(1), 7410.
21. Mok, D. H., Li, H., Zhang, G., Lee, C., Jiang, K., & Back, S. (2023). Data-driven discovery of electrocatalysts for CO2 reduction using active motifs-based ML. *Nature Communications, 14*(1), 7303.
22. Williams, T., McCullough, K., & Lauterbach, J. A. (2019). Enabling catalyst discovery through ML and high-throughput experimentation. *Chemistry of Materials, 32*(1), 157–165.
23. Evans, J. D., & Coudert, F. X. (2017). Predicting the mechanical properties of zeolite frameworks by ML. *Chemistry of Materials, 29*(18), 7833–7839.
24. Gundry, L., Kennedy, G., Bond, A. M., & Zhang, J. (2023). Establishing zone regions in cyclic voltammetry using unsupervised ML. *Journal of Electroanalytical Chemistry, 942*, 117551.
25. Zhu, S., Sun, X., Gao, X., Wang, J., Zhao, N., & Sha, J. (2019). Equivalent circuit model recognition of electrochemical impedance spectroscopy via ML. *Journal of Electroanalytical Chemistry, 855*, 113627.
26. Xu, Y., Verma, D., Sheridan, R. P., Liaw, A., Ma, J., Marshall, N. M., et al. (2020). Deep dive into ML models for protein engineering. *Journal of Chemical Information and Modeling, 60*(6), 2773–2790.
27. Heid, E., McGill, C. J., Vermeire, F. H., & Green, W. H. (2023). Characterizing uncertainty in ML for chemistry. *Journal of Chemical Information and Modeling, 63*(13).

28. Ozates, M. E., Karabulut, D., Salami, F., Wolf, S. I., & Arslan, Y. Z. (2023). ML-based prediction of joint moments based on kinematics in patients with cerebral palsy. *Journal of Biomechanics, 155*, 111668.
29. Burton, W. S., II, Myers, C. A., & Rullkoetter, P. J. (2021). ML for rapid estimation of lower extremity muscle and joint loading during activities of daily living. *Journal of Biomechanics, 123*, 110439.
30. Said, S., Yang, Z., Clauser, P., Ruiter, N. V., Baltzer, P. A. T., & Hopp, T. (2023). Estimation of the biomechanical mammographic deformation of the breast using ML models. *Clinical Biomechanics, 110*, 106117.
31. Li, Y., Mache, M. A., & Todd, T. A. (2020). Automated identification of postural control for children with autism spectrum disorder using a ML approach. *Journal of Biomechanics, 113*, 110073.
32. Halilaj, E., Rajagopal, A., Fiterau, M., Hicks, J. L., Hastie, T. J., & Delp, S. L. (2018). ML in human movement biomechanics: Best practices, common pitfalls, and new opportunities. *Journal of Biomechanics, 81*, 1–11.
33. Bao, Z., Bufton, J., Hickman, R. J., Aspuru-Guzik, A., Bannigan, P., & Allen, C. (2023). Revolutionizing drug formulation development: The increasing impact of ML. *Advanced Drug Delivery Reviews, 115108.*
34. Wang, N., Zhang, Y., Wang, W., Ye, Z., Chen, H., Hu, G., & Ouyang, D. (2023). How can ML and multiscale modeling benefit ocular drug development? *Advanced Drug Delivery Reviews*, 114772.
35. Chen, C., Yaari, Z., Apfelbaum, E., Grodzinski, P., Shamay, Y., & Heller, D. A. (2022). Merging data curation and ML to improve nanomedicines. *Advanced Drug Delivery Reviews, 183*, 114172.
36. He, S., Leanse, L. G., & Feng, Y. (2021). Artificial intelligence and ML assisted drug delivery for effective treatment of infectious diseases. *Advanced Drug Delivery Reviews, 178*, 113922.
37. Jariwala, N., Putta, C. L., Gatade, K., Umarji, M., Rahman, S. N. R., Pawde, D. M., et al. (2023). Intriguing of pharmaceutical product development processes with the help of artificial intelligence and deep/ML or artificial neural network. *Journal of Drug Delivery Science and Technology, 104751.*
38. Hussein, S., Kandel, P., Bolan, C. W., Wallace, M. B., & Bagci, U. (2019). Lung and pancreatic tumor characterization in the deep learning era: novel supervised and unsupervised learning approaches. *IEEE Transactions on Medical Imaging, 38*(8), 1777–1787.
39. Melki, L., Tourni, M., & Konofagou, E. E. (2021). Electromechanical wave imaging with ML for automated isochrone generation. *IEEE Transactions on Medical Imaging, 40*(9), 2258–2271.
40. Wei, L., Yang, Y., Nishikawa, R. M., & Jiang, Y. (2005). A study on several machine-learning methods for classification of malignant and benign clustered microcalcifications. *IEEE Transactions on Medical Imaging, 24*(3), 371–380.
41. Suri, M., & Parmar, V. (2015). Exploiting intrinsic variability of filamentary resistive memory for extreme learning machine architectures. *IEEE Transactions on Nanotechnology, 14*(6), 963–968.
42. Gómez, J. T., Kuestner, A., Simonjan, J., Unluturk, B. D., & Dressler, F. (2022). Nanosensor location estimation in the human circulatory system using ML. *IEEE Transactions on Nanotechnology, 21*, 663–673.
43. Kaplan, R., Yavits, L., & Ginosar, R. (2018). PRINS: Processing-in-storage acceleration of ML. *IEEE Transactions on Nanotechnology, 17*(5), 889–896.
44. Wang, Y., Yu, H., Ni, L., Huang, G. B., Yan, M., Weng, C., et al. (2015). An energy-efficient nonvolatile in-memory computing architecture for extreme learning machine by domain-wall nanowire devices. *IEEE Transactions on Nanotechnology, 14*(6), 998–1012.
45. Li, H., Wang, J., Xu, Q., Dai, L., Liao, W., Yuan, R., et al. (2023). High-strength medium-entropy alloy designed by precipitation-strengthening mechanism via ML. *Materials Science and Engineering: A, 882*, 145443.

46. Liu, W., Wang, C., Liang, C., Chen, J., Tan, H., Yang, J., et al. (2023). Optimal design of γ'-strengthened high-entropy alloys via ML multilayer structural model. *Materials Science and Engineering: A, 871*, 144852.
47. Dong, Z., You, K., Yuan, H., Wu, Y., Jia, Y., Wang, G., et al. (2022). ML assisted design of high-strength Sn-3.8 Ag-0.7 Cu alloys with the co-additions of Bi and In. *Materials Science and Engineering: A, 861*, 144257.
48. Zhang, S., Zhu, G., Fan, Y., Zeng, X., & Wang, L. (2023). A ML study of grain boundary damage in Mg alloy. *Materials Science and Engineering: A, 867*, 144721.
49. Luo, Y. W., Zhang, B., Feng, X., Song, Z. M., Qi, X. B., Li, C. P., et al. (2021). Pore-affected fatigue life scattering and prediction of additively manufactured Inconel 718: An investigation based on miniature specimen testing and ML approach. *Materials Science and Engineering: A, 802*, 140693.
50. Qiao, L., Ramanujan, R. V., & Zhu, J. (2022). ML discovery of a new cobalt free multi-principal-element alloy with excellent mechanical properties. *Materials Science and Engineering: A, 845*, 143198.
51. Wang, D., Yu, L., Li, X., Lu, Y., Niu, C., Fan, P., et al. (2024). Intelligent quantitative recognition of sulfide using ML-based ratiometric fluorescence probe of metal-organic framework UiO-66-NH2/Ppix. *Journal of Hazardous Materials, 464*, 132950.
52. Huang, J., Fan, G., Liu, C., & Zhou, D. (2023). Predicting soil available cadmium by ML based on soil properties. *Journal of Hazardous Materials, 460*, 132327.
53. Kida, M., Pochwat, K., & Ziembowicz, S. (2024). Assessment of ML-based methods predictive suitability for migration pollutants from microplastics degradation. *Journal of Hazardous Materials, 461*, 132565.
54. Kourehpaz, P., & Molina Hutt, C. (2022). ML for enhanced regional seismic risk assessments. *Journal of Structural Engineering, 148*(9), 04022126.
55. Chen, M., Mangalathu, S., & Jeon, J. S. (2022). ML–based seismic reliability assessment of bridge networks. *Journal of Structural Engineering, 148*(7), 06022002.
56. Feng, D. C., Wang, W. J., Mangalathu, S., & Taciroglu, E. (2021). Interpretable XGBoost-SHAP machine-learning model for shear strength prediction of squat RC walls. *Journal of Structural Engineering, 147*(11), 04021173.
57. Manzini, N., Orcesi, A., Thom, C., Brossault, M. A., Botton, S., Ortiz, M., & Dumoulin, J. (2022). ML models applied to a GNSS sensor network for automated bridge anomaly detection. *Journal of Structural Engineering, 148*(11), 04022171.
58. Solhmirzaei, R., Salehi, H., & Kodur, V. (2022). Predicting flexural capacity of ultrahigh-performance concrete beams: ML–based approach. *Journal of Structural Engineering, 148*(5), 04022031.
59. Althoff, D., Filgueiras, R., & Rodrigues, L. N. (2020). Estimating small reservoir evaporation using ML models for the Brazilian savannah. *Journal of Hydrologic Engineering, 25*(8), 05020019.
60. Yilmaz, A. G., & Muttil, N. (2014). Runoff estimation by ML methods and application to the Euphrates Basin in Turkey. *Journal of Hydrologic Engineering, 19*(5), 1015–1025.
61. Ersoy, M., Keskin, M. E., & Gürfidan, R. (2023). Rainfall forecasting with hybrid and ML models based on hyperparameter optimization. *Journal of Hydrologic Engineering, 28*(11), 04023031.
62. Bhattacharya, B., Price, R. K., & Solomatine, D. P. (2007). ML approach to modeling sediment transport. *Journal of Hydraulic Engineering, 133*(4), 440–450.
63. Pei, S., Liu, H., Zhu, Y., Zhang, C., Zhao, M., Fu, G., et al. (2021). Identifying flow patterns in water pipelines using complex network theory. *Journal of Hydraulic Engineering, 147*(6), 04021019.
64. Song, Y., Shen, C., & Liu, X. (2023). A surrogate model for shallow water equations solvers with deep learning. *Journal of Hydraulic Engineering, 149*(11), 04023045.
65. Froehlich, D. C. (2020). Neural network prediction of alluvial stream bedforms. *Journal of Hydraulic Engineering, 146*(12), 04020084.

66. Liu, T., Wang, Z., Long, Z., Zeng, J., Wang, J., & Zhang, J. (2022). Direct shear strength prediction for precast concrete joints using the ML method. *Journal of Bridge Engineering, 27*(5), 04022026.
67. Bedriñana, L. A., Sucasaca, J., Tovar, J., & Burton, H. (2023). Design-oriented machine-learning models for predicting the shear strength of prestressed concrete beams. *Journal of Bridge Engineering, 28*(4), 04023009.
68. Figueiredo, E., Moldovan, I., Santos, A., Campos, P., & Costa, J. C. (2019). Finite element–based machine-learning approach to detect damage in bridges under operational and environmental variations. *Journal of Bridge Engineering, 24*(7), 04019061.
69. Cao, R., Zuo, X., Agrawal, A. K., El-Tawil, S., & Wong, W. (2023). Evaluating the performance of protection beams subject to overheight vehicular impacts using analytical and ML–based methods. *Journal of Bridge Engineering, 28*(12), 04023091.
70. Guezzaz, A., Asimi, Y., Azrour, M., & Asimi, A. (2021). Mathematical validation of proposed ML classifier for heterogeneous traffic and anomaly detection. *Big Data Mining and Analytics, 4*(1), 18–24.
71. Tekouabou, S. C. K., Hartini, S., Rustam, Z., Silkan, H., & Agoujil, S. (2021). Improvement in automated diagnosis of soft tissues tumors using ML. *Big Data Mining and Analytics, 4*(1), 33–46.
72. Yu, N., Li, Z., & Yu, Z. (2018). Survey on encoding schemes for genomic data representation and feature learning – From signal processing to ML. *Big Data Mining and Analytics, 1*(3), 191–210.
73. Siemuri, A., Selvan, K., Kuusniemi, H., Valisuo, P., & Elmusrati, M. E. (2022). A systematic review of ML techniques for GNSS use cases. *IEEE Transactions on Aerospace and Electronic Systems, 99*, 1–42.
74. Linty, N., Farasin, A., Favenza, A., & Dovis, F. (2018). Detection of GNSS ionospheric scintillations based on ML decision tree. *IEEE Transactions on Aerospace and Electronic Systems, 55*(1), 303–317.
75. Ibrahim, S. K., Ahmed, A., Zeidan, M. A. E., & Ziedan, I. E. (2018). ML methods for spacecraft telemetry mining. *IEEE Transactions on Aerospace and Electronic Systems, 55*(4), 1816–1827.
76. Carvalho, O., Cueva, R. Y. D. L. C., & Barradas Filho, A. O. (2022). Nowcasting of amplitude ionospheric scintillation based on ML techniques. *IEEE Transactions on Aerospace and Electronic Systems, 58*(6), 4917–4927.
77. Qin, W., & Dovis, F. (2021). Situational awareness of chirp jamming threats to GNSS based on supervised ML. *IEEE Transactions on Aerospace and Electronic Systems, 58*(3), 1707–1720.
78. Herrmann, A., Stephenson, M. A., & Schaub, H. (2023). Single-agent reinforcement learning for scalable earth-observing satellite constellation operations. *Journal of Spacecraft and Rockets*, 1–19.
79. Johnson, K. B., Ferguson, D., & Nix, A. (2023). Machine-learning-based rotating detonation engine diagnostics: evaluation for application in experimental facilities. *Journal of Propulsion and Power*, 1–14.
80. Ispir, A. C., Saracoglu, B. H., Magin, T., & Coussement, A. (2023). A methodology for estimating hypersonic engine performance by coupling supersonic reactive flow simulations with ML techniques. *Aerospace Science and Technology, 140*, 108501.
81. Cao, R., Lu, Y., & He, Z. (2022). System identification method based on interpretable ML for unknown aircraft dynamics. *Aerospace Science and Technology, 126*, 107593.
82. Krummenacher, G., Ong, C. S., Koller, S., Kobayashi, S., & Buhmann, J. M. (2017). Wheel defect detection with ML. *IEEE Transactions on Intelligent Transportation Systems, 19*(4), 1176–1187.
83. Liu, T., Yang, Y., Huang, G. B., Yeo, Y. K., & Lin, Z. (2015). Driver distraction detection using semi-supervised ML. *IEEE Transactions on Intelligent Transportation Systems, 17*(4), 1108–1120.

84. Gupta, B. B., Gaurav, A., Marín, E. C., & Alhalabi, W. (2022). Novel graph-based ML technique to secure smart vehicles in intelligent transportation systems. *IEEE Transactions on Intelligent Transportation Systems, 99*, 1–9.
85. Mao, T., Mihăiţă, A. S., Chen, F., & Vu, H. L. (2021). Boosted genetic algorithm using ML for traffic control optimization. *IEEE Transactions on Intelligent Transportation Systems, 23*(7), 7112–7141.
86. Alghorani, Y., Chekkouri, A. S., Chekired, D. A., & Pierre, S. (2020). Improved S-AF and S-DF relaying schemes using ML based power allocation over cascaded Rayleigh fading channels. *IEEE Transactions on Intelligent Transportation Systems, 22*(12), 7508–7520.
87. Panda, J. P. (2023). ML for naval architecture, ocean and marine engineering. *Journal of Marine Science and Technology, 28*(1), 1–26.
88. Ray, T., Gokarn, R. P., & Sha, O. P. (1996). Neural network applications in naval architecture and marine engineering. *Artificial Intelligence in Enginering, 10*(3), 213–226.
89. Yonghui, S., Guoyuan Li, X., Skulstad, C. R., Jinshan, X., Honghai, L., & Houxiang, Z. (2019). An efficient neuralnetwork based approach to automatic ship docking. *Ocean Engineering, 191*, 106514.
90. Luo, W., Moreira, L., & Guedes Soares, C. (2014). Manoeuvring simulation of catamaran by using implicit models based on support vector machines. *Ocean Engineering, 82*, 150–159.
91. Huang, T., Yong, Y., & Helong, S. (2019). A model for vessel trajectory prediction based on long short-term memory neural network. *Journal of Marine Engineering & Technology*, 1–10.
92. Parkes, A. I., Sobey, A. J., & Hudson, D. A. (2018). Physics-based shaft power prediction for large merchant ships using neural networks. *Ocean Engineering, 166*, 92–104.
93. Christos, G., Iraklis, L., & Gerasimos, T. (2019). Machine learning models for predicting ship main engine fuel oil consumption: A comparative study. *Ocean Engineering, 188*, 106282.
94. Gao, M., & Shi, G.-Y. (2020). Ship collision avoidance anthropomorphic decision-making for structured learning based on ais with seq-cgan. *Ocean Engineering, 217*, 107922.
95. Lou, R., Lv, Z., Dang, S., Su, T., & Li, X. (2021). Application of ML in ocean data. *Multimedia Systems*, 1–10.
96. Chen, Y. S., Jiang, H. L., Li, C. Y., Jia, X. P., & Ghamisi, P. (2016). Deep feature extraction and classification of hyperspectral images based on convolutional neural networks. *IEEE Transactions on Geoscience and Remote Sensing, 54*, 6232–6251.
97. Chang, W., Liu, B., & Zhang, Q. (2019). *Oil slick extraction from hyperspectral images using a modified stacked auto-encoder network.* In Proceedings of the SPIE – The International Society for Optical Engineering, Guangzhou, China, 10–13 May 2019.
98. Hu, Y. B., Zhang, J., Ma, Y., An, J. B., Ren, G. B., Li, X. M., & Yang, J. G. (2019). Hyperspectral coastal wetland classification based on a multi-object convolutional neural network model and decision fusion. *IEEE Geoscience and Remote Sensing Letters, 16*, 1110–1114.
99. Hu, Y. B., Zhang, J., Ma, Y., An, J. B., Ren, G. B., Li, X. M., & Sun, Q. P. (2019). Deep learning classification of coastal wetland hyperspectral image combined spectra and texture features: A case study of Yellow River Estuary Wetland. *Acta Oceanologica Sinica, 38*, 142–150.
100. Zhu, X., Li, Y., Zhang, Q., & Liu, B. (2019). Oil film classification using deep learning-based hyperspectral remote sensing technology. *ISPRS International Journal of Geo-Information, 8*, 181.
101. Yang, J., Ma, Y., Hu, Y., Jiang, Z., Zhang, J., Wan, J., & Li, Z. (2022). Decision fusion of deep learning and shallow learning for marine oil spill detection. *Remote Sensing, 14*(3), 666.
102. Ryazanov, I., Nylund, A. T., Basu, D., Hassellöv, I. M., & Schliep, A. (2021). Deep learning for deep waters: An expert-in-the-loop ML framework for marine sciences. *Journal of Marine Science and Engineering, 9*(2), 169.
103. Davis, J., Edgar, T., Graybill, R., Korambath, P., Schott, B., Swink, D., & Wetzel, J. (2015). Smart manufacturing. *Annual Review of Chemical and Biomolecular Engineering, 6*, 141–160.

104. Lee, J., Lapira, E., Bagheri, B., & Kao, H. (2013). Recent advances and trends in predictive manufacturing systems in big data environment. *Manufacturing Letters, 1*, 38–41.
105. Lang, S. (2007). *Durchgängige Mitarbeiterinformation zur Steigerung von Effizienz und Prozesssicherheit in der Produktion* (Dissertation). Universitat Erlangen-Nurnberg, Bamberg: Meisenbach Verlag.
106. Wuest, T. (2015). *Identifying product and process state drivers in manufacturing systems using supervised machine learning* (Springer theses). Springer.
107. Wuest, T., Weimer, D., Irgens, C., & Thoben, K. D. (2016). ML in manufacturing: advantages, challenges, and applications. *Production & Manufacturing Research, 4*(1), 23–45.
108. Chinnam, R. B. (2002). Support vector machines for recognizing shifts in correlated and other manufacturing processes. *International Journal of Production Research, 40*, 4449–4466.
109. Rai, R., Tiwari, M. K., Ivanov, D., & Dolgui, A. (2021). ML in manufacturing and industry 4.0 applications. *International Journal of Production Research, 59*(16), 4773–4778.
110. Glaeser, A., Selvaraj, V., Lee, S., Hwang, Y., Lee, K., Lee, N., Lee, S., & Min, S. (2021). Applications of deep learning for fault detection in industrial cold forging. *International Journal of Production Research, 59*, 4826–4835.
111. Paturi, U. M. R., & Cheruku, S. (2021). Application and performance of ML techniques in the manufacturing sector from the past two decades: A review. *Materials Today: Proceedings, 38*, 2392–2401.
112. Ademujimi, T. T., Brundage, M. P., & Prabhu, V. V. (2017). A review of current ML techniques used in manufacturing diagnosis. In *Advances in production management systems*. The Path to intelligent, collaborative and sustainable manufacturing: IFIP WG 5.7 international conference, APMS 2017, Hamburg, Germany, September 3–7, 2017, Proceedings, Part I (pp. 407–415). Springer.
113. Stulp, F., & Sigaud, O. (2015). Many regression algorithms: one unified model: A review. *Neural Networks, 69*, 60–79.
114. Schölkopf, B., & Smola, A. J. (2002). *Learning with kernels: Support vector machines, regularization, optimization and beyond*. MIT Press.
115. Bengio, Y., & LeCun, Y. (2007). Scaling learning algorithms towards AI. *Large-Scale Kernel Machines, 34*(5), 1–41.
116. Bengio, Y. (2009). Learning deep architectures for AI. *Foundations and Trends® in Machine Learning, 2*(1), 1–127.
117. Lee, J. H., Shin, J., & Realff, M. J. (2018). ML: Overview of the recent progresses and implications for the process systems engineering field. *Computers & Chemical Engineering, 114*, 111–121.
118. Hinton, G. E., Osindero, S., & Teh, Y.-W. (2006). A fast learning algorithm for deep belief nets. *Neural Computation, 18*(7), 1527–1554.
119. Silver, D., Huang, A., Maddison, C. J., Guez, A., Sifre, L., Van Den Driessche, G., et al. (2016). Mastering the game of Go with deep neural networks and tree search. *Nature, 529*(7587), 484–489.
120. Erhan, D., Bengio, Y., Courville, A., Manzagol, P.-A., Vincent, P., & Bengio, S. (2010). Why does unsupervised pre-training help deep learning? *Journal of Machine Learning Research, 11*(February), 625–660.
121. Khalil, K., Eldash, O., Kumar, A., & Bayoumi, M. (2020). Machine learning-based approach for hardware faults prediction. *Regular Papers, IEEE Transactions on Circuits and Systems, I*, 1–13.
122. Cai, B., Liu, Y., & Xie, M. (2017). A dynamic-bayesian-network-based fault diagnosis methodology considering transient and intermittent faults. *IEEE Transactions on Automation Science and Engineering, 14*(1), 276–285.
123. Liu, H., Zhou, J., Zheng, Y., Jiang, W., & Zhang, Y. (2018). Fault diagnosis of rolling bearings with recurrent neural network-based autoencoders. *ISA Transactions, 77*, 167–178.

124. Huang, T., Zhang, Q., Tang, X., Zhao, S., & Lu, X. (2022). A novel fault diagnosis method based on CNN and LSTM and its application in fault diagnosis for complex systems. *Artificial Intelligence Review*, 1–27.
125. Topuz, B., & Alp, N. Ç. (2023). ML in architecture. *Automation in Construction, 154*, 105012.
126. Feng, K., Lu, W., & Wang, Y. (2019). Assessing environmental performance in the early building design stage: An integrated parametric design and ML method. *Sustainable Cities and Society, 50*, 101596.
127. Zhang, J., Yin, X., Zhang, G., Gu, Y., & Fan, X. (2020). Prediction method of physical parameters based on linearized rock physics inversion. *Petroleum Exploration and Development, 47*(1), 59–67.
128. Zhang, D., Chen, Y., & Meng, J. (2018). Synthetic well logs generation via recurrent neural networks. *Petroleum Exploration and Development, 45*(4), 629–639.
129. Teixeira, A., & Secchi, A. (2019). Machine learning models to support reservoir production optimization. *IFAC Papers OnLine, 52*(1), 498–501.
130. Anifowose, F. A., Labadin, J., & Abdulraheem, A. (2017). Hybrid intelligent systems in petroleum reservoir characterization and modeling: The journey so far and the challenges ahead. *Journal of Petroleum Exploration and Production Technology, 7*(1), 251–263.
131. Noshi, C. I., & Schubert, J. J. (2018). The role of machine learning in drilling operations; a review. In *SPE/AAPG eastern regional meeting*. Society of Petroleum Engineers.
132. Nande, S. (2018). *Application of machine learning for closure pressure determination*. Society of Petroleum Engineers.
133. Saghir, F., Gilabert, H., & Boujonnier, M. (2018). *Edge analytics and the future of upstream automation*. Society of Petroleum Engineers.
134. Zliobaite, I., Pechenizkiy, M., & Gama, J. (2016). An overview of concept drift applications. In N. Japkowicz & J. Stefanowski (Eds.), *Big data analysis: New algorithms for a new society* (Studies in big data) (Vol. 16). Springer.
135. Sircar, A., Yadav, K., Rayavarapu, K., Bist, N., & Oza, H. (2021). Application of ML and artificial intelligence in the oil and gas industry. *Petroleum Research, 6*(4), 379–391.
136. Bhattacharya, S. (2021). A brief review of popular machine learning algorithms in geosciences. In *Briefs in petroleum geoscience & engineering*. Springer.
137. Abrahart, R. J., See, L. M., & Solomatine, D. P. (2008). *Practical hydroinformatics: Computational intelligence and technological developments in water applications*. Springer.
138. Jin, M., Cao, P., & Short, M. P. (2019). Predicting the onset of void swelling in irradiated metals with machine learning. *Journal of Nuclear Materials, 523*, 189–197.
139. Kautz, E. J., Hagen, A. R., Johns, J. M., & Burkes, D. E. (2019). A machine learning approach to thermal conductivity modeling: A case study on irradiated uranium-molybdenum nuclear fuels. *Computational Materials Science, 161*, 107–118.
140. Morgan, D., Pilania, G., Couet, A., Uberuaga, B. P., Sun, C., & Li, J. (2022). ML in nuclear materials research. *Current Opinion in Solid State and Materials Science, 26*(2), 100975.
141. Rezaeianjouybari, B., & Shang, Y. (2020). Deep learning for prognostics and health management: State of the art, challenges, and opportunities. *Measurement, 163*, 107929.
142. Nguyen, T. N., Downar, T., & Vilim, R. (2020). A probabilistic model-based diagnostic framework for nuclear engineering systems. *Annals of Nuclear Energy, 149*, 107767.
143. Wang, H., Peng, M., Xu, R., Ayodeji, A., & Xia, H. (2020). Remaining useful life prediction based on improved temporal convolutional network for nuclear power plant valves. *Frontiers in Energy Research, 8*, 584463.
144. Ma, J., & Jiang, J. (2011). Applications of fault detection and diagnosis methods in nuclear power plants: A review. *Progress in Nuclear Energy, 53*, 255–266.
145. Hu, G., Zhou, T., & Liu, Q. (2021). Data-driven ML for fault detection and diagnosis in nuclear power plants: A review. *Frontiers in Energy Research, 9*, 663296.
146. Welsh, R. (2019). Defining artificial intelligence. *SMPTE Motion Imaging Journal, 128*(1), 26–32.

147. Go, H., Kang, M., & Suh, S. C. (2020). ML of robots in tourism and hospitality: interactive technology acceptance model (iTAM)–cutting edge. *Tourism Review, 75*(4), 625–636.
148. Feng, Z., Rong, P., Cao, P., Zhou, Q., Zhu, W., Yan, Z., Liu, Q., & Wang, W. (2018). Machine learning-based quantitative texture analysis of CT images of small renal masses: Differentiation of angiomyolipoma without visible fat from renal cell carcinoma. *European Radiology, 28*, 1625–1633.
149. Kocak, B., Durmaz, E. S., Ates, E., Kaya, O. K., & Kilickesmez, O. (2019). Unenhanced CT texture analysis of clear cell renal cell carcinomas: A machine learning-based study for predicting histopathologic nuclear grade. *AJR American Journal of Roentgenology, 212*, W1–W8.
150. Lin, F., Cui, E.-M., Lei, Y., & Luo, L.-P. (2019). CT-based machine learning model to predict the Fuhrman nuclear grade of clear cell renal cell carcinoma. *Abdominal Radiology, 44*, 2528–2534.
151. Ma, R., Vanstrum, E. B., Lee, R., Chen, J., & Hung, A. J. (2020). ML in the optimization of robotics in the operative field. *Current Opinion in Urology, 30*(6), 808.
152. Esteva, A., Robicquet, A., Ramsundar, B., Kuleshov, V., DePristo, M., Chou, K., Cui, C., Corrado, G., Thrun, S., & Dean, J. (2019). A guide to deep learning in healthcare. *Nature Medicine, 25*, 24–29.
153. Wang, Z., Zi, B., Ding, H., You, W., & Yu, L. (2018). Hybrid grey prediction model-based autotracking algorithm for the laparoscopic visual window of surgical robot. *Mechanism and Machine Theory, 123*, 107–123.
154. Sun, Y., Pan, B., Fu, Y., & Cao, F. (2020). Development of a novel intelligent laparoscope system for semi-automatic minimally invasive surgery. *The International Journal of Medical Robotics and Computer Assisted Surgery, 16*, 879.
155. Rivas-Blanco, I., López-Casado, C., Pérez-del-Pulgar, C. J., García-Vacas, F., Fraile, J. C., & Muñoz, V. F. (2018). Smart cable-driven camera robotic assistant. *IEEE Transactions on Human-Machine Systems, 48*, 183–196.
156. Osa, T., Sugita, N., & Mitsuishi, M. (2018). Online trajectory planning and force control for automation of surgical tasks. *IEEE Transactions on Automation Science and Engineering, 15*, 675–691.
157. Baek, D., Hwang, M., Kim, H., & Kwon, D. (2018). *Path planning for automation of surgery robot based on probabilistic roadmap and reinforcement learning*. In 2018 15th International conference on ubiquitous robots (UR), Honolulu, HI, pp. 342–347.
158. Alambeigi, F., Wang, Z., Liu, Y.-H., Taylor, R. H., & Armand, M. (2018). Toward semi-autonomous cryoablation of kidney tumors via model-independent deformable tissue manipulation technique. *Annals of Biomedical Engineering, 46*, 1650–1662.
159. Ivanov, S. H., & Webster, C. (2017). *Designing robot-friendly hospitality facilities*. In Proceedings of the Scientific Conference "Tourism. Innovations. Strategies" 13–14 October, Bourgas, pp. 74–81.
160. Soori, M., Arezoo, B., & Dastres, R. (2023). Artificial intelligence, ML and deep learning in advanced robotics, A review. *Cognitive Robotics*.
161. Baduge, S. K., Thilakarathna, S., Perera, J. S., Arashpour, M., Sharafi, P., Teodosio, B., Shringi, A., & Mendis, P. (2022). Artificial intelligence and smart vision for building and construction 4.0: machine and deep learning methods and applications. *Automation in Construction, 141*, 104440.
162. Sardar, P., Abbott, J. D., Kundu, A., Aronow, H. D., Granada, J. F., & Giri, J. (2019). Impact of artificial intelligence on interventional cardiology: from decision-making aid to advanced interventional procedure assistance. *Cardiovascular Interventions, 12*, 1293–1303.
163. Dalzochio, J., Kunst, R., Pignaton, E., Binotto, A., Sanyal, S., Favilla, J., & Barbosa, J. (2020). Machine learning and reasoning for predictive maintenance in industry 4.0: Current status and challenges. *Computers in Industry, 123*, 103298.
164. Abdar, M., Khosravi, A., Islam, S. M. S., Acharya, U. R., & Vasilakos, A. V. (2022). The need for quantification of uncertainty in artificial intelligence for clinical data analysis: increasing

the level of trust in the decision-making process. *IEEE Systems, Man, Cybernetics Magazine, 8*, 28–40.
165. Begoli, E., Bhattacharya, T., & Kusnezov, D. (2019). The need for uncertainty quantification in machine-assisted medical decision making. *Nature Machine Intelligence, 1*, 20–23.
166. Zohuri, B., Moghaddam, M., & Mossavar-Rahmani, F. (2022). Business resilience system integrated artificial intelligence system. *International Journal of Theoretical and Computational Physics, 3*, 1–7.
167. Dimiduk, D. M., Holm, E. A., & Niezgoda, S. R. (2018). Perspectives on the impact of machine learning, deep learning, and artificial intelligence on materials, processes, and structures engineering. *Integrated Materials Manufacturing Innovation, 7*, 157–172.
168. Badage, A. (2018). Crop disease detection using ML: Indian agriculture. *International Research Journal of Engineering and Technology, 5*(9), 866–869.
169. Ferentinos, K. P. (2018). Deep learning models for plant disease detection and diagnosis. *Computers and Electronics in Agriculture, 145*, 311–318.
170. Ahmad, A., Saraswat, D., & El Gamal, A. (2023). A survey on using deep learning techniques for plant disease diagnosis and recommendations for development of appropriate tools. *Smart Agricultural Technology, 3*, 100083.
171. Su, H. Y., & Hong, H. H. (2021). An intelligent data-driven learning approach to enhance online probabilistic voltage stability margin prediction. *IEEE Transactions on Power Systems, 36*(4), 3790–3793.
172. Taghipourbazargani, N., Dasarathy, G., Sankar, L., & Kosut, O. (2022). A machine learning framework for event identification via modal analysis of PMU data. *IEEE Transactions on Power Systems.*
173. Wazirali, R., Yaghoubi, E., Abujazar, M. S. S., Ahmad, R., & Vakili, A. H. (2023). State-of-the-art review on energy and load forecasting in microgrids using artificial neural networks, machine learning, and deep learning techniques. *Electric Power Systems Research, 225*, 109792.
174. Le, V., Yao, X., Miller, C., & Tsao, B. H. (2020). Series DC arc fault detection based on ensemble machine learning. *IEEE Transactions on Power Electronics, 35*(8), 7826–7839.
175. Liao, M., Li, H., Wang, P., Sen, T., Chen, Y., & Chen, M. (2022). Machine learning methods for feedforward power flow control of multi-active-bridge converters. *IEEE Transactions on Power Electronics, 38*(2), 1692–1707.
176. Li, W., Limoge, D. W., Zhang, J., Sauer, D. U., & Annaswamy, A. M. (2021). Estimation of potentials in lithium-ion batteries using machine learning models. *IEEE Transactions on Control Systems Technology, 30*(2), 680–695.
177. Devasia, S. (2017). Iterative machine learning for output tracking. *IEEE Transactions on Control Systems Technology, 27*(2), 516–526.
178. Matschek, J., Bethge, J., & Findeisen, R. (2023). Safe machine-learning-supported model predictive force and motion control in robotics. *IEEE Transactions on Control Systems Technology.*
179. Ko, K., Lee, J. K., & Shin, H. (2020). Variability-aware machine learning strategy for 3-D NAND flash memories. *IEEE Transactions on Electron Devices, 67*(4), 1575–1580.
180. HNI, University of Paderborn. (n.d.). https://www.hni.uni-paderborn.de/
181. Huang, K., Wei, K., Li, F., Yang, C., & Gui, W. (2023). LSTM-MPC: A deep learning based predictive control method for multimode process control. *IEEE Transactions on Industrial Electronics, 70*(11), 11544–11554.
182. Mokhtari, S., Abbaspour, A., Yen, K. K., & Sargolzaei, A. (2021). A machine learning approach for anomaly detection in industrial control systems based on measurement data. *Electronics, 10*(4), 407. https://doi.org/10.3390/electronics10040407
183. Wang, Q., Spronck, P., & Tracht, R. (2003, November). *An overview of genetic algorithms applied to control engineering problems.* In Proceedings of the 2003 international conference on machine learning and cybernetics (IEEE Cat. No. 03EX693) (Vol. 3, pp. 1651–1656). IEEE.

184. Paul Ayeng'o, S., Axelsen, H., Haberschusz, D., & Sauer, D. U. (2019). A model for direct-coupled PV systems with batteries depending on solar radiation, temperature and number of serial connected PV cells. *Solar Energy, 183*, 120–131.
185. Sohani, A., et al. (2022). Using machine learning in photovoltaics to create smarter and cleaner energy generation systems: A comprehensive review. *Journal of Cleaner Production, 364*, 132701.
186. Data for Validating Models for PV Module Performance – Metadata. 10.21948/1811521.
187. Green, M. A., et al. (2019). Solar cell efficiency tables (version 54). *Progress in Photovoltaics: Research and Applications, 27*(7), 565–575.
188. Gupta, P., & Singh, R. (2021). PV power forecasting based on data-driven models: a review. *International Journal of Sustainable Engineering, 14*(6), 1733–1755.
189. Gaboitaolelwe, J., Zungeru, A. M., Yahya, A., Lebekwe, C. K., Vinod, D. N., & Salau, A. O. (2023). Machine learning based solar photovoltaic power forecasting: A review and comparison. *IEEE Access, 11*, 40820–40845.
190. Voyant, C., Notton, G., Kalogirou, S., Nivet, M. L., Paoli, C., Motte, F., & Fouilloy, A. (2017). Machine learning methods for solar radiation forecasting: A review. *Renewable Energy, 105*, 569–582.
191. Sharma, S. (2017, September 23). *Epoch vs Batch size vs Iterations*. Towards Data Science.
192. Yu, H., & Wilamowski, B. M. (2011). Levenberg–marquardt training. In B. M. Wilamowski & J. D. Irwin (Eds.), *The industrial electronics handbook* (2nd ed.). CRC Press.

Shallow Learning vs. Deep Learning in Finance, Marketing, and e-Commerce

Ismail A. Mageed, Ashiq H. Bhat, and Sayyed A. Edalatpanah

1 Introduction: Shallow Learning and Deep Learning

Three primary categories are included in the machine learning discipline of shallow learning (SL) [1]: supervised learning, unsupervised learning, and reinforcement learning. While unsupervised learning concentrates on data reduction and clustering using unlabeled data, supervised learning employs labeled data to predict outcomes. In the construction industry, common unsupervised learning techniques include Principal Component Analysis, K-means, and spectral clustering, while supervised learning algorithms like KNN, Perceptron, and Logistic Regression are frequently used. Figure 1 in the given context illustrates the application of KNN and Support Vector Machines (SVM) for text classification. KNN is a supervised learning algorithm that assigns unlabeled samples to categories based on many of their closest neighbors. Deep learning (DL) is an advanced form of artificial neural network that addresses the limitations of traditional neural networks. It achieves this by automatically extracting features and utilizing network structures like Convolutional Neural Networks (CNNs) and Recurrent Neural Networks (RNNs). These advancements make DL more capable of handling complex and nonlinear functions, enhancing its effectiveness in various applications.

I. A. Mageed (✉)
Faculty of Engineering and Digital Technologies, School of Computer Science, AI, and Electronics, University of Bradford, Bradford, UK
e-mail: iamoham@bradford.ac.uk

A. H. Bhat
Department of Biostatistics, Rajendra Memorial Research, Institute of Medical Sciences, Patna, India

S. A. Edalatpanah
Department of Research Center, Ayandegan Institute of Higher Education, Tonenkabon, Iran

Ö. F. Ertuğrul et al. (eds.), *Shallow Learning vs. Deep Learning*, The Springer Series in Applied Machine Learning, https://doi.org/10.1007/978-3-031-69499-8_3

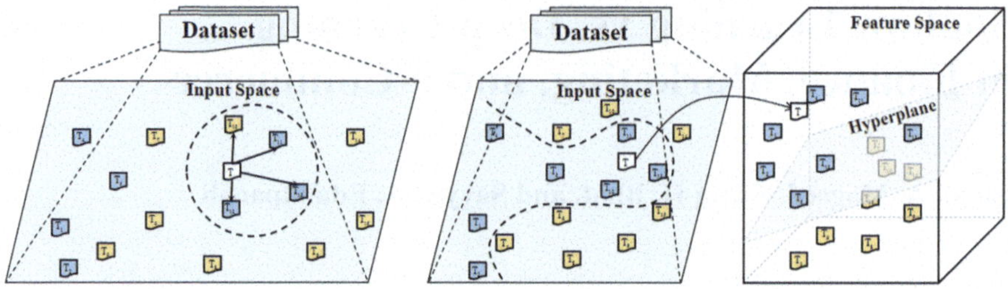

Fig. 1 The KNN structure, depicted on the left side of Fig. 1, corresponding to $k = 4$, represents the K-Nearest Neighbors algorithm used for text classification. In the given context, the structure mentioned consists of nodes that represent different texts. Nodes with distinct outlines indicate various classifications. On the right side, the SVM structure refers to Support Vector Machines, which is an algorithm used for text classification [2]

The shortcomings of conventional neural networks are addressed by DL [3], an enhanced type of artificial neural network. It utilizes techniques like backpropagation to train multilayer networks and has shown effectiveness in various domains, particularly in image classification tasks. DL employs network topologies, for example, CNNs and RNNs, to extract significant features and data patterns.

To eliminate the necessity for human feature extraction, deep neural networks (DNNs) [4] are trained using algorithms that automatically extract features from input and learn representations. DNNs are more effective than SL models in mapping complex and nonlinear functions due to their deeper architecture. The availability of large datasets and affordable processing units like GPUs has facilitated the widespread adoption of deep learning, particularly in image recognition tasks, where CNNs are used to extract features and classify images.

Figure 2 illustrates the functioning of a CNN in processing incoming data to create hierarchical representations for accurate recognition of higher-level objects. The red squares in the highlighted area of the image indicate a condensed representation of the pixel values. This method, which enables CNN to automatically extract features from the input and create predictions based on those characteristics, revolutionizes image recognition tasks. CNNs are made to progressively extract more sophisticated information from input data within the context of DL until they can correctly identify images. The procedure entails using labeled training data to train the network, using independent testing data to assess the network's performance, and utilizing the trained network to forecast fresh, unseen data. This iterative process allows CNNs to learn and improve their ability to recognize and classify images. Image recognition by a CNN refers to the process of using a CNN for images' analysis and classification. CNNs are specialized deep learning models that are designed for image-processing tasks. They could automatically extract features from images and use those features to make predictions or classifications. This approach refers to a method that has significantly transformed the field of image recognition. It has achieved this by removing the requirement for manual feature engineering, which involves manually selecting and designing features for image

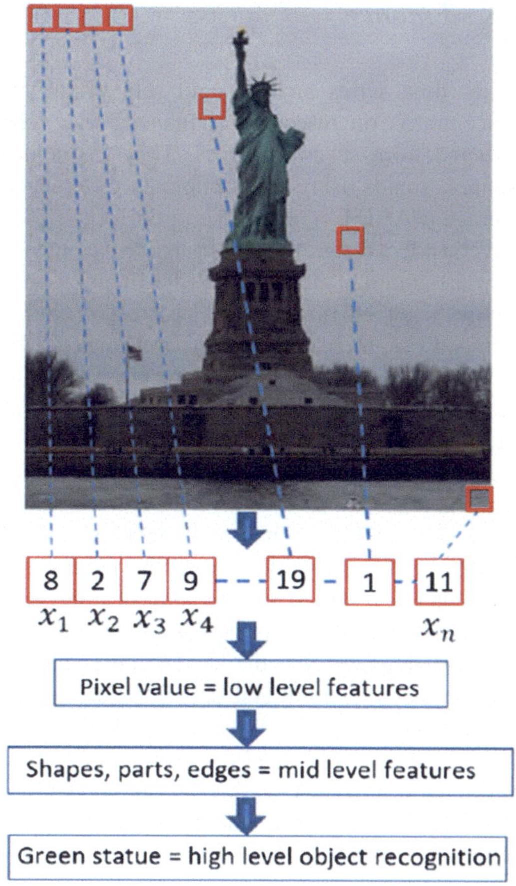

Fig. 2 CNN image recognition [4]

analysis. Instead, this approach automatically learns and extracts features from images, leading to improved accuracy in classifying images.

The remaining sections of this chapter are as follows: Sect. 2 examines applications of shallow and deep learning techniques in finance, marketing, and e-commerce. Section 3 offers several difficult open problems. Section 4 provides concluding thoughts and future research directions.

2 Method: Shallow Learning vs Deep Learning in Finance, Marketing, and e-Commerce

This section is devoted to overviewing some real-world applications, namely, finance, marketing, and e-commerce.

2.1 *Finance*

Using time series analysis and self-similar business modeling, the authors [5] concentrated on researching financial risk forecast analysis using DL and data gathered from several sources. They [5] achieved this by analyzing self-similar business trends using the fractionally differenced autoregressive integrated moving average (FARIMA) model. To optimize the DL model, they [5] employed a kernel that considers the probability of light entering and leaving a specific color. After that, they use simulations to verify the model's performance in risk forecast analysis. Figure 3 (c.f., [5]) shows the process of obtaining data from various sources for financial risk forecast analysis based on DL. This highlights the need to consider factors such as noise, fading characteristics, and error control codes in the transmission of data through communication channels. The proposed methodology incorporates fractional neural networks, which combine fractional order theory with artificial neural networks, to leverage memory and heredity properties for improved analysis.

Examining Fig. 3 in further detail reveals the sources of data collection and emphasizes the difficulties caused by fading and noise in communication routes. It highlights how important error control codes are for identifying and fixing information transmission faults brought on by distortion when the data is being transmitted. The text indicates that the sender must continuously convey control information relating to modulation and coding to the receiver in the context of financial risk forecast analysis due to the changing conditions of the communication channel. This is required to identify and fix faults brought on by gearbox distortion. The control information guarantees efficient communication between the sender and recipient, and the modulation and coding format are modified in accordance with the channel conditions.

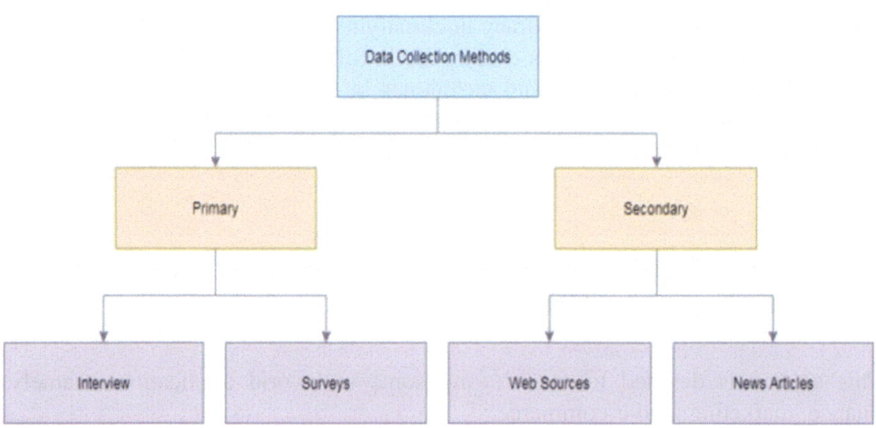

Fig. 3 The gathered data from various sources

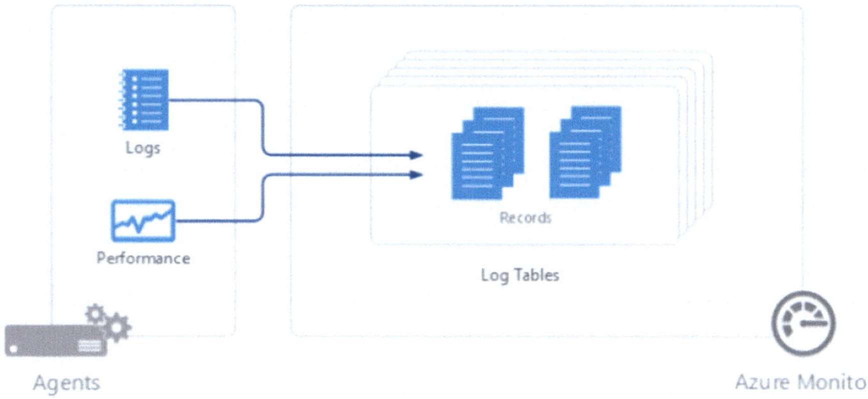

Fig. 4 DL structure model [5]

Notably, the focus on feature learning and the depth of the model structure are the primary differences between DL and SL. While SL is generally used for prediction and classification tasks and depends on human knowledge to extract sample characteristics, DL concentrates on extracting complicated data features through multilayer perceptrons [6, 7]. Deep models like multilayer perceptrons are better at approximating complex functions and describing data features than SL models like SOFM neural networks and single hidden layer BP networks. However, there are difficulties in training DL models since their non-convex objective function frequently exhibits several local extrema. The DL structure model is shown in Fig. 4, as indicated [8, 9].

Large dataset analysis with DL can advance the ability of financial prediction. Its ability to identify intricate nonlinear patterns in data, however, raises questions about the notion of market efficiency [10] and may allow for the costing of assets with little mistakes. This raises questions about whether existing theoretical models can compete with DL models, and whether new tests of market efficiency will be needed in the future.

SL models can be seen as involving a low-dimensional auxiliary variable (Z) and a prediction rule made up of functions in the context of data reduction approaches [10]. The prediction rule ($\widehat{Y}$) is determined by applying these functions to the auxiliary variable (Z), which is obtained through another set of functions applied to the original input data (X). This formulation aligns with DL architectures and highlights the challenge of finding the appropriate auxiliary variable (Z) and estimating its relationship to the original data.

Principal component analysis (PCA), linear discriminant analysis (LDA), reduced rank regression (RRR), project pursuit regression (PPR), and logistic regression are examples of SL algorithms. While PCA and PPR are dimension-reduction techniques that transform the input data into lower-dimensional representations, RRR and LDA are linear regression techniques used for classification and prediction applications. These techniques operate independently of the desired output and may discard valuable information, but sliced inverse regression (SIR) partially addresses this limitation by incorporating both input and output data to estimate the layer function.

DL is a powerful approach that addresses limitations in traditional SL methods [10] by simultaneously estimating multiple functions (f_1, f_2) using a large dataset containing both input (X) and output (Y) information.

Predicting the direction of the stock market is one of the hardest things to do in finance. Despite their development, shallow computational approaches like Support Vector Machine and Backpropagation Network have not produced satisfactory results [11]. A DL model called the stacked denoising autoencoder (SDAE) is used in [11] to reliably predict the daily CSI 300 index from the Shanghai and Shenzhen Stock Exchanges in China. The experiment shows that when it comes to CSI 300 index prediction, the DL model performs better than alternative techniques.

SVM uses a high-dimensional feature space, which may contain hundreds or thousands of dimensions, to store the input data for stock market direction prediction. This is a serious drawback of SVM [11]. Furthermore, storing the huge matrices needed for SVM results in high memory and computational expenses [11]. SVM is regarded as an SL [11] technique that has limited learning capacity and is prone to noise interference in financial time series data, much as ANN.

The SDAE (Stacked Denoising Autoencoder) [11], which is a DL method capable of handling more complexity compared to shallow layer methods like BPNN and SVM, demonstrates superior performance, showcasing the power of DL techniques.

Furthermore, the SDAE outperforms SVM in terms of cumulative return, as shown in Fig. 5 (c.f., [11]). This is a significant consideration for many investors. In simpler terms, compared to the other approaches, SDAE has a lesser likelihood of

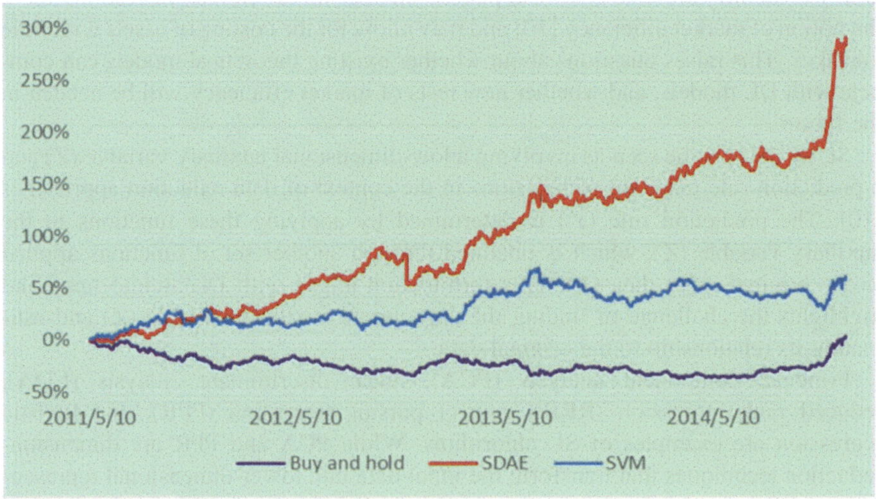

Fig. 5 The SDAE and SVM cumulative return

experiencing severe losses. More intriguingly, Fig. 5 shows the cumulative return of two models—SDAE and SVM—when attempting to forecast the CSI 300 index's direction. The total profit or loss produced by these models across several phases or time periods is represented by the cumulative return. According to the findings [11], the trading strategy based on the SDAE model often beats the market and generates statistically significant excess returns, suggesting that it may be useful for anticipating market movements.

2.2 *Marketing*

Homogeneity-oriented behavior analysis (HOBA), a unique framework for behavior analysis, was suggested in [12] for credit card transactions. Furthermore, they produce improved fraud detection findings and better transaction behavior prediction by using DL methodologies instead of traditional machine learning techniques. Some of the study's contributions include enhanced feature variables for predicting fraudulent activity, the application of HOBA to analyze past behavior for fraud detection, and practical implications for credit card issuers to efficiently identify fraudulent transactions and reduce fraud losses.

In the context of fraud detection [12], previous feature engineering frameworks have used the recency, frequency, monetary (RFM) framework, a marketing technique which analyzes consumer behavior based on the timing, frequency, and monetary value of their purchases. However, this framework overlooks the inherent heterogeneity in credit card transactions, where different types of transactions exhibit distinct behaviors. For instance, cash withdrawals pose a higher risk as fraudsters are more likely to target stolen cards for cash withdrawals.

Because of the amount of transactions, it is not practical to manually detect fraudulent credit [12] card transactions. As a result, ML techniques are frequently used to automatically identify fraudulent transactions. Because some prior transactions may not have been correctly labeled, supervised learning [12] in fraud detection faces difficulties. Furthermore, a major obstacle is the extremely uneven character of credit card transaction statistics, with less than 0.1% of transactions being fraudulent.

On the contrary, unsupervised approaches typically have greater false alarm rates since anomalous behavior is frequently linked to real transactions. Consequently, supervised learning techniques are the primary area of fraud detection research to improve performance and reduce false alarm rates. By utilizing the HOBA feature set, the DBN classifier outperforms the benchmark models by a significant margin. Results also indicate [12] that all data mining techniques benefit from the use of HOBA-based feature engineering to detect fraudulent transactions. This shows that the improved performance of the proposed strategy can be attributed to the combination of HOBA feature engineering and deep learning techniques. The work [12] describes how four elements—aggregation time, aggregation characteristic, transaction behavior measure, and aggregation statistics—may be used to build a feature

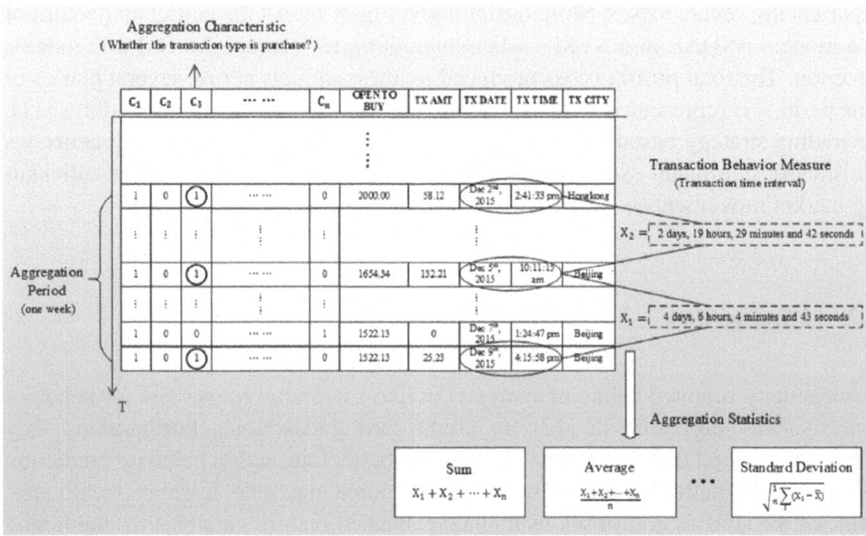

Fig. 6 A schematic diagram illustrating the application of the transaction aggregation strategy using the HOBA principle

variable in the given scenario. The flowchart, which is shown in Fig. 6 (c.f., [12]), demonstrates how to extract historical transactions with attributes, calculate the intervals between consecutive transactions, and use the average value of these intervals as a feature variable. It offers a graphic summary of the procedures that make up the aggregation technique for transaction behavior analysis.

This study's [13] goal was to investigate the factors influencing users' intentions to use wearable payments. The study focuses on how perceived attractiveness, technological preparedness, mobile usefulness, and mobile usability affect consumers' behavioral intentions. The study used DL and a dual-stage analysis to look at 307 replies. Apart from the relationship between behavioral intention and mobile ease of use, it was found that all correlations were validated.

The study's findings [13] offered useful data that companies that process payments and manufacturers of intelligent wearables might utilize to develop strategies and advertising campaigns that would successfully encourage consumers to embrace wearable payments. Furthermore, a more comprehensive knowledge of consumer acceptance of wearable payments is provided by the incorporation of the Fashion Theory, Technology Readiness Theory, and Mobile Technology Adoption Model into the conceptual framework used in this study, which allows for the assessment of component importance.

The correlations between TRI (Technology Readiness Index), PA (Perceived Usefulness), MTAM (Mobile Technology Acceptance Model), and BI (Behavioral Intention) to use wearable payment have been illustrated in a study model that the researchers constructed based on the given hypotheses. This model provides a visual

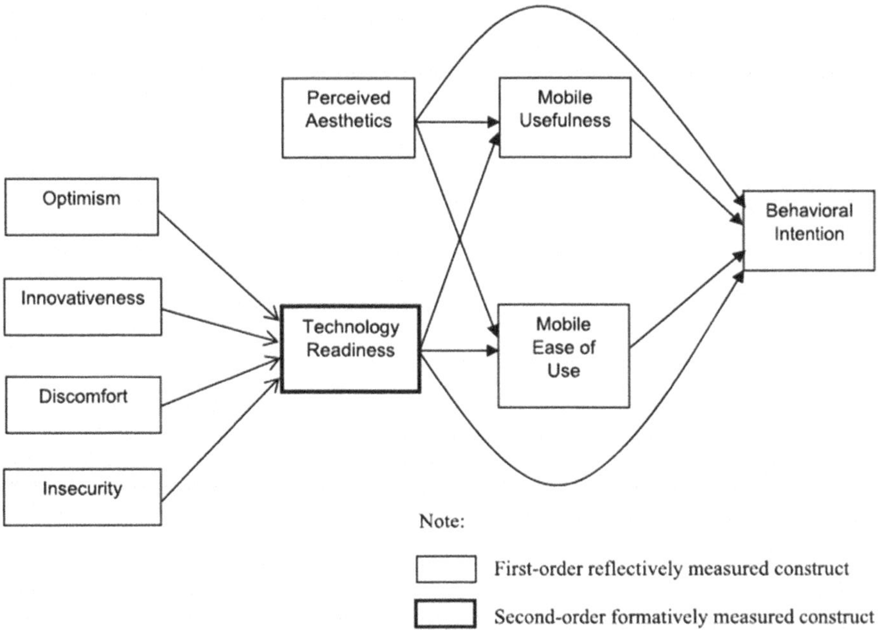

Fig. 7 The study model is a graphical representation demonstrating the relationships among the Technology Readiness Index (TRI), Perceived Usefulness (PU), Perceived Ease of Adoption (PEOU), and Behavioral Intention (BI) to adopt wearable payment systems. It presents a structure for understanding the various ways in which these factors influence consumers' inclinations to use wearable payments

depiction of how these parameters are related to the adoption of wearable payments, as seen in Fig. 7 (c.f., [13])

Theoretically, a second analysis was conducted by comparing the deep ANN architecture previously used to a shallow ANN architecture with a single hidden layer [13]. According to the findings [13], the deep ANN architecture produced higher R2 values and lower root mean square error (RMSE) values, indicating improved prediction ability and learning quality. This shows that the study's ANN models' output neurons can be better explained, and that deeper learning is made possible by the deep ANN architecture.

2.3 E-Commerce

In the context of E-commerce [14], tasks like search and recommendation are crucial for addressing the problem of information overload and catering to the diverse needs of users. Personalization plays a significant role in these tasks, but current techniques

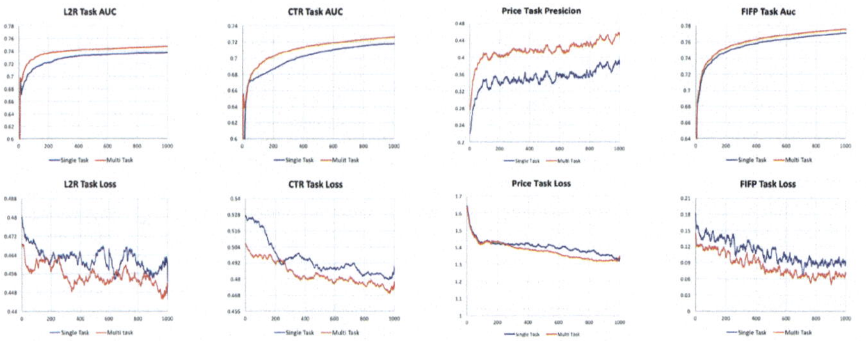

(a) the comparison of L2R task (b) the comparison of CTR task (c) the comparison of PPP task (d) the comparison of FIFP task

Fig. 8 Comparison of four distinct tasks (L2R, CTR, PPP, and FIFP) including single-task and multitask learning. The x-coordinate represents the learning iterations, and the y-coordinate represents the evaluation metric. Contiguous batch points smooth the AUC (Area Under the Curve) and Precision Curves, making performance comparisons between the two learning modalities possible (c.f., [14])

often treat each task separately, resulting in suboptimal performance due to the lack of shared user information across tasks.

The authors [14] proposed a method called Deep User Perception Network (DUPN) to enhance personalization in e-commerce tasks. They use user behavior sequences, such as clicks, bookmarks, or purchases, and model them using Long Short-Term Memory (LSTM) and attention mechanisms. By sharing and learning user representations across multiple tasks, DUPN improves the effectiveness of reflecting user interests and can be transferred to new tasks, as demonstrated through offline and online experiments. The authors also deploy DUPN in large-scale operational tasks in Taobao and provide detailed implementations to address practical issues in real-world applications.

The splittable network topology of DUPN [14] provides an additional benefit. In real-time systems like Taobao search, the computational requirements can be minimized without sacrificing efficiency by splitting the integrated network into two sections: one for creating user representations and tasks based exclusively on those representations, like PPP (Price Preference Task), and the other for tasks involving both user and item representations. This justifies the reason [14] behind dividing the network architecture used in the Taobao search system into two parts. The first part handles most of the calculations and has a fixed input under a certain query. Tasks like CTR prediction and item ranking scores are carried out by the second part, less complex and computationally demanding portion, using both user and object representations.

Figure 8 (c.f., [14]) illustrates the learning process and demonstrates two separate phases: a first phase where the AUC (Area Under the Curve) expands quickly and a second phase where the AUC grows gradually and steadily over a few days. This behavior is explained by the imbalance of features: sparse features take longer to learn because they appear less frequently than frequent features, which are learned more quickly since they are more common in training examples. Furthermore, the outcomes [14] show that multitask learning performs 1% better than single task learning across the board with faster convergence. With this method, the model may learn general representations that help with every activity and extract more usable information, which makes it more effective and practical for online real-time systems.

The growth of online businesses has a great need for ML and data mining techniques to understand user preferences and optimize sales. In the work [15], the accuracy and generalizability of a complete connection model and a CNN model are especially compared with regard to building a sales prediction model for online products utilizing DL methods.

Previous research by [16] used a shallow neural network for big data processing in the context of predicting goods sales on e-commerce platforms, but the prediction accuracy needed to be improved. Conversely, [16] created a crown model with a DL algorithm tailored for agricultural e-commerce sales data, allowing for the classification and prediction of online sales of agricultural products.

The DL algorithm has special benefits for predicting online goods sales. Longer training times are possible because of its solution to the gradient disappearance problem, which also offers strong generalization capabilities in huge data settings. With its layer-by-layer methodology, DL can extract more useful information from large amounts of data while also increasing prediction accuracy by enabling the extraction of higher-level characteristics. Justifiably, the authors [15] were able to create an index system for variables impacting the sales of online products and employ DL algorithms to create flexible models for sales prediction for different kinds of online products. It is worth noting that L2R, C2R, PPP, and FIFP are, respectively, referred to Learning to Rank, Click Through Rate Prediction, Price Preference Prediction, and Fashion Icon Following Prediction (c.f., [14]).

3 Results

DL theory [15] offers several advantages. Firstly, it utilizes distributed representation, which helps overcome the limitations of traditional machine learning algorithms by reducing computation and avoiding overfitting. Secondly, the depth structure of deep learning enables effective expression of functions, allowing it to learn complex functions that other algorithms struggle with. Lastly, unsupervised pretraining, such as with deep belief networks and stack self-encoder methods, sets an initial stage for fine-tuning the depth structure, minimizing variance, and introducing bias to promote effective unsupervised training.

The following open problems are adapted from [12, 17], respectively:

- There is growing interest in and potential for the adaption of 2D CNN implementations in a variety of financial application sectors. Further exploration of this technique in different problems is recommended as there are still ample opportunities for its application. The potential benefits and possibilities of using 2D CNN in finance are still challenging open problems.

- Graph CNN is a model that has shown potential in financial applications, although it has not been extensively utilized. By transforming financial data into graphs and incorporating sentiment analysis, researchers can explore the possibilities of constructing different models and improving the performance of financial applications. This approach opens opportunities for building more effective financial models by leveraging the representation and analysis of graph structures.
- Recently developed DL models, such as GAN (Generative Adversarial Networks) and Capsule Networks, have shown promise in various non-financial studies. Nevertheless, as of right now, these models have not been applied to any financial applications. For financial scholars and practitioners, investigating these novel models as well as cutting-edge paradigms like transfer learning and one-shot learning may present new prospects.
- Financial text mining is becoming more and more important in the field of financial analysis. Researchers can combine data semantics, expand data models like Stock2Vec, and add ensemble models based on Natural Language Processing (NLP) to increase the accuracy of already available models. Furthermore, by creatively fusing spatial and temporal data, hybrid models that blend several deep learning layers—like cascaded CNN-LSTM blocks—can produce superior outcomes.
- The authors' observations indicate that most studies favor hybrid models—which integrate multiple deep learning layers—over standalone models. Although it is anticipated that this tendency will continue, researchers are urged to use more flexible and unusual models to get better outcomes. To combine spatial and temporal information, hybrid models that incorporate cascaded CNN-LSTM blocks have the potential to be a breakthrough for researchers trying to enhance their models' performance. It is anticipated that conventional fields in financial research, such as risk assessment, portfolio management, and algorithmic trading, would continue to dominate. However, there is increasing attention toward emerging areas that offer new research opportunities and have significant potential impact on the financial world. These areas are gaining attention due to their forecasted influence and include topics such as cryptocurrencies, blockchain technology, and other innovative models.
- In the context of model development in the financial field, there is growing interest in adapting 2D CNN implementations and exploring their potential in various financial applications. Additionally, the use of graph CNN models and innovative paradigms like transfer learning and one-shot learning show promise for building better-performing financial applications. Financial research will probably continue to be dominated by algorithmic trading, portfolio management, and risk assessment in terms of implementation, but newer fields like blockchain technology and cryptocurrencies are drawing interest due to their potential influence on the financial industry.

- Portfolio management is an area that can benefit from innovative models, particularly in the context of robo-advisory systems that rely on high-performing automated decision support. DL models are well-suited for this purpose, and their utilization is expected to increase in the coming years. This tendency could result in the combination of optimization and learning models, improving system performance, and possibly establishing new avenues for research in the area.
- Researchers studying soft computing and DL have paid little attention to the financial derivatives market. For experts in the finance sector, these products, however, provide several benefits, including the ability to perform leveraged transactions with reduced risk and hedge assets. DL research in areas such as option pricing, futures trading, arbitrage trading, and option strategy optimization may be quite beneficial, even though there aren't many studies in this area.
- One industry that hasn't fully benefited from advances in ML research is high-frequency trading (HFT). Trading systems that incorporate machine learning models must avoid adding more latency because of the requirement for quick transaction processing. To tackle this issue, scholars may investigate deep learning models that are realized on specialized hardware such as Field Programmable Gate Arrays (FPGAs) or Graphic Processing Units (GPUs). Even though hardware is sometimes disregarded in research, there can be prospects in this area. Additionally, cryptocurrency and blockchain studies are expected to receive more attention in applied DL research for finance.
- The authors [12] acknowledged limitations in their work, specifically the lack of assessment regarding the computational cost of their proposed HOBA feature engineering framework. This framework constructs a larger variable set compared to RFM, which may impact the computational efficiency of the fraud detection model. These are the top challenging open issues. Further research is needed to evaluate the computational demands of real-time fraud detection systems and explore the application of advanced machine learning methods, including combinations of deep learning and traditional data mining techniques, in fraud detection.

There are strong motivations [18] behind the significant success of DL in various application domains. This highlights [18] that this success is not solely attributed to marketing or sociological factors but also to genuine advancements in performance. DL has enabled efficient solutions to previously challenging AI problems, which were considered beyond the capabilities of existing computational tools. More specifically [18], no distinct and unique innovation can adequately account for the extraordinary success of DL when considering the evolution of neural networks and the shift from "shallow" to "deep" designs. The reasons behind its effectiveness lie in a combination of factors, including empirical improvements, heuristic search, and the accumulation of knowledge and techniques over time. DL's achievements can be attributed to a collective effort rather than a single groundbreaking innovation.

The current leading [18] factors that are believed to have driven the development of deep neural networks, such as advancements in computing hardware and insights from neuroscience, are considered insufficient to fully justify the achievements of deep learning. Mathematics is seen as a promising field for providing some explanatory insights, but a complete mathematical explanation for the success of DL is still lacking. Furthermore [18], knowledge of deep neural networks is restricted to a particular class of networks, and rather than a strong theoretical base, DL advancement primarily depends on empirical findings and heuristics.

4 Concluding Remarks and Future Research Directions

The use of SL and DL techniques in real-world fields including marketing, e-commerce, and finance is covered in this chapter. It examines the advantages and disadvantages of using each method in these applications, providing insights into their effectiveness and limitations. Furthermore, the work addresses the provision of insights on selecting the most appropriate approach and contrasting the effectiveness of different methods. Notably, the inclusion of observations, unresolved open problems, and potential future research directions are provided.

References

1. Xu, Y., Zhou, Y., Sekula, P., & Ding, L. (2021). Machine learning in construction: From shallow to deep learning. *Developments in the Built Environment, 6*, 100045.
2. Li, Q., Peng, H., Li, J., Xia, C., Yang, R., Sun, L., et al. (2022). A survey on text classification: From traditional to deep learning. *ACM Transactions on Intelligent Systems and Technology (TIST), 13*(2), 1–41.
3. Suzuki, M., Pennartz, C. M., & Aru, J. (2023). How deep is the brain? The shallow brain hypothesis. *Nature Reviews Neuroscience*, 1–14.
4. Shrestha, A., & Mahmood, A. (2019). Review of deep learning algorithms and architectures. *IEEE Access, 7*, 53040–53065.
5. Yang, H., & Hu, X. (2022, January). *Financial risk forecast analysis based on deep learning considering the data collection from different sources*. In 2022 4th International Conference on Smart Systems and Inventive Technology (ICSSIT) (pp. 957–961). IEEE.
6. Golbayani, P., Florescu, I., & Chatterjee, R. (2020). A comparative study of forecasting corporate credit ratings using neural networks, support vector machines, and decision trees. *The North American Journal of Economics and Finance, 54*, 101251.
7. He, P., Niu, H., Sun, Z., & Li, T. (2020). Accounting index of COVID-19 impact on Chinese industries: A case study using big data portrait analysis. *Emerging Markets Finance and Trade, 56*(10), 2332–2349.
8. Sun, Y., Chen, L., Sun, H., & Taghizadeh-Hesary, F. (2020). Low-carbon financial risk factor correlation in the belt and road PPP project. *Finance Research Letters, 35*, 101491.
9. Patel, M. M., Tanwar, S., Gupta, R., & Kumar, N. (2020). A deep learning-based cryptocurrency price prediction scheme for financial institutions. *Journal of Information Security and Applications, 55*, 102583.
10. Heaton, J. B., Polson, N. G., & Witte, J. H. (2016). *Deep learning in finance*. arXiv preprint arXiv:1602.06561.
11. Lv, S., Hou, Y., & Zhou, H. (2019). *Financial market directional forecasting with stacked denoising autoencoder*. arXiv preprint arXiv:1912.00712.

12. Zhang, X., Han, Y., Xu, W., & Wang, Q. (2021). HOBA: A novel feature engineering methodology for credit card fraud detection with a deep learning architecture. *Information Sciences, 557*, 302–316.
13. Lee, V. H., Hew, J. J., Leong, L. Y., Tan, G. W. H., & Ooi, K. B. (2020). Wearable payment: A deep learning-based dual-stage SEM-ANN analysis. *Expert Systems with Applications, 157*, 113477.
14. Ni, Y., Ou, D., Liu, S., Li, X., Ou, W., Zeng, A., & Si, L. (2018, July). *Perceive your users in depth: Learning universal user representations from multiple e-commerce tasks*. In Proceedings of the 24th ACM SIGKDD international conference on knowledge discovery & data mining (pp. 596–605).
15. Yin, X., & Tao, X. (2021). Prediction of merchandise sales on E-commerce platforms based on data mining and deep learning. *Scientific Programming*, 1–9.
16. Liu, J., He, S., & Zhu, Q. (2017). Research on online agricultural product sales forecasting model based on deep learning. *Computer Application Research, 34*(8), 2291–2293.
17. Ozbayoglu, A. M., Gudelek, M. U., & Sezer, O. B. (2020). Deep learning for financial applications: A survey. *Applied Soft Computing, 93*, 106384.
18. Plebe, A., & Grasso, G. (2019). The unbearable shallow understanding of deep learning. *Minds and Machines, 29*, 515–553.

Shallow Learning vs. Deep Learning in Social Applications

Ismail A. Mageed, Ashiq H. Bhat, and Jihad Alja'am

1 Introduction

1.1 Shallow Learning (SL) and Deep Learning (DL)

Guided learning, unsupervised learning, and reinforcement learning are the three primary categories of shallow learning (SL) [1]. While unsupervised learning concentrates on data reduction and clustering issues using unlabeled data, supervised learning employs patterns from labeled data to predict results. In the construction industry, unsupervised learning techniques like Principal Component Analysis, K-means, and spectral clustering are common, while supervised learning algorithms like kNN, Perceptron, and Logistic Regression are frequently employed. Figure 1 illustrates the development of the K-Nearest Neighbors (KNN) method and Support Vector Machines (SVM) for text classification. By identifying the category with the greatest number of samples among its closest neighbors, the KNN algorithm classifies an unlabeled sample. Figure 1's left side depicts the KNN structure, with each node denoting a text and nodes with various shapes denoting distinct categories.

DL [3] is an advanced form of artificial neural network that overcomes the limitations of traditional neural networks. It was made possible by the introduction

I. A. Mageed (✉)
Faculty of Engineering and Digital Technologies, School of Computer Science, AI, and Electronics, University of Bradford, Bradford, UK
e-mail: iamoham@bradford.ac.uk

A. H. Bhat
Department of Biostatistics, Rajendra Memorial Research, Institute of Medical Sciences, Patna, India

J. Alja'am
Liverpool J. Mores University, Liverpool, UK
e-mail: M.Aljaam@ljmu.ac.uk

Ö. F. Ertuğrul et al. (eds.), *Shallow Learning vs. Deep Learning*, The Springer Series in Applied Machine Learning, https://doi.org/10.1007/978-3-031-69499-8_4

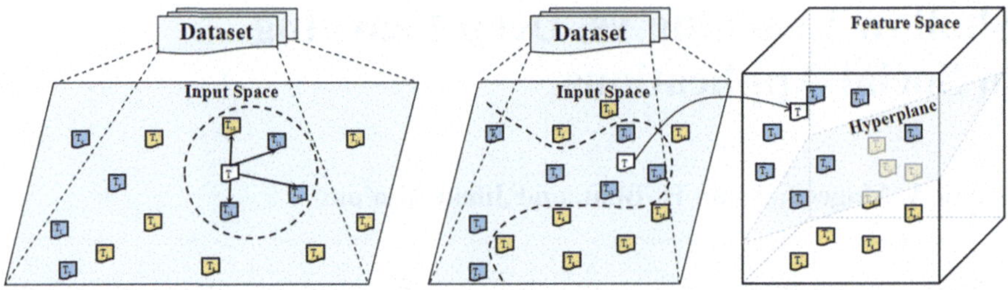

Fig. 1 The KNN structure (left) with $k = 4$, and the SVM structure (right). Every node depicts a text, while nodes with various outlines stand for various classifications [2]

of backpropagation, which allowed for the training of multilayer networks. With models like LeNet and AlexNet, DL has demonstrated efficacy in various fields, chief among them being picture classification. To extract significant characteristics and patterns from data, DL uses network topologies such as convolutional neural networks (CNNs) and recurrent neural networks (RNNs).

Using algorithms, deep neural networks (DNNs) [4], are trained to automatically extract features from data and learn representations; human feature extraction is not necessary. DNNs are more effective at mapping complex and nonlinear functions than SL models, which have fewer layers. The proliferation of huge data and the accessibility of less expensive processing units, such as GPUs, have enabled a shift toward deep learning and improved machine learning, with ramifications for a wide range of applications.

Figure 2 depicts CNN in action, using incoming data to build hierarchical representations to accurately recognize higher-level items. A condensed representation of the pixel values in the highlighted portion is shown in the picture by the red squares. Through layers, CNNs gradually extract more complex information until they can identify the image. Obtaining training and testing data, training the network, and using the test data to make predictions are the steps involved in putting neural networks into practice. Image recognition by a CNN refers to the process of using a CNN to analyze and classify images. CNNs are deep learning models specifically designed for image processing tasks, capable of automatically extracting features from images and making predictions based on those features. This approach has revolutionized image recognition tasks by eliminating the need for manual feature engineering and achieving high accuracy in classifying images.

The following is this chapter's road map: Opinion mining, sentiment analysis, and social network analysis are among the social applications covered in Sect. 2. Section 3 examines applications of shallow and deep learning techniques in the social domain. Section 4 offers several difficult open problems. Section 5 provides concluding thoughts and future research directions.

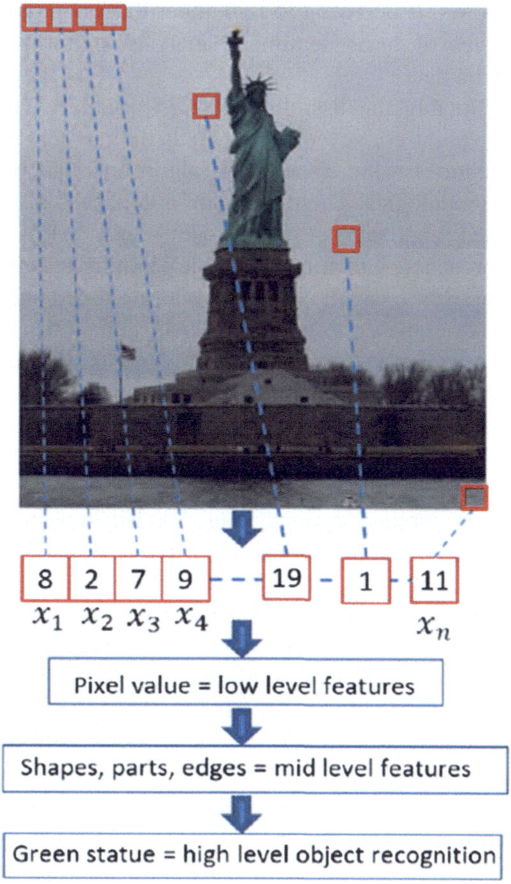

Fig. 2 CNN image recognition [4]

2 Methods: Social Applications

This section is devoted to overviewing some social applications, namely, sentiment analysis (SA), opinion mining (OM), and social network analysis (SNA).

2.1 Sentiment Analysis

SA [5] is a field of ML that focuses on understanding consumer sentiment through statistical analysis and natural language processing. It gathers subjective information from textual sources, such as opinions, and classifies it as neutral, negative, or favorable. This method has several uses, including anticipating behavior in people and figuring out how customers feel about goods or services based on textual reviews. It is crucial to remember that count-based metrics and conventional techniques of basic sentiment analysis are now the only ways to analyze social media data.

Notably, the discipline of SA [5] entails text analysis to ascertain the sentiment or viewpoint conveyed within it. It divides text into positive, negative, and neutral categories using a variety of algorithms and Natural Language Processing (NLP) methodologies. The three main approaches to sentiment analysis systems are rule-based approaches, which use preset rules to identify sentiment; automated approaches, which use machine learning techniques to categorize text; and hybrid approaches, which combine both rule-based and automatic methods.

A rule-based technique in SA [6] quantifies the number of positive and negative phrases in a text to assess the text's mood using a predetermined set of rules. This approach uses stemming, tokenization, and part-of-speech tagging, among other linguistic methods. Compared to rule-based systems, automatic approaches—which classify text into positive, negative, or neutral categories using machine learning techniques—offer greater flexibility and accuracy.

On another separate note, SA [6] is a subfield of data mining that involves analyzing and extracting subjective information from text. It focuses on classifying text based on its sentiment polarity (positive, negative, or neutral) or associating it with emotional states. SA also includes the task of identifying subjectivity/objectivity in text, which can be more challenging due to contextual dependencies and varying definitions of subjectivity.

More potentially, SA [6] has diverse applications across various domains. It is commonly used in business analytics to analyze consumer product reviews, predict sales performance, rank products and merchants based on sentiments expressed in reviews, and even predict box-office revenues. SA is also applied in stock market analysis, where sentiments from sources like social media are leveraged to gain insights into market trends and make informed investment decisions.

SA involves several key subtasks [6]. First, opinion identification is performed to identify words or phrases expressing opinions. Then, feature extraction is carried out to identify important features for analysis, such as part-of-speech, term presence, negations, and opinion words. Ultimately, SA is done to categorize the text's polarity as good, negative, or neutral, and the findings are summarized. Figure 3 provides an illustration of these subtasks.

Elections are essential for choosing leaders and determining a nation's course [7]. SA is the process of examining language, including tweets, to determine the thoughts and feelings of the users. To find out what people's opinions were of Mr. Yogi Adityanathth following the UP elections, live tweets were gathered and examined.

SA is a challenging task [7] in language technology that aims to determine the positive or negative sentiments, opinions, emotions, or subjectivity expressed in natural language text. It can be complicated, as different human annotators may disagree on the categorization of the text due to personal interpretation and cultural factors. Short texts, like comments or tweets on social networks, can pose additional difficulties in interpretation if not written clearly.

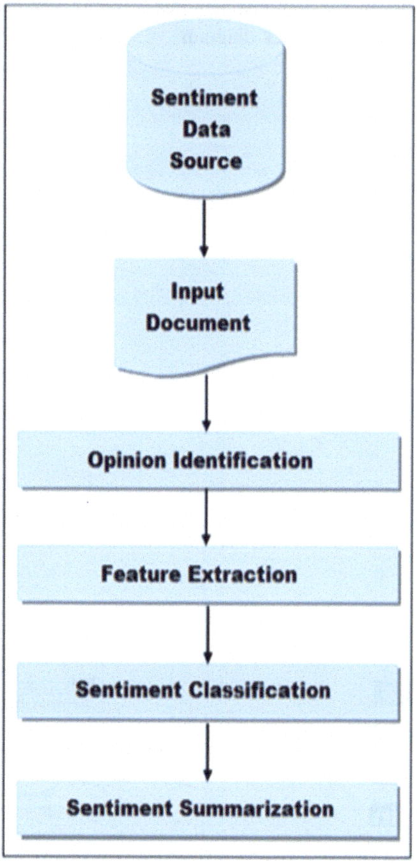

Fig. 3 Sentiment analysis framework [6]

Figure 4 (c.f., [4]) depicts an integrated signature verification system with three stages: preprocessing, feature extraction, and verification. During the preprocessing phase, unwanted components are removed from the signature. During the feature extraction phase, local, global, and transit features are recovered from the signature image, and a decision tree SVM approach is applied for verification. The method matches individual scores using a hybrid technique to enhance signature quality and detect advanced frauds.

2.2 *Opinion Mining*

The study of OM [8] focuses on categorizing texts' polarity as positive, negative, or neutral. It takes opinions out of unstructured texts by applying natural language processing algorithms. The major goal is to comprehend and evaluate people's attitudes toward different things or subjects.

Fig. 4 Integrated signature verification block diagram

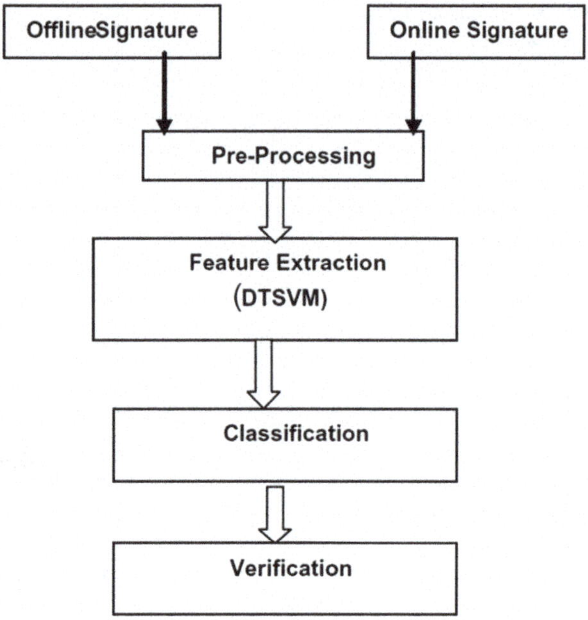

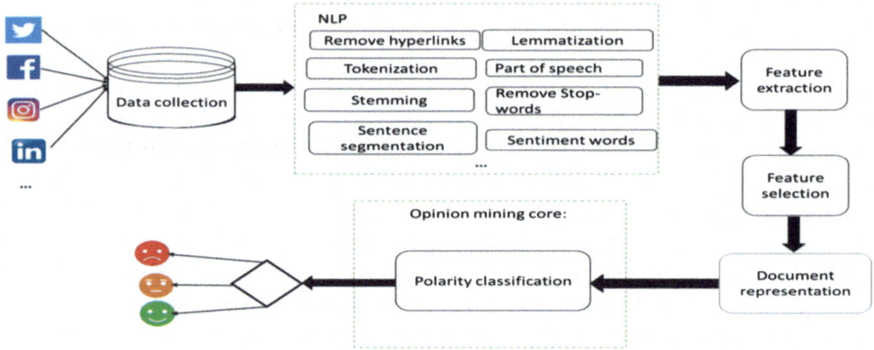

Fig. 5 The flow of OM process (c.f., [8])

A good or negative mood, attitude, feeling, or assessment of a thing or one of its aspects is referred to as an opinion. Regular opinions (like "Camera1 is good") and comparison instances (like "Camera1 is better than Camera2") are two different types of opinions. Additionally, they can be implicit (more difficult to recognize) or explicit (easier to detect and classify). While most research focuses on explicit opinions, limited work has been done on implicit opinions. An outline of the opinion mining procedure is shown in Fig. 5. To process the input text, it requires several components including document representation, feature engineering, polarity computation, and natural language processing (NLP) techniques. Opinions can be analyzed at multiple levels in the context of opinion mining analysis.

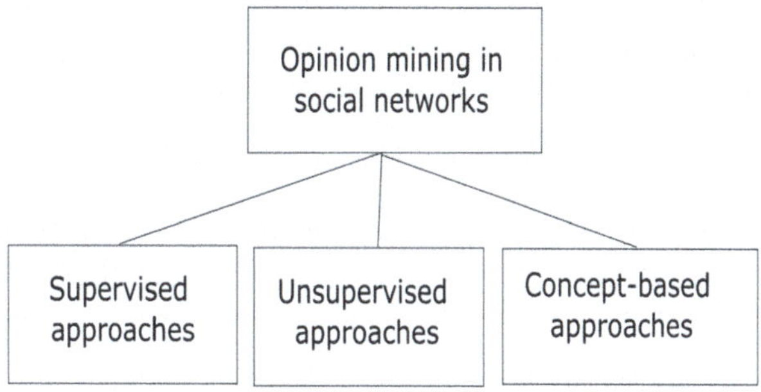

Fig. 6 OM approaches

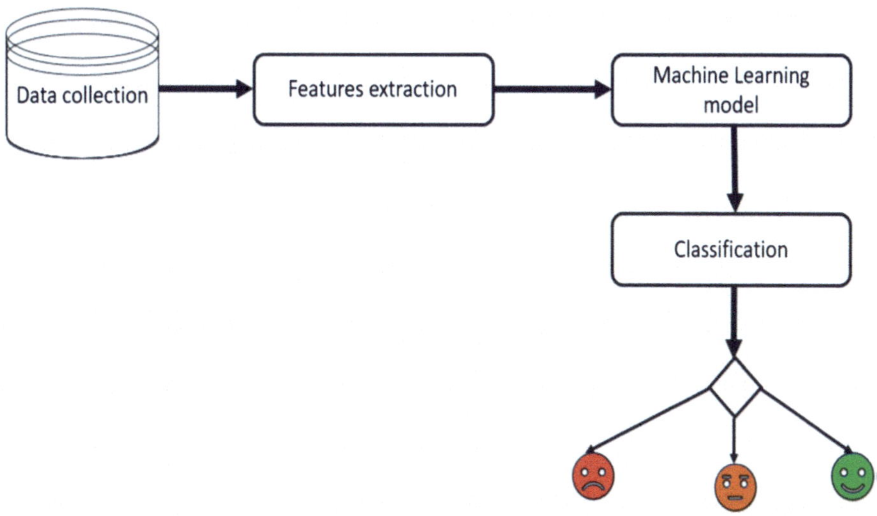

Fig. 7 Generic ML workflow for OM

Document-level, sentence-level, entity-level, and aspect-level analysis are some of these levels. Sentence-level analysis examines opinions encompassed inside individual sentences, whereas document-level analysis focuses on viewpoints reflected within a work as a whole. Entity-level analysis examines opinions related to specific entities or objects, and aspect-level analysis delves into opinions about specific aspects or features of those entities. Understanding these different levels is crucial before delving into the details of each step in OM analysis.

OM can be categorized into three different approaches [8], namely, supervised machine learning, unsupervised lexicon-based approaches, and concept-based approaches. This is depicted by Fig. 6 (c.f., [8]).

Several ML techniques have been developed to overcome the problem in OM. According to the level of analysis—document, sentence, or entity—these methods can be divided into several categories and adhere to a general workflow, as seen in Fig. 7 (c.f., [8]). This classification helps in understanding and applying supervised machine-learning techniques for opinion-mining tasks.

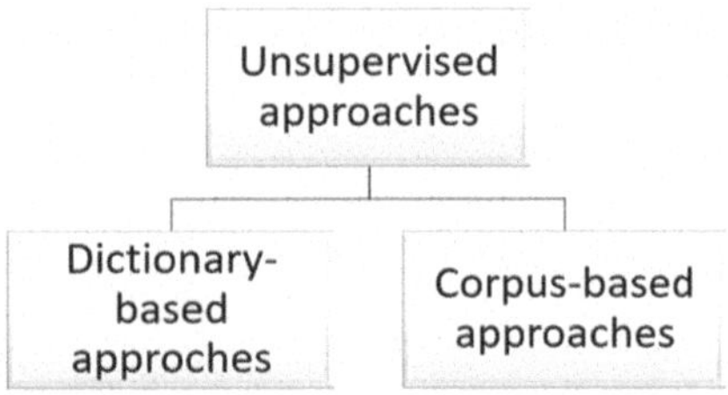

Fig. 8 The unsupervised approaches for OM

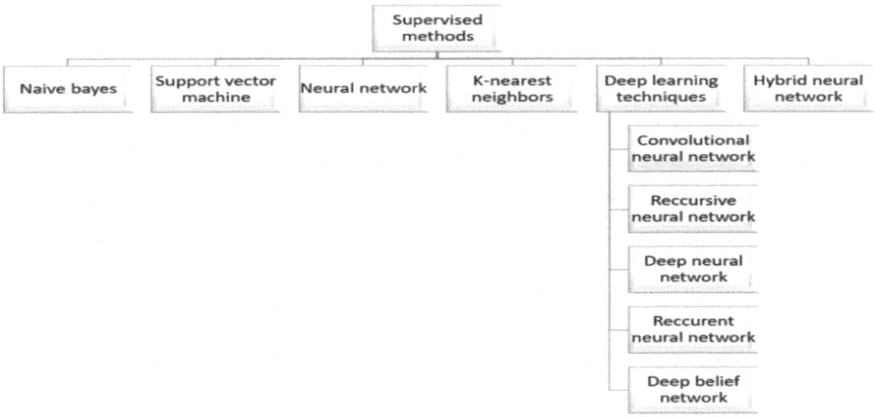

Fig. 9 ML algorithms for OM

In OM, there are various unsupervised approaches that have been proposed. These approaches can be categorized into two types: dictionary-based and corpus-based methods. Dictionary-based methods rely on predefined word lists, while corpus-based methods analyze the overall context of the text. This categorization helps researchers understand and compare different techniques used in OM—see Fig. 8 (c.f., [8]).

Regarding algorithms in general, Fig. 9 (c.f., [8]) shows the ML algorithms used for OM. For this job, various techniques are used in the literature, including Naïve Bayes, maximum entropy classification (EM), support vector machines (SVM), artificial neural networks (ANN), and rule-based machine learning classifiers. To find the best method for sentiment analysis and opinion classification, several algorithms are contrasted.

The text refers to the general structure of the OMLML method, as proposed by [9]. OMLML stands for Opinion Mining using Latent Mixture Models. This method outlines the overall framework or design of the approach used for opinion mining, which involves analyzing and extracting opinions or sentiments from textual data using latent mixture models (Fig. 10).

Ontologies are used by concept-based approaches in OM [8], often referred to as ontology-based approaches, to supply machine- and human-understandable knowledge about areas. An ontology provides a semantic basis for

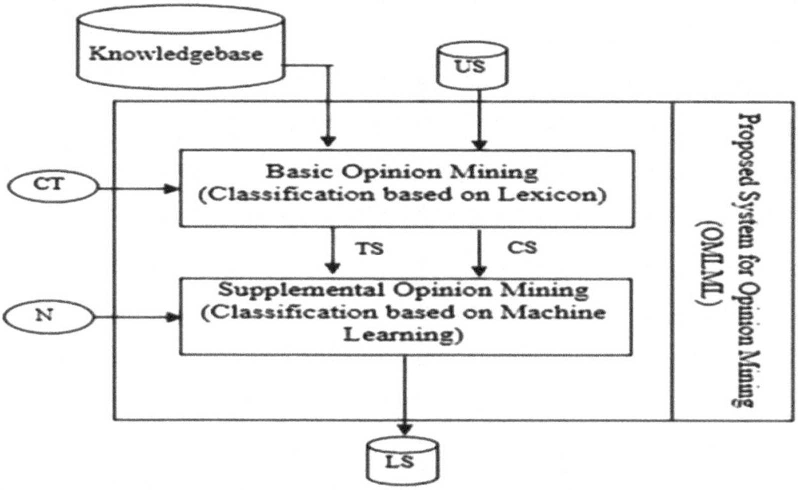

Fig. 10 The OMLML method's general structure

machine-understandable descriptions of digital content by conceptualizing a domain and integrating concepts and relationships among them. Opinion mining algorithms can create deeper and richer sources of knowledge by using ontologies to extract and classify text segments containing concepts linked to a certain subject, like movies.

Stated differently, the concept-based approaches in opinion mining represent a relatively new class that concentrates on extracting and classifying text segments associated with certain domains utilizing ontologies, such as movies. These approaches leverage the conceptualization of a domain and the relationships among concepts to provide a deeper understanding of the content. Figure 11 illustrates the flow of the concept-based OM process.

2.3 *Social Network Analysis*

One fundamental method for analyzing social networks by looking at the connections between social entities is social network analysis (SNA) [10]. It goes beyond traditional statistical techniques and helps understand dependencies, behaviors, and the overall impact of social entities on the network.

In numerous practical applications, data is depicted as an interconnected network of objects [8]. Analyzing the links between these objects will help to extract more information, which makes data analysis more difficult. One popular type of network that is constructed with relational data is a social network. In a graph model, social elements like people, groups, and organizations are represented as vertices and their relationships as edges. Figure 12 provides an example of a network graph (c.f., [10]).

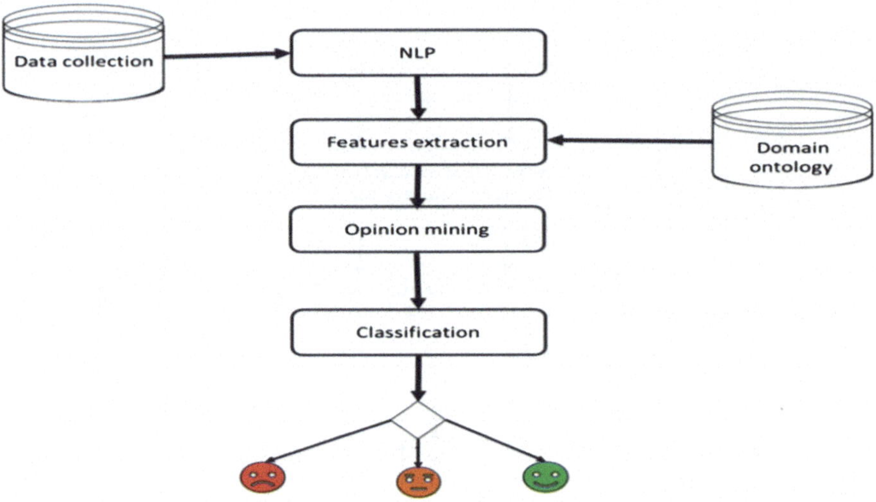

Fig. 11 The flow of concept-based OM

Fig. 12 A schematic of a network representation

According to SNA, a network's nodes—people or organizations—are characterized by the connections that bind them together [10]. These linkages can stand for several kinds of relationships, such as those based on resemblance or being brought together. In a product network, for instance, the links may signify resemblance or being purchased in tandem with a group of products. Like this, different sets of nodes may have unique associations with one another; examples of this include heterogeneous networks, which include user-product networks.

[10] brought attention to the increased complexity of network analysis because of the volume of data and the need for developments in high-velocity enormous scale

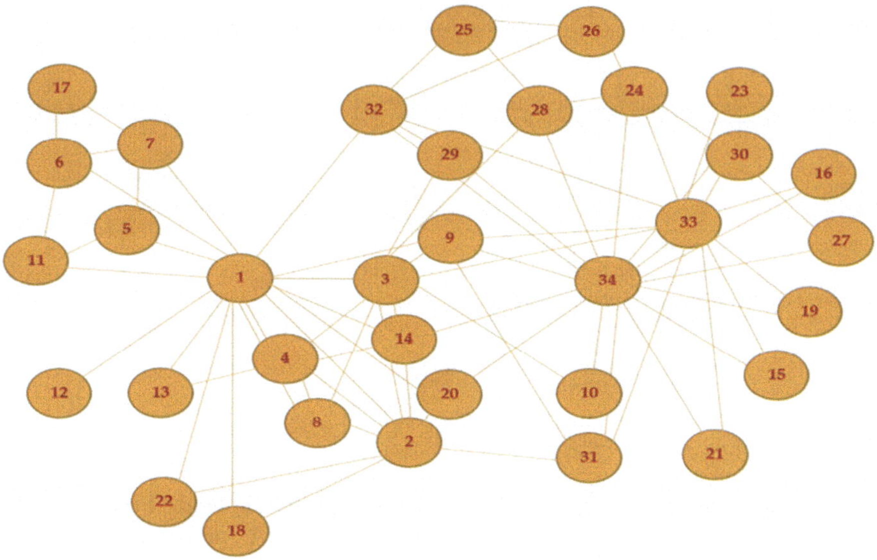

Fig. 13 Zachary's Network of Karate Clubs [12]

network processing and mining. More potentially, an exposition of the challenges and recent developments in designing algorithms for efficient computations of SNA measures was undertaken [10]. On another separate note, the emerging trends and applications in fields such as social reputation, smart cities, bioinformatics, and other inter-disciplinary fields were investigated by [10].

From an alternative perspective, SNA can be defined as the study of nodes, or individuals or things, and the relationships that they have within social networks [11]. By treating the relationships between living and non-living entities as an information network, it seeks to learn more about these structures. The field of social network analysis (SNA) has progressed from examining small-scale networks, like karate clubs (see Fig. 13), to examining large-scale networks seen on online social platforms, which may include millions of interconnected nodes.

The use of online social networks [11], such as Facebook, Twitter, Instagram, and LinkedIn, has grown significantly due to the widespread use of the Internet and advancements in technology. This has led to a massive amount of data being generated as millions of people connect and share their views and emotions on these platforms. As a result, computational advancements and social network analysis methods have emerged to analyze this valuable data.

Online social networks (OSNs) are complex structures consisting of platforms like email services and social network services. These networks are composed of nodes and links that facilitate interactions, collaborations, and information sharing. OSNs have evolved significantly, allowing for the application of various SNA methods, such as Stance Detection, Anomaly Detection, Topic Detection, Role Detection, Phishing Detection, Irony Detection, Opinion Leader Detection, and

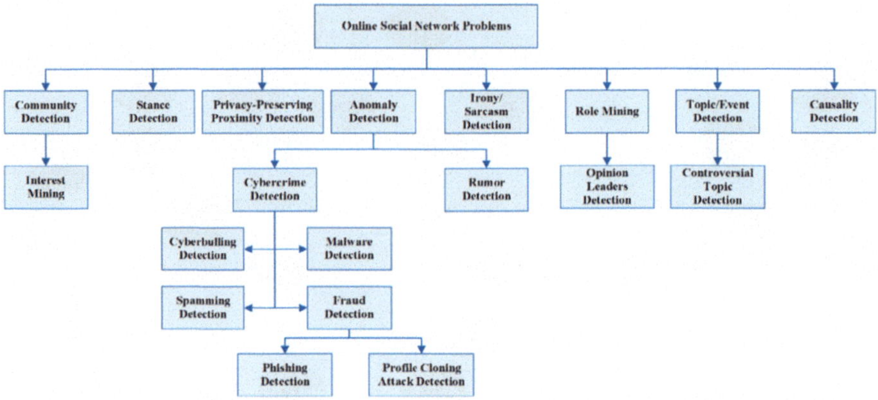

Fig. 14 Issues with online social network analysis

Rumor Detection, which have diverse applications in different fields. Figure 14 (c.f., [11]) provides a visual representation of these SNA problems for better understanding.

Social network structures [11] play a crucial role in facilitating the sharing and dissemination of information. Consequently, social SNA problems have evolved significantly over time, from being applied to small groups to now being applicable to large social networks. This expansion has led to the development of numerous SNA methods, which have the potential to contribute significantly across various domains.

Undoubtedly, researching social networks [11] and the innovations brought about by the Internet revolution has historical relevance. The application of SNA methods to large networks has created opportunities and produced new quantitative and qualitative results. These highlight the significance of SNA in several application areas, including OSNs and community, anomaly, topic, event, and stance detection.

This chapter is organized as follows: Social applications, such as SA, OM, and SNA, are reviewed in Sect. 2. Section 3 provides a comparison of the performance of shallow and deep learning methods in these areas. Some suggested challenging open problems are provided in Sect. 4. Concluding remarks and future research directions are provided in Sect. 5.

3 Results: Applications of Shallow and Deep Learning Techniques in the Social Domain

The current section overviews a comparison between shallow learning (SL) and deep learning (DL) in the social domain, for example, in SA, OM, and SNA.

3.1 Shallow Learning vs. Deep Learning in Sentiment Analysis

Scholars across other fields have shown interest in the use of Deep Convolution Neural Networks (CNNs) in computer vision and image analysis. NgramCNN is a neural network architecture that was created for SA of lengthy text documents by the inventors of [13]. NgramCNN shows promise for SA of variable-length texts by outperforming shallow convolution networks or deeper recurrent networks in terms of accuracy by using pretrained word embeddings and a straightforward single-layer classifier.

The authors [13] introduced NgramCNN, a deep neural architecture that utilizes the speed and effectiveness of Convolutional Neural Networks (CNNs) to extract important features from n-grams in different types of text. In principle, they [13] followed a similar approach that has been successful in image analysis, focusing on complex features and a simple classifier. To represent text features, they [13] use pretrained word embeddings from Google News, which offer excellent generalization and reduced dimensionality. For feature extraction and selection, the architecture uses repeated convolution and pooling layers, using localized max-pooling on text regions rather than global pooling. Each document's emotional category is predicted using a single-layer classifier. If larger datasets are available, the architecture can be enlarged both horizontally and vertically by concatenating and stacking additional convolutions. Experiments verify NgramCNN's efficacy by contrasting it with baseline RNN, shallow CNN, and linear models.

The study's findings [13] showed that the NgramCNN architecture performs better than alternative techniques in terms of training time and prediction accuracy. This implies that NgramCNN is a more efficient method for categorizing and evaluating textual data, including evaluations of smartphones, movies, and song lyrics. Figure 15 shows a schematic of NgramCNN (c.f., [13]).

SA is a technique [14] used to determine the sentiment expressed in a document or sentence, either through lexicon-based approaches or ML methods. While DL models have shown superior performance in automated feature engineering and classification, their application to Persian SA has been limited compared to English. Based on this, the authors [14] developed a novel DL architecture for Persian SA and compared their performance with traditional shallow classifiers.

People across the world are using social media more and more to voice their opinions on a variety of subjects. One method for automatically categorizing these viewpoints as favorable or negative is sentiment analysis. Although sentiment analysis in Persian has been the subject of many research, this work focuses on creating a novel architecture for deep learning algorithms to analyze feelings in datasets of movie and hotel reviews.

Persian movie reviews and hotel review datasets were used for testing and validation using both shallow and deep learning algorithms to evaluate the efficacy of the suggested approach [14]. The outcomes demonstrated that deep learning techniques performed better than conventional shallow ML strategies. More

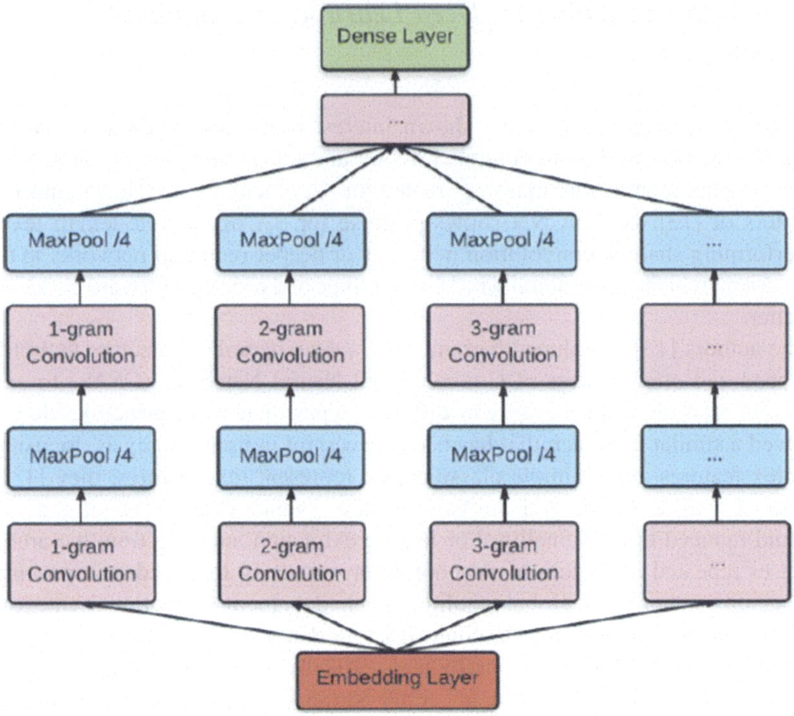

Fig. 15 NgramCNN structure

specifically, for the movie dataset, the stacked-bidirectional-LSTM yielded a maximum accuracy of up to 95.61%, whereas for the hotel dataset, the 2D-CNN fared better with an accuracy of 89.76%. The effectiveness of deep learning classifiers with various layers was also assessed in the study.

A significant amount of data is produced by the World Wide Web [15] in the form of user perspectives, feelings, and opinions on social networks, forums, and review websites. These expressions can be categorized using the SA approach into good, negative, or neutral categories. Sentiment analysis has combined with deep learning techniques to make use of their autonomous learning capabilities; however, the problem is the shortage of labeled data. The application of deep learning models, such as convolutional and deep neural networks, to resolve a range of sentiment analysis issues, such as sentiment categorization and product review analysis, was covered by the writers [15].

Fundamentally, [15] discussed several studies on SA using deep learning models, highlighting their effectiveness in achieving more efficient and accurate SA compared to shallow models. Deep learning networks, with their multiple hidden layers, outperform Support Vector Machines (SVMs) and normal neural networks, and they can be trained in both supervised and unsupervised ways. While DL models offer advantages such as automatic feature extraction and flexibility in handling different problem statements, they also have limitations, including the need for large datasets and expensive training resources.

3.2 *Shallow Learning vs. Deep Learning in Opinion Mining*

Promising outcomes in OM have been observed when ensemble approaches and recent deep learning developments are coupled. Nevertheless, there hasn't been much research done on this method for the Vietnamese language. Based on experimental results, the authors of [16] presented an ensemble approach that combines classical (shallow) and DL algorithms to increase the sentiment classification accuracy for Vietnamese sentiment datasets.

[16] presented an ensemble model for Vietnamese OM (see Fig. 16) that blends shallow and deep learning approaches. Through testing on three sentiment datasets, the authors [16] discovered that the ensemble models improve the precision of individual algorithms in identifying Vietnamese sentiment. Interestingly, ensembles with a mix of shallow and deep learning algorithms typically perform better than those with only DL models.

With the goal of integrating e-commerce with agricultural product marketing, cutting out middlemen, and encouraging the expansion of rural e-commerce outlets, China launched its "Internet plus agriculture" project in 2015. To overcome the obstacles that rural e-commerce faces, the authors [17] investigated the use of deep learning and data mining approaches to optimize e-commerce marketing in rural areas.

Online data processing requires more than standard shallow models when processing transaction data related to agricultural products. The study [17] suggested using DL theory, which is excellent at processing large amounts of data as well as images and voices, to optimize the model to address this. A novel customer value

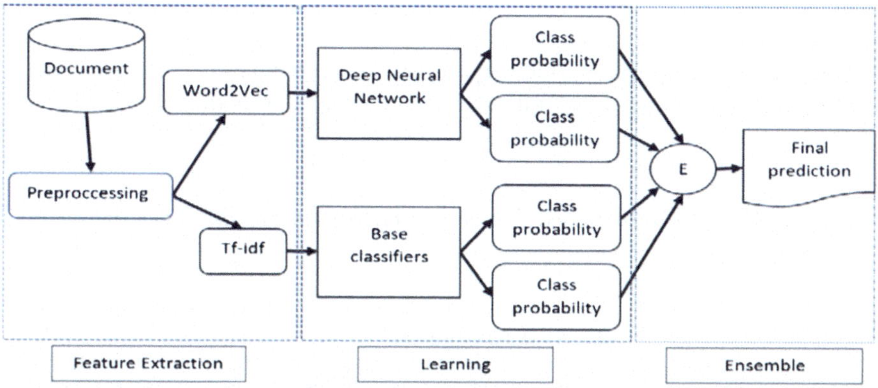

Fig. 16 The ensemble model (c,f., [16])

Fig. 17 Sales platform's proportion (c.f., [17])

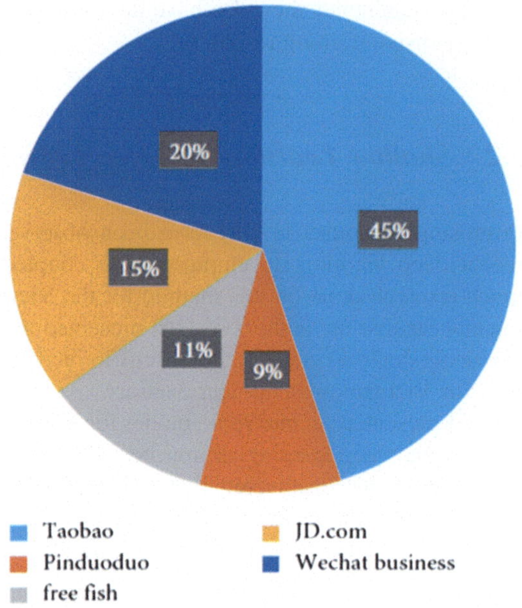

evaluation model was created to improve marketing optimization in agricultural goods e-commerce by fusing data mining and DL technologies.

The distribution of preferences among online shop farmers in a county is shown in Fig. 17, whereby over 45% of farmers choose Taobao, 9% choose Pinduoduo, 11% choose Xianyu, 15% choose Jingdong, and the remaining farmers choose WeChat. Additionally, it states that most of the farmers have been running their Internet stores for 1 to 3 years, while a few have been doing so for more than 3 years. It is also mentioned that most workers at online retailers are related, which results in customer service, packaging, and warehousing duties being completed internally. Furthermore, the reputation of online stores differs depending on educational attainment, with just 16% of them being at the crown level and 50% at the diamond level.

The improvement effect of DL, compared to a benchmark classifier of shallow learning, is shown in Fig. 18. This suggests that DL techniques have demonstrated a positive impact in terms of performance when compared to traditional SL methods.

3.3 Shallow Learning vs. Deep Learning in Social Network Analysis

An exposition of the increasability in online social media usage during COVID-19, which led to the spread of both accurate and inaccurate information, was undertaken by [18]. Notably, the author [18] highlighted that the development of a Hate Speech Detection (HSD) system to address the rise of degrading, racist, and offensive posts

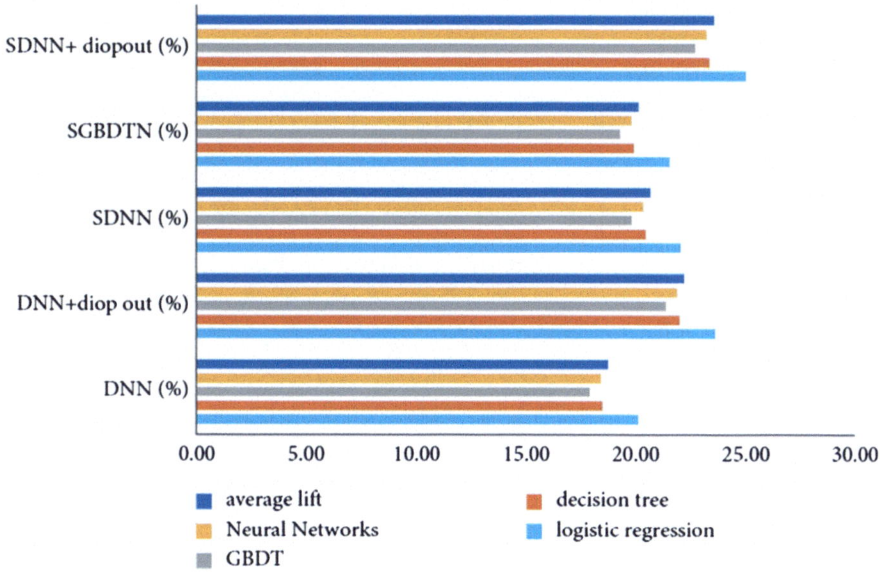

Fig. 18 Percentage of the effect of relative improvement [17]

targeting Asian people as responsible for the outbreak. Using SL and DL techniques, the study in [18] carried out a textual-based analysis, optimizing various models for COVID-19-related hate speech datasets and contrasting their efficacy according to F-score, accuracy, precision, and sensitivity standards. The RNN model demonstrated the highest accuracy values, and it is anticipated that the methods employed in this study will be applicable to additional COVID-19-related social media and network issues.

Hate speech (HS) has been a significant issue on social networks even before the COVID-19 pandemic. It encompasses derogatory, critical, and insulting content targeting individuals or groups. While social media allows freedom of expression, there are important boundaries between expression and hate speech, as exemplified by the blaming of East Asian societies for the COVID-19 outbreak, as illustrated by Fig. 19 (c.f., [19]).

A combination of SNA techniques was applied [18] to analyze HSD in online social networks during the COVID-19 outbreak. The study's contribution [18] includes the use of DL models alongside traditional supervised learning models to improve the performance of HSD related to COVID-19. It was shown by [18] that DL networks consistently outperformed traditional machine learning SL approaches in terms of essential measurement metrics. This suggests that DL models were more effective in achieving higher performance results compared to SL models for the automatic detection of HS in online social networks.

DL has gained attention for its powerful processing capabilities in tasks like speech, image, and text processing. Traditional tools and technologies struggle to analyze the vast amount of data from digital social media (SM), making DL a

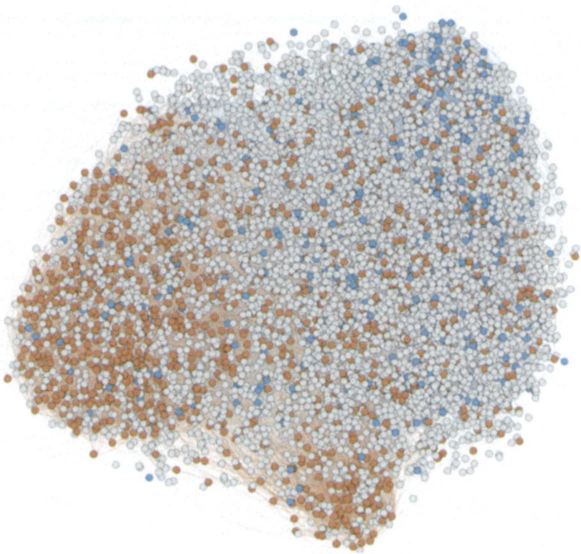

Fig. 19 The distribution of HS (represented by orange), counter HS (represented by blue), and neutral content (represented by grey) on social media platforms. This analysis provides insights into the prevalence and nature of HS during the pandemic, which can help in developing effective strategies to address and prevent the spread of hate speech online

suitable solution. The author's [20] focus was on discussing DL architectures and their applications in SM analytics (SMA), highlighting the problems faced in SMA and the DL-based solutions, along with research challenges and future trends in DL.

SM data is growing rapidly, necessitating the need for advanced techniques to extract meaningful patterns and features for knowledge discovery. Traditional learning methods often employ shallow-structured architectures, but there is a shift toward using DL techniques [20], which automatically learn hierarchical representations for classification tasks. DL offers potential benefits [20] for analyzing SM data due to its ability to uncover complex relationships and has been successfully applied in various domains such as image processing, speech recognition, and collaborative filtering. Figure 20 displays the taxonomy of DL for social media (c.f., [20]).

4 Discussion

In summary, social media platforms pose several difficulties for deep learning techniques [20], but they also provide chances for useful data visualizations across a range of fields, including business, sentiment, user behavior, and anomaly detection. The need for resources, increasing productivity, cutting computing expenses, and figuring out effective data representations from various social data sources are some of the issues that still need to be resolved [20].

The open problems are adapted from [21].

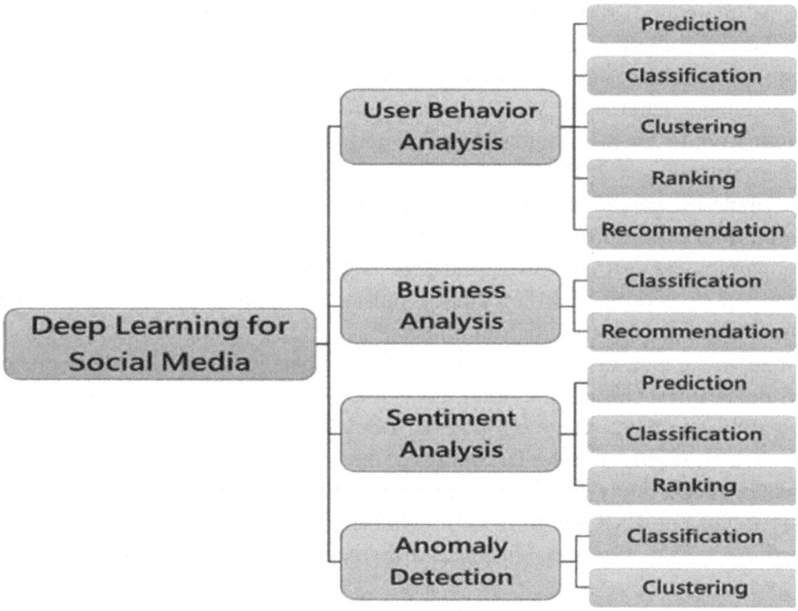

Fig. 20 Taxonomy of DL methods for social media

4.1 Some Challenging Open Problems

- Traditional bullying is defined as intentional harm repeatedly inflicted on a victim who struggles to defend themselves. Cyberbullying extends this definition to aggressive behavior using electronic platforms against a victim who cannot easily defend themselves. However, applying this definition to online environments poses challenges in measuring repetitive behavior and power imbalance. It is important to build ML models that accurately recognize cyberbullying participation. Some academics have proposed to simplify the idea of cyberbullying, describing it as simply the use of electronic communication technology to bully others. This poses a challenging open problem for DL developers to tackle these socially based issues by selecting the appropriate social domain.
- Many cyberbullying prediction studies collect data by tracking posts with specific keywords, which can introduce sampling bias and limit the prediction model to only those posts. It is critical to incorporate a variety of posts that demonstrate cyberbullying when developing machine learning models and to make sure the data gathered accurately reflects all potential social media platform activity. Potential bias in data collection should be avoided by researchers since it might cause overfitting and impair the model's capacity to generate accurate predictions on unobserved data.
- Enhancing the predictive power of ML models for cyberbullying on social media platforms is mostly dependent on features. Previous studies have focused on

extracting features such as age and gender from user profiles, but these may not be available for all users. Other important features, such as words related to aggression, neuroticism, and sociability, can be derived from the content of posts and engagement patterns on social media platforms, providing comprehensive discriminative features for better classifier performance.

- When selecting a ML algorithm for a specific dataset, it is challenging to determine the best classifier. Researchers can use a literature review on machine learning for cyberbullying detection, previous research findings, and performance comparisons of comprehensive datasets as guides to narrow down the selection. However, a variety of ML methods must be evaluated to determine the best classifier for accurate prediction models.
- In real-world datasets, it is common to have imbalanced classes, where one class has many instances and the other has a small number. This is also the case with cyberbullying data, where the number of cyberbullying posts is much less than non-cyberbullying posts. To improve the performance of machine learning models in accurately classifying the examples, approaches such as cost-sensitive adjustment and SMOTE (Synthetic Minority Oversampling Technique) can be used to create synthetic instances or modify the weights of the classes, respectively.
- The selection of suitable evaluation metrics for ML models predicting cyberbullying on social media is contingent upon the characteristics of the labeled data. When there is an imbalance in the data, such as when the number of non-cyberbullying posts is much higher than the number of cyberbullying posts, metrics like AUC (Area Under the Curve) can be used to evaluate the model's performance more thoroughly than metrics like accuracy. To ensure correct evaluation of machine learning algorithms in handling imbalanced datasets, careful consideration of the evaluation metric is essential.

5 Concluding Remarks and Future Research Directions

The current chapter discusses the exploration of shallow and deep learning techniques in the social domain. It specifically focuses on applications like sentiment analysis, opinion mining, and social network analysis, comparing the performance of SL and DL methods within these areas. Clearly, DL was shown to outperform SL in these social domains. On a very potential note, some challenging open problems are highlighted, to strengthen the establishment of more innovative explorations from experts in the field to add more to the knowledge. Future research pathways involve investigating the proposed open problems and exploring more mathematical insights to enhance the overall DL performance in dealing with other emerging social domain issues, fundamentally, through the provision of more solutions to the needed improvement to the current state of the art in the field.

References

1. Xu, Y., Zhou, Y., Sekula, P., & Ding, L. (2021). Machine learning in construction: From shallow to deep learning. *Developments in the Built Environment, 6*, 100045.
2. Li, Q., Peng, H., Li, J., Xia, C., Yang, R., Sun, L., et al. (2022). A survey on text classification: From traditional to deep learning. *ACM Transactions on Intelligent Systems and Technology (TIST), 13*(2), 1–41.
3. Suzuki, M., Pennartz, C. M., & Aru, J. (2023). How deep is the brain? The shallow brain hypothesis. *Nature Reviews Neuroscience*, 1–14.
4. Shrestha, A., & Mahmood, A. (2019). Review of deep learning algorithms and architectures. *IEEE Access, 7*, 53040–53065.
5. Saju, B., Jose, S., & Antony, A. (2020, July). Comprehensive study on sentiment analysis: Types, approaches, recent applications, tools, and APIs. In *2020 advanced computing and communication technologies for high performance applications (ACCTHPA)* (pp. 186–193). IEEE.
6. Tedmori, S., & Awajan, A. (2019). Sentiment analysis main tasks and applications: A survey. *Journal of Information Processing Systems, 15*(3), 500–519.
7. Baid, P., & Chaplot, N. (2019). Sentiment analysis of live tweets after elections. In *Emerging trends in expert applications and security: Proceedings of ICETEAS 2018* (pp. 307–314). Springer Singapore.
8. Messaoudi, C., Guessoum, Z., & Ben Romdhane, L. (2022). Opinion mining in online social media: A survey. *Social Network Analysis and Mining, 12*(1), 25.
9. Keyvanpour, M., Karimi Zandian, Z., & Heidarypanah, M. (2020). OMLML: A helpful opinion mining method based on lexicon and machine learning in social networks. *Social Network Analysis and Mining, 10*, 1–17.
10. Tabassum, S., Pereira, F. S., Fernandes, S., & Gama, J. (2018). Social network analysis: An overview. *Wiley Interdisciplinary Reviews: Data Mining and Knowledge Discovery, 8*(5), e1256.
11. Can, U., & Alatas, B. (2019). A new direction in social network analysis: Online social network analysis problems and applications. *Physica A: Statistical Mechanics and its Applications, 535*, 122372.
12. Zachary, W. W. (1977). An information flow model for conflict and fission in small groups. *Journal of Anthropological Research, 33*(4), 452–473.
13. Çano, E., & Morisio, M. (2018, April). *A deep learning architecture for sentiment analysis*. In Proceedings of the international conference on geoinformatics and data analysis (pp. 122–126).
14. Dashtipour, K., Gogate, M., Adeel, A., Larijani, H., & Hussain, A. (2021). Sentiment analysis of persian movie reviews using deep learning. *Entropy, 23*(5), 596.
15. Ain, Q. T., Ali, M., Riaz, A., Noureen, A., Kamran, M., Hayat, B., & Rehman, A. (2017). Sentiment analysis using deep learning techniques: A review. *International Journal of Advanced Computer Science and Applications, 8*(6).
16. Nguyen, H. Q., & Nguyen, Q. U. (2018, November). An ensemble of shallow and deep learning algorithms for Vietnamese sentiment analysis. In *2018 5th NAFOSTED conference on information and computer science (NICS)* (pp. 165–170). IEEE.
17. Yang, H., Zheng, Z., & Sun, C. (2022). E-commerce marketing optimization of agricultural products based on deep learning and data mining. *Computational Intelligence and Neuroscience, 2022*.
18. Baydogan, C. (2022). Deep-Cov19-Hate: A textual-based novel approach for automatic detection of hate speech in online social networks throughout COVID-19 with shallow and deep learning models. *Tehnički vjesnik, 29*(1), 149–156.
19. Ziems, C., He, B., Soni, S., & Kumar, S. (2020). *Racism is a virus: Anti-Asian hate and counter hate in social media during the covid-19 crisis.*

20. Hayat, M. K., Daud, A., Alshdadi, A. A., Banjar, A., Abbasi, R. A., Bao, Y., & Dawood, H. (2019). Towards deep learning prospects: insights for social media analytics. *IEEE Access, 7*, 36958–36979.
21. Al-Garadi, M. A., Hussain, M. R., Khan, N., Murtaza, G., Nweke, H. F., Ali, I., et al. (2019). Predicting cyberbullying on social media in the big data era using machine learning algorithms: Review of literature and open challenges. *IEEE Access, 7*, 70701–70718.

Shallow Learning vs. Deep Learning in Image Processing

Muhammed Fatih Akil and Ömer Faruk Ertuğrul

1 Introduction

Machine learning (ML)—a method where a computer discovers rules to execute a data processing task, given training examples—can generally be divided into two categories: Shallow learning (SL) and deep learning (DL) methods [1]. Deep learning uses many successive layered representations of data (i.e., hundreds of convolutions/filters), while shallow learning typically uses one or two layered representations of the data [2]. Deep learning has shown great promise for tackling many tasks such as image recognition, natural language processing, speech recognition, superhuman Go playing, and autonomous driving [2–4]. Shallow ML heavily relies on such well-defined features, and therefore its performance is dependent on a successful extraction process. Multiple feature extraction techniques have emerged over time that are applicable to different types of data. For example, when analyzing time-series data, it is common to apply techniques to extract time-domain features (e.g., mean, range, skewness) and frequency-domain features (e.g., frequency bands) [5]; for image analysis, suitable approaches include histograms of oriented gradients (HOG) [6], scale-invariant feature transform (SIFT) [7], and the Viola-Jones method [8], and in NLP, it is common to use term frequency-inverse document frequency (TF-IDF) vectors [9], part-of-speech (POS) tagging, and word shape features [10].

DL and SL methods have been extensively used in image processing, specifically in medical image analysis. Until 2017, SL methods, like SVM, ANN, and k-NN, were much more popular than DL [11]. After GPU was developed, DL methods became intensively used. SL and DL algorithms are important ways in medical imaging to predict the symptoms of early disease. According to [12], SL and DL techniques have been used for breast cancer classification and achieved 98.33% and 98.95%, respectively. Moreover, Rana and Brushan applied DL vs. SL methods to

M. F. Akil (✉) · Ö. F. Ertuğrul
Department of Electrical and Electronics Engineering, Batman University, Batman, Turkey

Ö. F. Ertuğrul et al. (eds.), *Shallow Learning vs. Deep Learning*, The Springer Series in Applied Machine Learning, https://doi.org/10.1007/978-3-031-69499-8_5

MRI images for classification of tumors. They achieve a 97.6% accuracy rate for DL, specifically CNN, and 96.9% with Random Forest (a SL method). Besides, Meherwar mentions [13] SL methods have been successfully used for medical predictions, specifically, Naïve Bayes to predict diabetes and attain 95% accuracy and SVM to estimate heart disease and gain 94% accuracy rate. In addition, Ertuğrul and Akıl used DL to detect hemorrhage types [14] and SL methods to estimate the mismatch between diffusion-weighted images and perfusion-weighted images after stroke [15].

In this chapter, deep learning and shallow learning algorithms are compared for the classification of heartbeats extracted from ECG images. Because the ECG is a common, inexpensive, and noninvasive tool for measuring the electrical activity of the heart [16], it is used to identify heart-related cardiovascular diseases [16, 17]. According to the World Health Organization, cardiovascular diseases (heart diseases) are the leading cause of death worldwide. They claim an estimated 17.9 million lives each year, accounting for 32% of all deaths worldwide. About 85% of all deaths from heart disease are due to heart attacks, also known as myocardial infarctions (MI) [18]. Many lives can be saved if an efficient diagnosis of cardiovascular disease is detected at an earlier stage [18]. Different techniques are used in the healthcare system to detect heart diseases, such as electrocardiogram (ECG), echocardiography (echo), cardiac magnetic resonance imaging, computed tomography, blood tests, etc. [19]. ECG is widely used by cardiologists and medical practitioners for monitoring cardiac health. The main problem with the manual analysis of ECG signals, similar to many other time-series data, lies in the difficulty of detecting and categorizing different waveforms and morphologies in the signal especially for practitioners at first step healthcare organizations.

Shallow learning for ECG classification is a well-established method. Countless studies have used random forest [20], support vector machine (SVM) [21], boosted regression trees [22], and many other algorithms to classify ECG signals. These algorithms typically work on algorithms that extract the numerical value of the raw ECG signals. With this numerical data, a shallow learning model can be built through methods such as kernel methods/decision boundaries [23], decision trees [22], and gradient boosting [24]. These models can then be used to predict.

The use of deep learning in ECG classifying is a more recent development, and it is gaining popularity [25]. Of the many types of deep learning approaches that exist, convolutional neural network (CNN) methods have been selected. Because it has achieved satisfactory results on image classification tasks. Hasan and Bhattacharjee [26] used a convolution neural network (CNN) to classify ECG databases. They achieved a 97.70% accuracy rate for the MIT-BIH database, a 99.71% accuracy rate for the St- Petersburg database, and a 98.24% accuracy rate for the PTB database with their proposed method. Of course, not only these researchers studied, but Sun et al. [27], Remya et al. [28], Hammad et al. [29], Banerjee and Mitra [30], Padhy and Dandapat [31], and Sahoo et al. [32] have also made various classification studies on the ECG dataset, especially on the PTB dataset.

Based on this explanation, a comparative study was conducted on classifying ECG heartbeat images with shallow and deep learning algorithms. The goal is to determine which classification performance is better between algorithms based on shallow learning and deep learning.

2 Materials and Method

2.1 Dataset

In this chapter, Kaggle's ecg-image_data has been used. This dataset had been generated from PhysioNet MIT-BIH Arrhythmia ECG Databases as a data source for labeled ECG records. The MIT-BIH dataset consists of ECG recordings from 47 different subjects recorded at the sampling rate of 360 Hz. Each beat is annotated by at least two cardiologists. Kachuee et al. used annotations in this dataset to create five different beat categories in accordance with the Association for the Advancement of Medical Instrumentation (AAMI) EC57 standard [33]. Table 1 is the summary of mappings between beat annotations in each category.

Heartbeat images of different categories are shown in Fig. 1.

3 Methodology

Shallow ML training and deep learning training have been conducted according to the process shown in Figs. 2 and 3.

Table 1 Summary of mappings between beat annotations and categories [33]

Category	Annotations	Number of training samples	Number of test s amples
N	_ Normal _ Left/right bundle branch block _ Atrial escape _ Nodal escape	13,626	18,926
S	_ Atrial premature _ Aberrant atrial premature _ Nodal premature _ Supra-ventricular premature	2223	556
V	Premature ventricular contraction _ Ventricular escape	5789	1447
F	Fusion of ventricular and normal	642	161
Q	_ Paced _ Fusion of paced and normal _ Unclassifiable	6431	1608

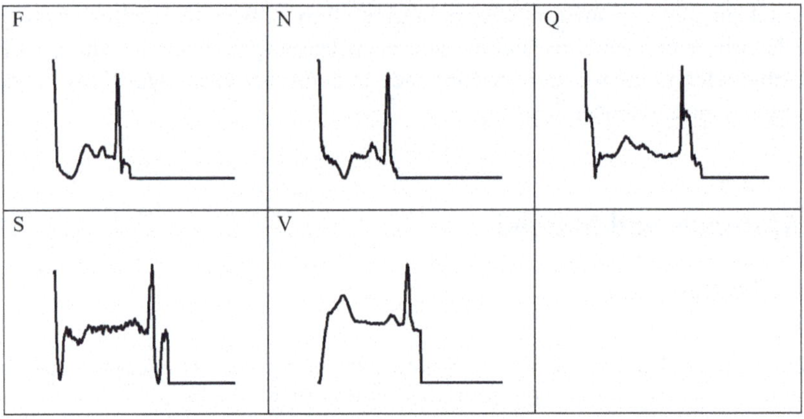

Fig. 1 Heartbeat images for each category

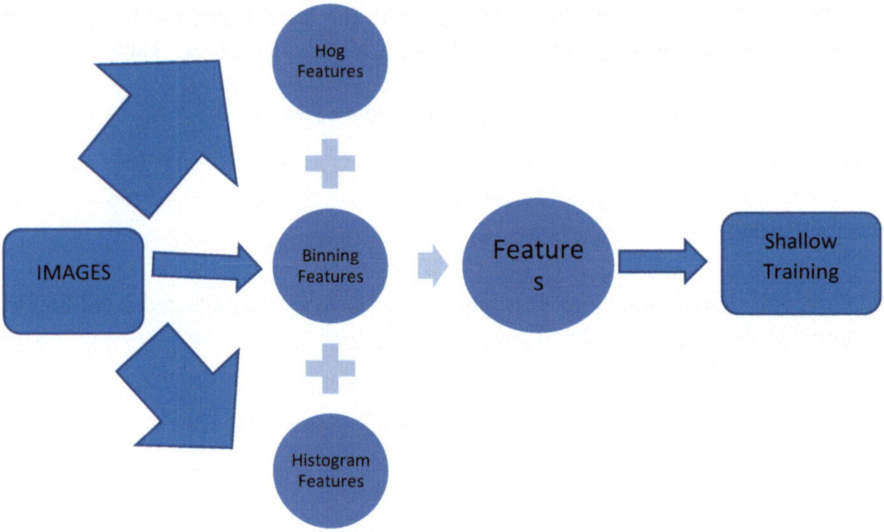

Fig. 2 Shallow ML process

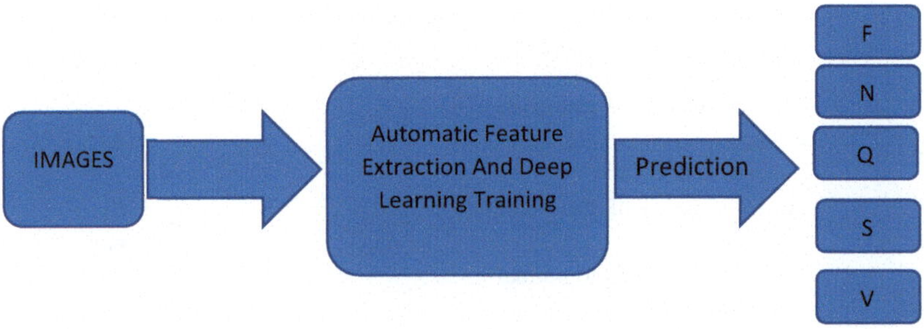

Fig. 3 Deep learning process

During the shallow process, different features have been trained separately, and a combination of feature extraction methods has been used in the training process to investigate features extraction methods effect on prediction results.

A. Feature Extraction.

In this study, three different feature extraction methods have been used for shallow ML training.

(a) Histogram of Oriented Gradients (HOG).

Histogram of oriented gradients (HOG) is a gradient-based feature extraction method, which was first proposed in the study by Dalal and Triggs [6]. This algorithm defines an image as a series of local histograms. Each of the local histograms is the distribution of the number of occurrences of gradients in specified directions that are calculated in a particular area defined as a cell on the image [34]. HOG extraction method has been explained in detail by Ertuğrul et al. in [34]. In this study, histogram divisions were created by calculating unsigned gradients over rectangular cells (R− HOG) [35]. The entire image was divided into 16 × 16 rectangular cells, and a local histogram was created from each cell. Then, four histogram divisions per cell (180° / 45° = 4) were combined with a total of 486 local histograms to acquire a length of (18 × 27) 486 units HOG vector. So, the 1944 feature has been extracted.

(b) Binning Feature.

The binning feature is used as a second feature extraction method. The simple dimension of images is reduced from (288, 432) to (32,32), and the image is flattened to generate 1024 features.

(c) Histogram.

Histograms are constructed by splitting the range of the data into equal-sized bins (called classes). Then, for each bin, the numbers of points from the data set that fall into each bin are counted. In this study, 256 features have been selected for histogram features.

B. Shallow Machine Learning Classifiers.

In this study, three different types of shallow ML classifier have been used to see classifiers performance and effect of choosing ML classifier methods.

(a) Decision Tree (DT) Classifier.

DTs [36] are powerful, efficient, and popular approaches in data mining and knowledge discovery (KD) to explore large and complex datasets to find useful patterns. DTs are very effective tools in various disciplines including data mining, machine learning, information extraction, text mining, and pattern recognition. There are different algorithms that are used for DTs. In this study, skilit library has been used for DTs. Skilit library is using optimized version of CART algorithm which is proposed by [37].

(b) Random Forest Classifier.

The random forest classifier consists of a combination of tree classifiers where each classifier is generated using a random vector sampled independently from the input vector, and each tree casts a unit vote for the most popular class to classify an input vector [38]. Breiman suggests that as the number of trees increases, the generalization error always converges even without pruning the tree, and overfitting is not a problem because of the Strong Law of Large Numbers. The number of features used at each node to generate a tree and the number of trees to be grown are two user-defined parameters required to generate a random forest classifier.

(c) XGBoost Classifier.

XGBoost, also called eXtreme Gradient Boosting, is a machine learning algorithm that is becoming widespread as it won many Kaggle data science competitions [39]. It provides satisfactory results in many applications such as disease prediction [40], diesel fuel brands identification [41], estimation of the tunnel boring machine advance rate [42] and prediction of concrete electrical resistivity for structural health monitoring [43]. XGBoost presented by Chen and Guestrin [44] is a gradient boosted decision tree model. XGBoost algorithm trains decision trees on training data sequentially. The algorithm adds a new decision tree at each iteration to the previous decision trees to improve the objective function's value.

C. Deep Learning Algorithm.

Convolutional neural network (CNN) models are developed for 2D image recognition [45]. A CNN comprises successive convolutional (filtering) and pooling (subsampling) layers with nonlinearity applied before or after pooling, potentially followed by one or more fully connected layers. In classification problems, the last layer of a CNN is commonly a softmax (multinomial logistic regression) layer. CNNs are trained using iterative optimization with the backpropagation algorithm [45]. Multiple convolutional and pooling layers can be incorporated into a model to make it "deeper" and increase its recognition ability to classify complex tasks. After a series of convolutional and pooling layers, the resulting feature map is flattened into a single list of vectors before being fed to the fully connected layers, establishing connections between output and input via learnable weights [45]. Model representation can be seen in Fig. 3. Dropout layers are placed in the model at a convolutional and fully connected layer to prevent the overfitting problem (Fig. 4).

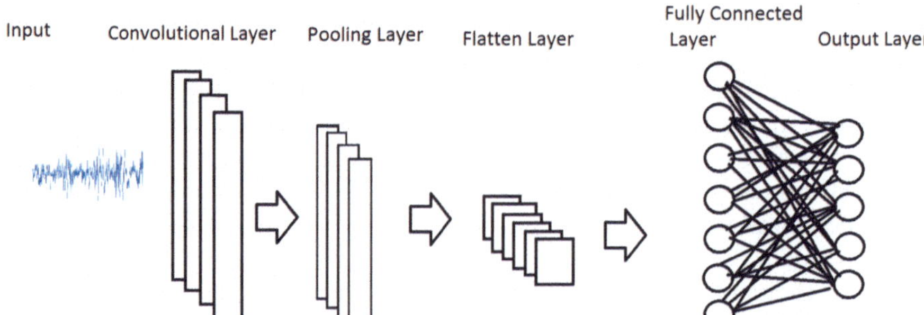

Fig. 4 CNN architecture demonstration

Table 2 CNN layers, layer outputs, and parameters

Layer	Output of layer	Number of parameters
Conv2D	288, 432, 32	832
Batch normalization	288, 432, 32	128
MaxPooling2D	144, 216, 32	0
Conv2D	144, 216, 64	51,264
Batch normalization	144, 216, 64	256
MaxPooling2D	72, 108, 64	0
Conv2D	72, 108, 64	102,464
Batch normalization	72, 108, 64	256
MaxPooling2D	36, 54, 64	0
Conv2D	36, 54, 90	144,090
Batch normalization	36, 54, 90	360
MaxPooling2D	18, 27, 90	0
Flatten	43,740	0
Dense	3032	132,622,712
Dropout	3032	0
Softmax	5	18,198

Table 3 Test sets accuracy rates

	HOG	Binning	Histogram	HOG + Binning	All features
DT	90.18	88.21	72.84	90.38	90.93
RF	**97.79**	**96.50**	71.00	97.81	97.94
XGB	97.70	96.06	**76.63**	**97.81**	**97.97**
Deep learning	**98.10**				

All of the layers of the model and detailed parameter representations of these layers are given in Table 2.

4 Experiment Results

The experimental results were collected separately for each feature and combination of best effective features. Besides, results were collected for deep learning. It can be seen that deep learning has higher accuracy. Moreover, XGB model with all features has higher accuracy among shallow learnings (Table 3).

The confusion matrix of best shallow ML for each feature and deep learning are demonstrated in Figs. 5, 6, 7, 8, 9, and 10. In these figures, category 0 is label F, 1 is label N, 2 is label Q, 3 is label S, and 4 is label V.

It can be seen from Fig. 5, as that RF training by HOG has a 97.79% accuracy rate, and most of the labels are predicted correctly. However, false predictions are confused with label N (category 1) as it is the most dominated category.

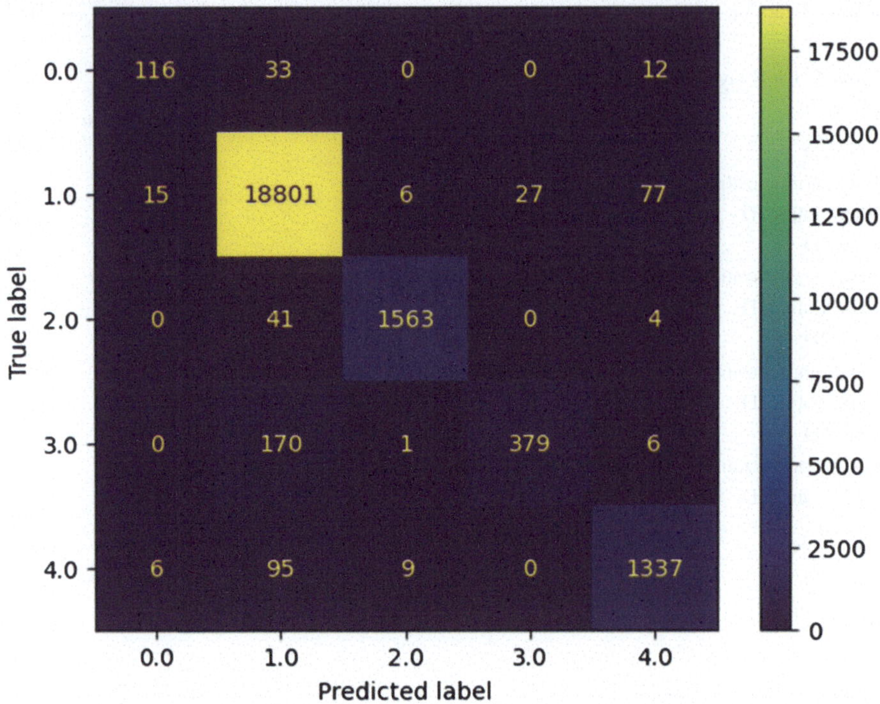

Fig. 5 Confusion Matrix for RF trained by HOG

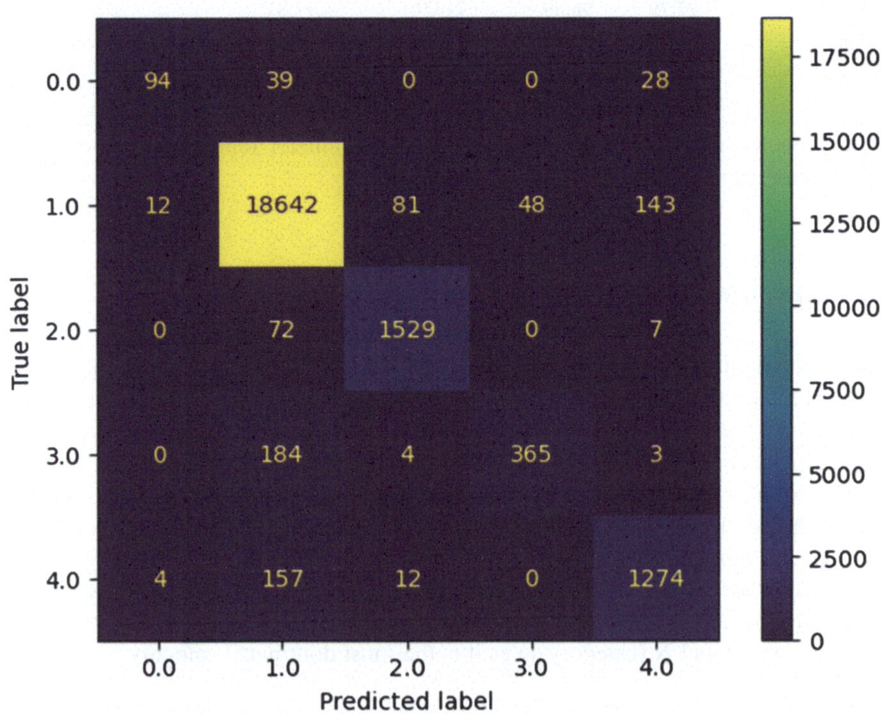

Fig. 6 Confusion Matrix for RF trained by Binning

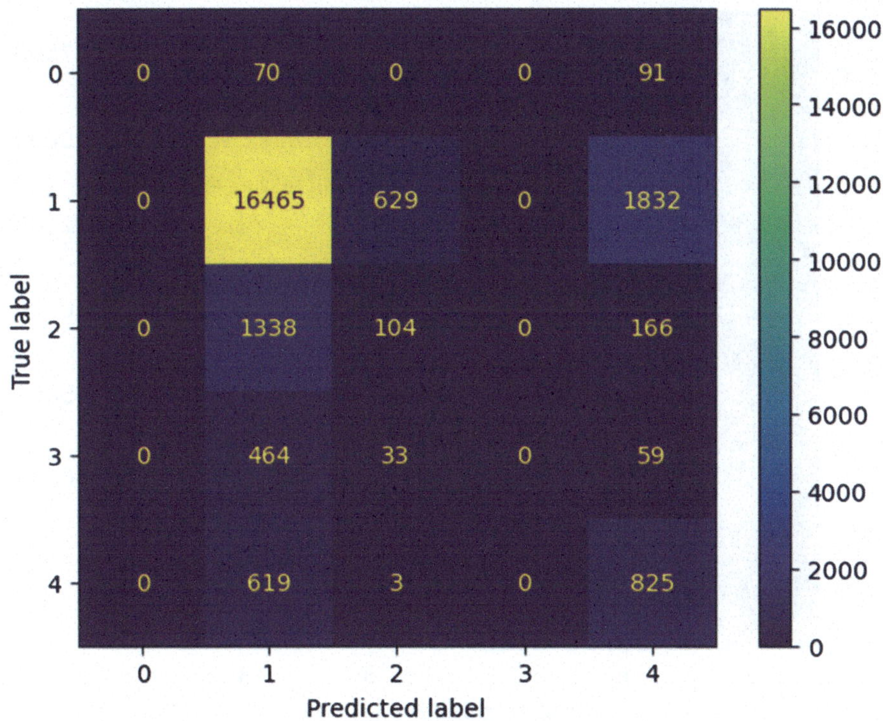

Fig. 7 Confusion Matrix for XGB trained by histogram

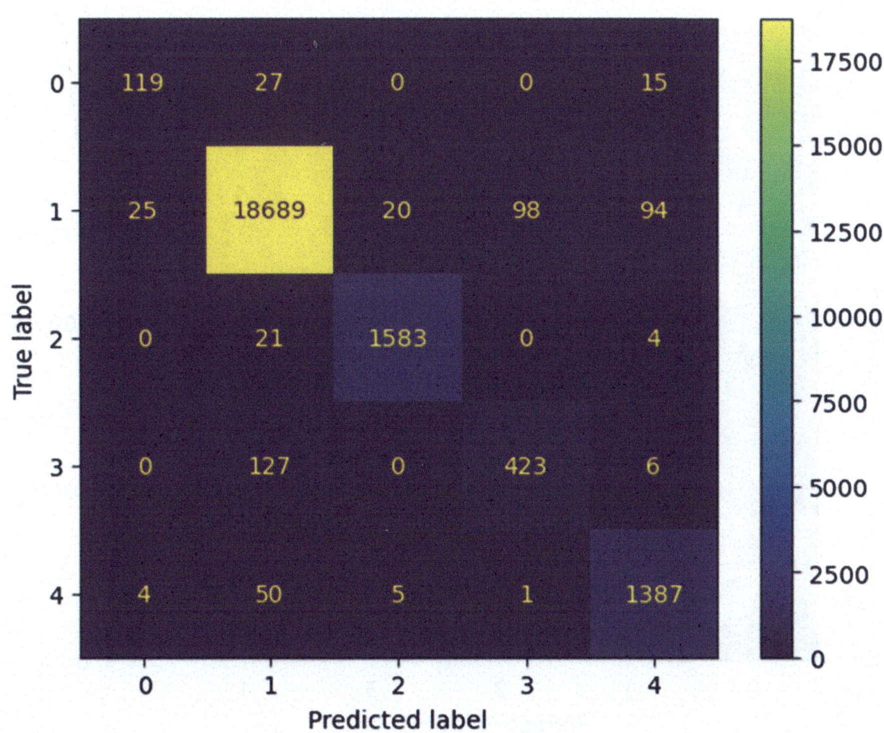

Fig. 8 Confusion Matrix for XGB trained by HOG + Binning

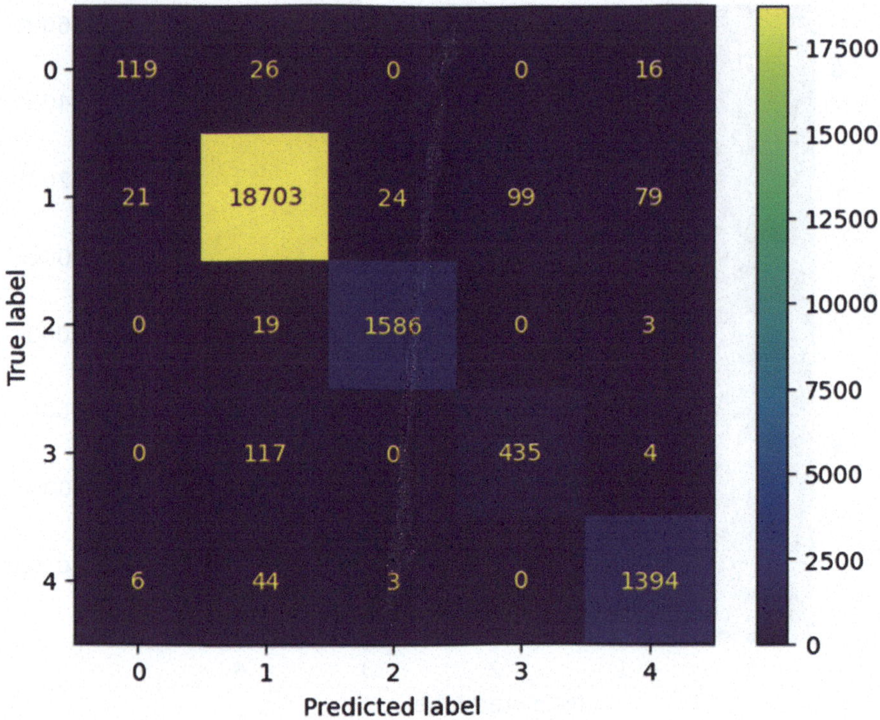

Fig. 9 Confusion Matrix for XGB trained by all features

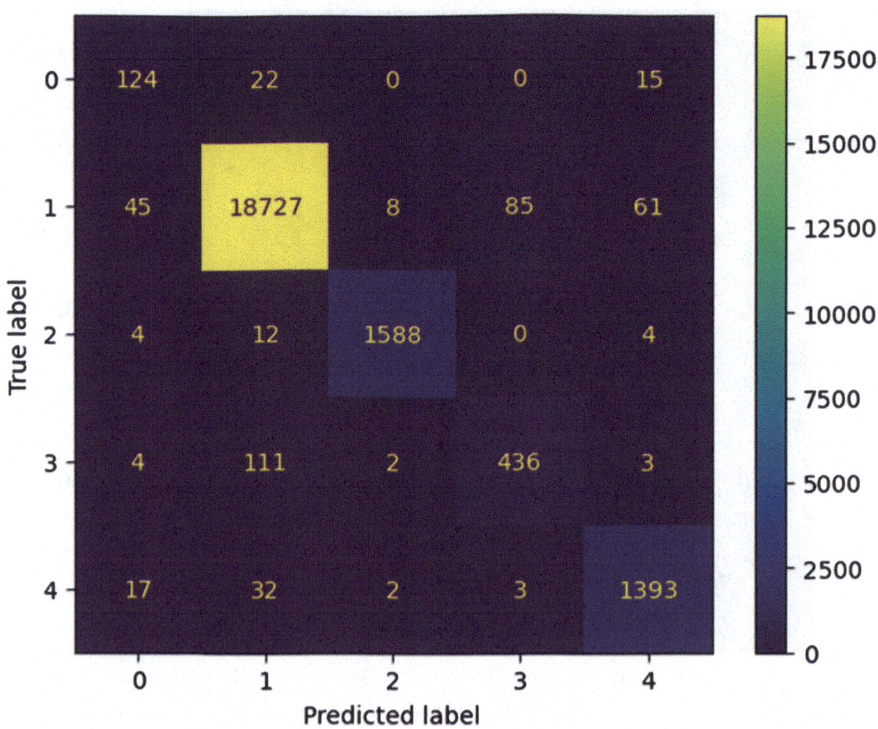

Fig. 10 Confusion Matrix for deep learning

In Fig. 6, Confusion Matrix of RF trained by Binning, it is clear that most false predictions are confused with label N again. Besides, label N is confused with label Q and V (category 2 and 4) which are second and third dominated categories.

It can be clearly seen from Fig. 7 that XGB trained by histogram has very low accuracy, and there are many confusions between labels. Although it seems it has 76.63% accuracy, most of the categories have less than 10% accuracy. For example, category 2 has 6.5% accuracy, and category 3 and category 0 have 0% accuracy.

In Fig. 8, as XGB trained by HOG + Binning has 97.81% accuracy, there are few confusions. Interestingly, although label S (category 3) is not dominated, label 1 is most confused with label 3.

Figure 9 demonstrates a very close pattern to Fig. 8 by increasing each category's accuracy except category 0.

In Fig. 10, it can be seen that the deep learning model's accuracy rate is the best in each category except category 4 (label V). Besides, the category 4 accuracy rate is very close to the shallow learning best result.

5 Discussion

Experiment results demonstrate deep learning has higher accuracy and has better confusion matrix patterns. This result supports previous research [1, 17, 19]. Besides, it is clear that shallow ML has comparable results. Well-defined features may achieve high enough accuracy. Histogram-based prediction demonstrates features that should be selected carefully as histogram feature achieves low accuracy and have a low effect on the best result among all features. Moreover, although it seems histogram-based prediction has a 76.63% accuracy rate, confusion matrix demonstrates that it has terrible performance as it has 0% accuracy rate for a category. Although histogram-based prediction has low performance, it positively contributes to overall accuracy. Random Forest and XGB have close accuracy rate; however, DT has low accuracy; moreover, while training HOG + Binning features accuracy rate was the same for RF and XGB, XGB had better confusion matrix which implies model selection is also very important for shallow ML. Besides, confusion matrix comparison indicates that even though proposed solutions may have high accuracy rate, this may show poor performance due to data imbalance. In Table 4, it can be seen that the performance level of the proposed approach in both DL and SL is comparable or better than the proposed method accuracy rate using MIT-BIH arrhythmia dataset. Although Rahhal et al.'s model seems to have a better accuracy rate than the proposed solution, they exclude label Q as it is classified as an unknown label (Table 4).

Table 4 Research using MIT-BIH arrhythmia dataset for heartbeat classification

	No. of category	Method	Accuracy rate
Hasan and Bhattacharjee [26]	4	CNN (DL)	97.70
Kiranyaz et al. [46]	5	1D CNN (DL)	98.21
Pasolli [47]	6	SVM (SL)	96.91
Khalaf [48]	5	SVM (SL)	98.60
Rahhal et al. [49]	4	DAE (DL)	99.7
Luo [50]	5	DAE (DL	97.5
Proposed DL	5	CNN (DL)	98.1
Proposed SL	5	XGB (SL)	97.97

6 Conclusion

In conventional multiple heartbeat classification schemes, the ECG images are employed in feature learning and classification. In this study, shallow ML and deep learning have been compared based on ECG images. SL algorithms generally depend upon the manual selection of features using domain-specific expertise. This process can reduce modeling complexity and has easier adaptation to smaller and less complicated datasets [51]. DL classification algorithms are designed independently to detect, extract, and learn features, with minimal-to-no human influence, from large datasets [51]. Manual feature extraction is a tedious task as it usually requires a lot of domain expertise within an application-specific engineering process. For this reason, it is seemingly time-consuming, labor-intensive, and inflexible [52]. Besides, SL models can be beneficial in case DL models are not applicable due to lack of datasets.

As ECG is very important for diagnosis and it is not always easy to find an expert in all stages of a health system, the proposed type of prediction algorithm is very helpful and may be life-saving. Many lives can be saved if an efficient diagnosis of cardiovascular disease is detected at an earlier stage [18]. Proposed solutions show that accuracy rates are high enough at least to indicate the need for further diagnosis.

References

1. DeLancey, E. R., Simms, J. F., Mahdianpari, M., Brisco, B., Mahoney, C., & Kariyeva, J. (2019, December). Comparing deep learning and shallow learning for large-scale wetland classification in Alberta, Canada. *Remote Sensing, 12*(1), 2. https://doi.org/10.3390/rs12010002
2. Kim, S. (2020, March). Deep learning with R, FrançoisChollet, Joseph J.Allaire, Shelter Island, NY: Manning. *Biometrics, 76*(1), 361–362. https://doi.org/10.1111/biom.13224
3. Guo, Y., Liu, Y., Oerlemans, A., Lao, S., Wu, S., & Lew, M. S. (2016, April). Deep learning for visual understanding: A review. *Neurocomputing, 187*, 27–48. https://doi.org/10.1016/j.neucom.2015.09.116

4. Voulodimos, A., Doulamis, N., Doulamis, A., & Protopapadakis, E. (2018). Deep learning for computer vision: A brief review. *Computational Intelligence and Neuroscience, 2018*, 1–13. https://doi.org/10.1155/2018/7068349
5. Goyal, D., & Pabla, B. S. (2015, August). Condition based maintenance of machine tools – A review. *CIRP Journal of Manufacturing Science and Technology, 10*, 24–35. https://doi.org/10.1016/j.cirpj.2015.05.004
6. Dalal, N., & Triggs, B., *Histograms of oriented gradients for human detection*. In 2005 IEEE Computer society conference on computer vision and pattern recognition (CVPR'05) (pp. 886–893). IEEE. https://doi.org/10.1109/CVPR.2005.177.
7. Lowe, D. G. (2004, November). Distinctive image features from scale-invariant keypoints. *International Journal of Computer Vision, 60*(2), 91–110. https://doi.org/10.1023/B:VISI.0000029664.99615.94
8. Viola, P., & Jones, M., *Rapid object detection using a boosted cascade of simple features*. In Proceedings of the 2001 IEEE Computer Society Conference on Computer Vision and Pattern Recognition. CVPR 2001 (pp. I-511–I-518). IEEE Computer Society. https://doi.org/10.1109/CVPR.2001.990517.
9. Salton, G., & Buckley, C. (1988, January). Term-weighting approaches in automatic text retrieval. *Information Processing and Management, 24*(5), 513–523. https://doi.org/10.1016/0306-4573(88)90021-0
10. Wu, M., Liu, F., & Cohn, T., *Evaluating the utility of hand-crafted features in sequence labelling* *, pp. 2850–2856. [Online]. https://spacy.io/. Accessed 9 Dec 2023.
11. Rana, M., & Bhushan, M. (2023, July). Machine learning and deep learning approach for medical image analysis: Diagnosis to detection. *Multimedia Tools and Applications, 82*(17), 26731–26769. https://doi.org/10.1007/S11042-022-14305-W/TABLES/5
12. Houssein, E. H., Emam, M. M., Ali, A. A., & Suganthan, P. N. (2021, April). Deep and machine learning techniques for medical imaging-based breast cancer: A comprehensive review. *Expert Systems with Applications, 167*, 114161. https://doi.org/10.1016/J.ESWA.2020.114161
13. Fatima, M., Pasha, M., Fatima, M., & Pasha, M. (2017, January). Survey of machine learning algorithms for disease diagnostic. *Journal of Intelligent Learning Systems and Applications, 9*(1), 1–16. https://doi.org/10.4236/JILSA.2017.91001
14. Ertuğrul, Ö. F., & Akıl, M. F. (2022, January). Detecting hemorrhage types and bounding box of hemorrhage by deep learning. *Biomedical Signal Processing and Control, 71*, 103085. https://doi.org/10.1016/J.BSPC.2021.103085
15. Fatih Akıl, M., & Ertugrul, Ö. F. (2023). Estimation of diffusion weight imaging and perfusion-weighted imaging volume by texture methods. *Arabian Journal for Science and Engineering*, 48. https://doi.org/10.1007/s13369-022-07536-4
16. Lopes, R. R., et al. (2021, April). Improving electrocardiogram-based detection of rare genetic heart disease using transfer learning: An application to phospholamban p.Arg14del mutation carriers. *Computers in Biology and Medicine, 131*, 104262. https://doi.org/10.1016/j.compbiomed.2021.104262
17. Abubaker, M. B., & Babayiğit, B. (2023, April). Detection of cardiovascular diseases in ECG images using machine learning and deep learning methods. *IEEE Transactions on Artificial Intelligence, 4*(2), 373–382. https://doi.org/10.1109/TAI.2022.3159505
18. Kaptoge, S., et al. (2019, October). World Health Organization cardiovascular disease risk charts: Revised models to estimate risk in 21 global regions. *The Lancet Global Health, 7*(10), e1332–e1345. https://doi.org/10.1016/S2214-109X(19)30318-3
19. Swathy, M., & Saruladha, K. (2022, March). A comparative study of classification and prediction of cardio-vascular diseases (CVD) using machine learning and deep learning techniques. *ICT Express, 8*(1), 109–116. https://doi.org/10.1016/j.icte.2021.08.021
20. Kung, B.-H., Hu, P.-Y., Huang, C.-C., Lee, C.-C., Yao, C.-Y., & Kuan, C.-H. (2021, June). An efficient ECG classification system using resource-saving architecture and random forest. *IEEE Journal of Biomedical and Health Informatics, 25*(6), 1904–1914. https://doi.org/10.1109/JBHI.2020.3035191

21. Kohli, N., Verma, N. K., & Roy, A. (2010, September). *SVM based methods for arrhythmia classification in ECG.* In 2010 International conference on computer and communication technology (ICCCT) (pp. 486–490). IEEE. https://doi.org/10.1109/ICCCT.2010.5640480.
22. Mohebbanaaz, Kumari, L. V. R., & Sai, Y. P. (2022, April). Classification of ECG beats using optimized decision tree and adaptive boosted optimized decision tree. *Signal Image Video Process, 16*(3), 695–703. https://doi.org/10.1007/s11760-021-02009-x
23. Zhang, Y., & Liao, S. (2020, November). *A kernel perspective for the decision boundary of deep neural networks.* In 2020 IEEE 32nd international conference on tools with artificial intelligence (ICTAI) (pp. 653–660). IEEE. https://doi.org/10.1109/ICTAI50040.2020.00105.
24. Shi, H., Wang, H., Huang, Y., Zhao, L., Qin, C., & Liu, C. (2019, April). A hierarchical method based on weighted extreme gradient boosting in ECG heartbeat classification. *Computer Methods and Programs in Biomedicine, 171*, 1–10. https://doi.org/10.1016/j.cmpb.2019.02.005
25. Liu, X., Wang, H., Li, Z., & Qin, L. (2021, September). Deep learning in ECG diagnosis: A review. *Knowledge Based System, 227*, 107187. https://doi.org/10.1016/j.knosys.2021.107187
26. Hasan, N. I., & Bhattacharjee, A. (2019, July). Deep learning approach to cardiovascular disease classification employing modified ECG signal from empirical mode decomposition. *Biomedical Signal Processing and Control, 52*, 128–140. https://doi.org/10.1016/j.bspc.2019.04.005
27. Sun, L., Lu, Y., Yang, K., & Li, S. (2012, December). ECG analysis using multiple instance learning for myocardial infarction detection. *IEEE Transactions on Biomedical Engineering, 59*(12), 3348–3356. https://doi.org/10.1109/TBME.2012.2213597
28. Remya, R. S., Indiradevi, K. P., & Babu, K. K. A. (2016). Classification of myocardial infarction using multi resolution wavelet analysis of ECG. *Procedia Technology, 24*, 949–956. https://doi.org/10.1016/j.protcy.2016.05.195
29. Hammad, M., Luo, G., & Wang, K. (2019, January). Cancelable biometric authentication system based on ECG. *Multimedia Tools and Applications, 78*(2), 1857–1887. https://doi.org/10.1007/s11042-018-6300-2
30. Banerjee, S., & Mitra, M. (2014, February). Application of cross wavelet transform for ECG pattern analysis and classification. *IEEE Transactions on Instrumentation and Measurement, 63*(2), 326–333. https://doi.org/10.1109/TIM.2013.2279001
31. Padhy, S., & Dandapat, S. (2017, January). Third-order tensor based analysis of multilead ECG for classification of myocardial infarction. *Biomedical Signal Processing and Control, 31*, 71–78. https://doi.org/10.1016/j.bspc.2016.07.007
32. Sahoo, S., Kanungo, B., Behera, S., & Sabut, S. (2017, October). Multiresolution wavelet transform based feature extraction and ECG classification to detect cardiac abnormalities. *Measurement, 108*, 55–66. https://doi.org/10.1016/j.measurement.2017.05.022
33. Kachuee, M., Fazeli, S., & Sarrafzadeh, M. (2018, June). *ECG heartbeat classification: A deep transferable representation.* In 2018 IEEE international conference on healthcare informatics (ICHI) (pp. 443–444). IEEE. https://doi.org/10.1109/ICHI.2018.00092.
34. Ertuğrul, Ö. F., Acar, E., Aldemir, E., & Öztekin, A. (2021, February). Automatic diagnosis of cardiovascular disorders by sub images of the ECG signal using multi-feature extraction methods and randomized neural network. *Biomedical Signal Processing and Control, 64*, 102260. https://doi.org/10.1016/j.bspc.2020.102260
35. Junior, O. L., Delgado, D., Goncalves, V., & Nunes, U. (2009, October). *Trainable classifier-fusion schemes: An application to pedestrian detection.* In 2009 12th International IEEE conference on intelligent transportation systems (pp. 1–6). IEEE. https://doi.org/10.1109/ITSC.2009.5309700.
36. Cohen, S., Rokach, L., & Maimon, O. (2007, September). Decision-tree instance-space decomposition with grouped gain-ratio. *Information Sciences, 177*(17), 3592–3612. https://doi.org/10.1016/j.ins.2007.01.016
37. Breiman, L., Friedman, J. H., Olshen, R. A., & Stone, C. J. (2017). *Classification and regression trees*. Routledge. https://doi.org/10.1201/9781315139470

38. Breiman, L. (1999, October). Prediction games and arcing algorithms. *Neural Computation, 11*(7), 1493–1517. https://doi.org/10.1162/089976699300016106
39. Ndayishimiyepas, P., Wilson, C., & Kimwele, M. (2022, November). *A hybrid model for predicting missing records in data using XGBoost*. In 2022 IEEE international symposium on product compliance engineering – Asia (ISPCE-ASIA) (pp. 1–5). IEEE. https://doi.org/10.1109/ISPCE-ASIA57917.2022.9971092.
40. Budholiya, K., Shrivastava, S. K., & Sharma, V. (2022, July). An optimized XGBoost based diagnostic system for effective prediction of heart disease. *Journal of King Saud University – Computer and Information Sciences, 34*(7), 4514–4523. https://doi.org/10.1016/j.jksuci.2020.10.013
41. Wang, S., et al. (2020, December). A new method of diesel fuel brands identification: SMOTE oversampling combined with XGBoost ensemble learning. *Fuel, 282*, 118848. https://doi.org/10.1016/j.fuel.2020.118848
42. Zhou, J., Qiu, Y., Zhu, S., Armaghani, D. J., Khandelwal, M., & Mohamad, E. T. (2021, October). Estimation of the TBM advance rate under hard rock conditions using XGBoost and Bayesian optimization. *Underground Space, 6*(5), 506–515. https://doi.org/10.1016/j.undsp.2020.05.008
43. Dong, W., Huang, Y., Lehane, B., & Ma, G. (2020, June). XGBoost algorithm-based prediction of concrete electrical resistivity for structural health monitoring. *Automation in Construction, 114*, 103155. https://doi.org/10.1016/j.autcon.2020.103155
44. Chen, T., & Guestrin, C. (2016, August). *XGBoost*. In Proceedings of the 22nd ACM SIGKDD international conference on knowledge discovery and data mining (pp. 785–794). ACM. https://doi.org/10.1145/2939672.2939785.
45. Krizhevsky, A., Sutskever, I., & Hinton, G. E. (2017, May). ImageNet classification with deep convolutional neural networks. *Communications of the ACM, 60*(6), 84–90. https://doi.org/10.1145/3065386
46. Kiranyaz, S., Ince, T., Hamila, R., & Gabbouj, M. (2015, August). *Convolutional neural networks for patient-specific ECG classification*. In 2015 37th annual international conference of the IEEE engineering in medicine and biology society (EMBC) (pp. 2608–2611). IEEE. https://doi.org/10.1109/EMBC.2015.7318926.
47. Pasolli, E., & Melgani, F. (2010, November). Active learning methods for electrocardiographic signal classification. *IEEE Transactions on Information Technology in Biomedicine, 14*(6), 1405–1416. https://doi.org/10.1109/TITB.2010.2048922
48. Khalaf, A. F., Owis, M. I., & Yassine, I. A. (2015, November). A novel technique for cardiac arrhythmia classification using spectral correlation and support vector machines. *Expert Systems with Applications, 42*(21), 8361–8368. https://doi.org/10.1016/J.ESWA.2015.06.046
49. Rahhal, M. M. A., Bazi, Y., Alhichri, H., Alajlan, N., Melgani, F., & Yager, R. R. (2016, June). Deep learning approach for active classification of electrocardiogram signals. *Information Sciences, 345*, 340–354. https://doi.org/10.1016/J.INS.2016.01.082
50. Luo, K., Li, J., Wang, Z., & Cuschieri, A. (2017). Patient-specific deep architectural model for ECG classification. *Journal of Healthcare Engineering, 2017*. https://doi.org/10.1155/2017/4108720
51. Kashou, A. H., May, A. M., & Noseworthy, P. A. (2023, July). Comparison of two artificial intelligence-augmented ECG approaches: Machine learning and deep learning. *Journal of Electrocardiology, 79*, 75–80. https://doi.org/10.1016/J.JELECTROCARD.2023.03.009
52. Liastuti, L. D., et al. (2022, December). Detecting left heart failure in echocardiography through machine learning: A systematic review. *Reviews in Cardiovascular Medicine, 23*(12), 402. https://doi.org/10.31083/j.rcm2312402

Shallow Learning Versus Deep Learning in Biomedical Applications

Mithat Önder, Ümit Şentürk, and Kemal Polat

1 Introduction

Alzheimer's disease (AD) and Parkinson's disease (PD) are two degenerative neurological conditions that impact different aspects of brain function and movement. AD primarily affects recall, cognition, and conduct, leading to a decline in cognitive abilities and retention. On the other hand, PD primarily affects the ability to move, as it involves the gradual decline of cells in the brain responsible for producing dopamine, a neurotransmitter that governs movement and coordinates bodily actions [1, 2]. It is crucial to identify PD at an early stage, as this allows for prompt implementation of treatment methods and strategies for controlling the condition [3]. Clinicians can achieve an early diagnosis of AD, even with the subtle onset that challenges diagnostic skills, by asking the appropriate questions, paying attention to input from family members, and utilizing sensitive office-based screening tools [4]. The diagnosis of AD and PD is accomplished through a comprehensive evaluation that encompasses reviewing medical history, conducting a physical examination, performing a neurological assessment, utilizing imaging techniques,

The original version of the chapter has been revised. A correction to this chapter can be found at https://doi.org/10.1007/978-3-031-69499-8_13

M. Önder · K. Polat (✉)
Faculty of Engineering, Department of Electrical and Electronics Engineering, Bolu Abant Izzet Baysal University, Bolu, Turkey
e-mail: mithatonder@ibu.edu.tr; kpolat@ibu.edu.tr

Ü. Şentürk
Faculty of Engineering, Department of Computer Engineering, Bolu Abant Izzet Baysal University, Bolu, Turkey
e-mail: umit.senturk@ibu.edu.tr

Ö. F. Ertuğrul et al. (eds.), *Shallow Learning vs. Deep Learning*, The Springer Series in Applied Machine Learning, https://doi.org/10.1007/978-3-031-69499-8_6

and administering cognitive tests [5]. Recently, shallow learning (SL) and deep learning (DL) methods have been used in the diagnosis of AD and PD [6]. For the diagnosis of PD, freezing of gait, electroencephalography (EEG) signals, speech recordings, and handwriting were used to create datasets. These datasets were then applied to various SL and DL methods to diagnose the disease. A voice dataset for PD classification was examined in [7], where Artificial Neural Network (ANN), Support Vector Machine (SVM), and K-Nearest Neighbor (KNN) were utilized for classification. The ANN achieved the highest accuracy rate of 95.89% in classification. The feature selection methods used in [8] include Minimum-Redundancy Maximum Relevance (mRMR) and Recursive Feature Elimination (RFE). Additionally, Logistic Regression (LR), Naive Bayes (NB), KNN, Random Forest (RF), Decision Tree (DT), SVM, and XGBoost have been utilized and investigated for voice-based PD detection. To identify PD, researchers in [9] utilized a combination of tremor and voice data. The classification algorithms KNN, SVM, and NB were employed in this study. To enhance the obtained voice recordings, data augmentation techniques including pitch shifting, time stretching, and noise injection were employed, as mentioned in [10]. The classifiers used for early detection of PD were RF, SVM, and KNN. In [11], researchers implemented the Medium Gaussian SVM, Medium KNN, and Boosted Trees classification algorithms to detect PD. The features of gait were used as input for the algorithms. In [12], researchers investigated the utilization of wearable on-body sensors for measuring movement. They relied on discrete wavelet transform to extract the main features. To evaluate the classification accuracy of these features, SVM with a linear kernel function and ANN were employed. The primary objective was to identify the freezing of gait. The classification of patients with PD, based on their resting-state EEG signal, was conducted in [13] using SVM and Convolutional Neural Network (CNN). In [14], researchers have introduced two hybrid deep neural networks that merge CNN and Long Short-Term Memory (LSTM) to enhance the accuracy of diagnosing PD by classifying EEG signals more effectively. The diagnosis of PD is automated in [15] by utilizing EEG signals that are preprocessed using a flexible analytic wavelet transform. Multiple classifiers, such as SVM, LR, RF, Radial Basis Functions (RBF), and KNN, are employed to determine PD. Researchers employed a unique hybrid feature extraction technique in [16]. By combining ensemble empirical mode decomposition, and vector autoregressive model, the method aimed to determine the optimal number of features. The proposed approach achieved the highest classification accuracy using ANN. In [17], the characteristics derived from distinct handwritten tasks were subjected to classification using various ML techniques. These techniques included SVM, AdaBoost (AB), bagged Random Forest, and two different forms of DL models known as LSTM and Bidirectional LSTM (BLSTM). The detection of PD in its early stages was facilitated by using handwriting datasets in conjunction with DT and an innovative fuzzy classifier method, as demonstrated in study [18]. The technique has yet to meet the standard requirements for clinical performance, particularly in terms of sensitivity and specificity. As a result, it cannot be considered a dependable method for screening AD. Hence, there is increasing interest in the detection of AD with magnetic resonance imaging (MRI), EEG-based SL, and DL models. The prediction of AD relies on the analysis of MRI images from three perspectives: axial, sagittal, and coronal, as mentioned in [19].

Accurate and effective AD classification is accomplished by utilizing ML techniques such as multilayer perceptron (MLP) and SVM. The objective of [20] is to diagnose levels of AD based on MRI scans using a unique ML approach with CNN. In [21], the authors utilize a supervised DL technique to address the scarcity of data in an unbalanced 2D MRI dataset, with the aim to predict AD. They make use of several pretrained CNN models for this purpose. In [22], a combination of cross-sectional neuroimaging, longitudinal CSF analysis, and cognitive performance for MRI was implemented in a proposed framework utilizing multimodal recurrent neural networks to accurately predict the progression from Mild Cognitive Impairment (MCI) to AD. The resulting accuracy of the predictions was 81%. In [23], a method called SVM-RFE was presented for the identification of AD patients. This technique involves extracting multiple features from MRI scans to train a model capable of detecting AD. Notably, the performance of the model incorporating multiple features was found to be superior to the one using a single feature. The discrimination of AD is explored in [24], where MCI and MRI imaging are utilized. Import Vector Machine (IVM), RELM, and SVM algorithms are applied for AD discrimination in this study. A methodology presented in [25] involves the application of MRI and MCI analysis as well as the examination of brain function using ML techniques including SVM, LR, ANN, NB, DT, RF, and KNN. The purpose of this approach is to detect cases of AD. The EEG signals collected from three electrodes in the central lobe were analyzed in [26]. To determine statistical characteristics, wavelet transforms were utilized. In order to classify AD, various classification algorithms such as Diagonal Discriminant Analysis, SVM, KNN, RF, LR, and NB were employed. By analyzing the spectral and complex characteristics of the EEG signal gathered in [27], the KNN classifier was utilized to distinguish AD. The research paper [28] presents a unique framework called Adazd-Net, which aims to automate the identification of AD using EEG signals. This framework is adaptive and explanatory, providing a comprehensive solution to the problem. Its key feature is the adaptive flexible analytic wavelet transform, which effortlessly adapts to any variations in EEG signals. A Finite Impulse Response (FIR) filter is utilized in [29] to process the EEG signals, with a cut-off frequency of 7 Hz and 16 Hz. The diagnostic algorithm employed for AD is SVM, which yielded an accuracy rate of 86%. In [30], the database contains recordings of EEG from 19 different channels. The continuous wavelet transform was used to preprocess the signal. The diagnosis of AD was performed using DL approaches based on a CNN model.

In this study, EEG signals were pre-processed to filter out the blink noises from the EEG signals. The researchers employed both shallow and deep learning techniques to accurately identify and diagnose of PD and AD at the same conditions. The PD dataset, EEG signals, was taken from the San Diego Resting State [31, 32]. The AD dataset was formed by merging the datasets from [33, 34]. In the pre-processing stage for both shallow learning and deep learning, a FIR filter is used to pass EEG signals through a band of 7–30 Hz, effectively eliminating unwanted noise and artifacts. To further refine the signals, we then applied Independent Component Analysis (ICA) to the output of the filter. After that, the EEG signals are divided into segmented. Later on, the raw data is directly provided to both shallow learning and deep learning algorithms for the purpose of diagnosing PD. In the diagnosis of AD and PD, the features of the raw data are extracted using the Common Spatial Patterns

(CSP) approach. The CSP feature extraction is a technique used in different brain-computer interface (BCI) applications to identify distinctive patterns in electroencephalography (EEG) or functional magnetic resonance imaging (fMRI) data. There are 32 channels in the Parkinson's EEG signal dataset, which is separated into 1-s epochs. On the other hand, the Alzheimer's EEG signal dataset consists of 19 channels and is also divided into 1-s epochs. Next, the dataset was subjected to fivefold cross-validation (FCW) for the SL modeling process. AdaBoost (AB), CatBoost, XGBoost, Decision Tree (DT), Gaussian Naïve Bayes (GNB), Linear Discriminant Analysis (LDA), and KNN algorithms were employed in the diagnosis of PD and AD through SL techniques. For the DL modeling process, the dataset was subjected to a hold-out method with 80% allocated for training and 20% for testing. DL has employed as Deep Neural Networks (DNN) for the diagnosis of PD. The DNN architecture used in this application consists of 4 hidden layers, with each layer containing 100 neurons. The DL in diagnosing AD involves the utilization of a 2-Dimensional Convolutional Neural Network (2D-CNN). To enhance accuracy, each frequency range is trained individually using the 2D-CNN algorithm. This algorithm consists of layers such as convolution, flattening, and dense layers. The main motivation of the study is that early diagnosis of PD and AD is possible with ML methods, and thus the quality of life of patients can be increased [35]. The primary contribution of the study is the comparison of the performance of SL and DL methods in diagnosing PD and AD using EEG signals, based on performance metrics.

2 Material and Methods

2.1 Methods

2.1.1 Overview

In Fig. 1, the block diagram of the classification of Parkinson's disease and Alzheimer's disease is shown. The EEG signals were used to diagnose Alzheimer's disease (AD) and Parkinson's disease (PD). The PD dataset was taken from the San Diego Resting State [31, 32], and the AD dataset was formed by merging the datasets from [33, 34]. The dataset for patients with AD included electroencephalography (EEG) signals recorded over a period of 8 s using 19 channels. In the pre-processing stage, EEG signals of PD and AD were then filtered within the frequency range of 0.5–45 Hz. A 32-channel EEG signals recorder was utilized to collect the dataset for PD. The recording lasted for 180 s, while the samples were collected at a frequency of 512 Hz. In order to remove signals that were not within the frequency range of 0.5–45 Hz, the datasets for PD and AD were subjected to filtering using both a Finite Impulse Response (FIR) filter and a Butterworth band-pass filter. In the case of the PD dataset, the Independent Component Analysis (ICA) was utilized to filter out artifacts from muscle and eye blink, which are known to be significant sources of

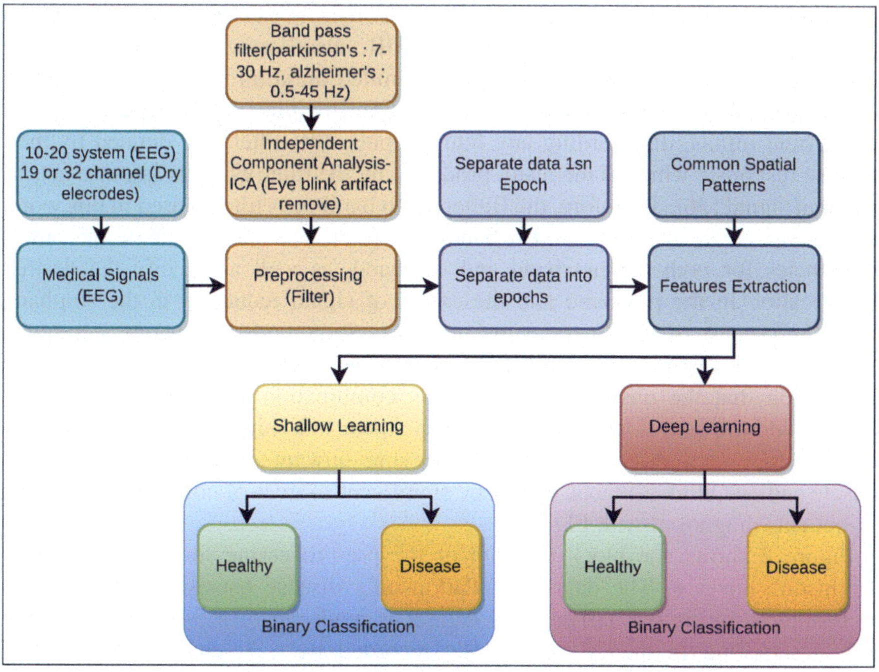

Fig. 1 The block diagram of the classification of Parkinson's disease and Alzheimer's disease

noise in EEG signals for patients with PD. During the segmentation stage, the EEG signals for PD and AD were divided into epochs of 1 s each. After the pre-processing and segmentation stages, the obtained AD dataset was used as raw data in the classification stage, but in PD, the dataset features are obtained by using the feature extraction method, and obtained dataset is used in the classification stage. In the classification stage, both shallow learning algorithms and deep learning algorithms were used to diagnose PD and AD. In the diagnosis of both diseases, AdaBoost (AB), CatBoost, XGBoost, Decision Tree (DT), Gaussian Naïve Bayes (GNB), Linear Discriminant Analysis (LDA), and K-Nearest Neighbors (KNN) algorithms were used as a shallow learning method. Fivefold cross-validation (FCW) for the SL modeling process is used before the classification. The DL modeling process utilized a hold-out technique for the dataset, with 80% designated for training and 20% for testing. The Deep Neural Network (DNN) was used as a deep learning method in the diagnosis of PD. The 2-Dimensional Convolutional Neural Network (2D-CNN) was used as a deep learning method in the diagnosis of AD.

2.1.2 Pre-processing

In order to analyze data more effectively, it is crucial to prioritize the pre-processing stage. At this stage, filtering takes on great importance. Filtering plays a vital role in obtaining reliable and easily interpretable outcomes, which are essential steps in diagnosing and classifying PD and AD. FIR, also known as Finite Impulse Response, is a digital filter that produces a limited duration of output response to an input signal. In simpler terms, the output of an FIR filter solely depends on past and current inputs, disregarding any future inputs. FIR filters are defined by their impulse response, which is the result produced by the filter when an impulse acts as the input signal [36]. Therefore, the Butterworth band-pass filter is used in this work. In order to create the Butterworth band-pass filter, one must indicate the cutoff frequencies for both the passband and stopband, as well as specify the desired amplification in the passband and the extent of signal reduction in the stopband [37]. The process of ICA involves using statistical methods to separate a complex signal into its distinct and unrelated components. This technique relies on the assumption that the observed signals are a combination of different underlying sources, and the goal is to accurately estimate these individual sources [38]. Consequently, ICA is highly beneficial in removing unwanted disturbances, such as muscle and eye blink artifacts, which are significant sources of noise in EEG signals for patients diagnosed with PD.

Figure 2 shows the block diagram of the pre-processing for classification of Parkinson's disease. The dataset of Parkinson's disease consists of 15 PD and 15 healthy controls (HC). The EEG dataset is made up of a 3-minute recording with a sampling frequency of 512 Hz, recorded by a 32-channel EEG signal recorder. Firstly, the EEG signals for the PD dataset were filtered using a 7–30 Hz bandpass filter using ICA. Secondly, the EEG signals are segmented into 1-second epochs. Thirdly, the raw dataset was created from 5849 epochs, each of which is a 512 * 32 matrix (512 datapoint, 32 channel).

Figure 3 shows the block diagram of the pre-processing for the classification of Alzheimer's disease. The dataset of Alzheimer's disease consists of 24 PD and 24 HC. The EEG dataset is made up of an 8-s recording with a sampling frequency of 128 Hz, recorded by a 19-channel EEG signal recorder. Firstly, the EEG signals for the PD dataset were filtered using a 0.5–45 Hz bandpass filter. Secondly, the EEG signals are segmented into 1-second epochs. Thirdly, the raw dataset was created from 384 epochs, each of which is a 128 * 19 matrix (128 datapoint, 19 channel).

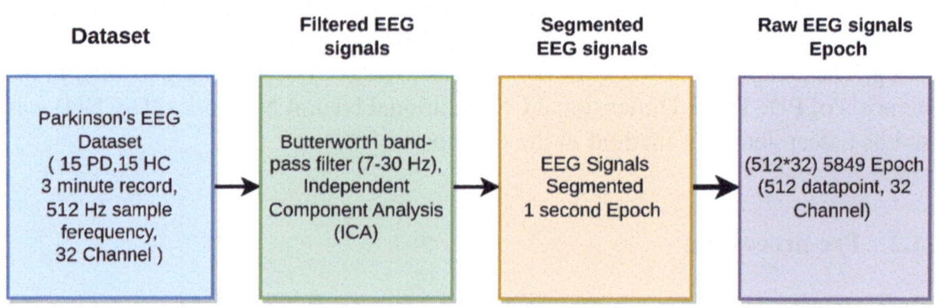

Fig. 2 The block diagram of the pre-processing for classification of Parkinson's disease

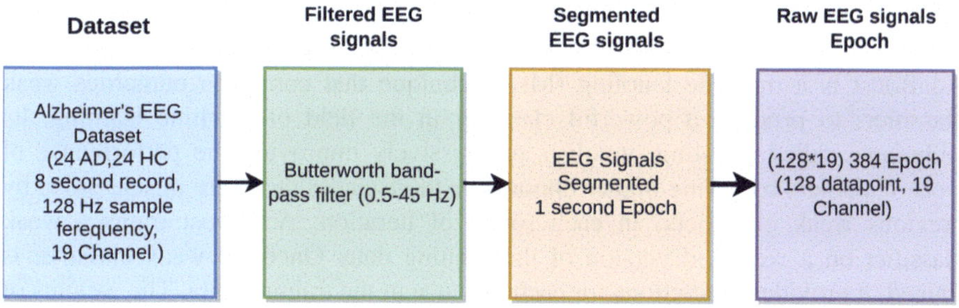

Fig. 3 The block diagram of the pre-processing for classification of Alzheimer's disease

2.1.3 Segmentation and Feature Extraction of EEG Signals

The process of EEG signal segmentation entails the partitioning of a continuous EEG signal, which is a recording of the brain's electrical activity, into discrete segments or epochs that hold significance. These epochs are comprised of a continuous stream of voltage values over time. Dismantling the continuous signal into smaller epochs aids in the examination of specific temporal occurrences or patterns in the brain's activity [39]. Segmenting the EEG signal is crucial because it enables the examination, identification of events, extraction of features, mitigation of noise, and modeling, thereby promoting a more profound comprehension of brain function and improved utilization of EEG data in both clinical and research environments. During the segmentation of EEG signals, epochs from various seconds were selected in each study. In [40], the comparative performance of the conventional 30-s epochs and the three consecutive and non-consecutive 6-s epochs is assessed. In [41], the segmentation of epochs into 1, 2, and 3 s epochs was carried out to guarantee values for the various segmentation methods. In this study, performance with different segmentations was not compared. Therefore, in this study, the EEG signal segmented 1-s epochs, shown in Figs. 2 and 3, for both the PD and AD datasets. The feature extraction entails the extraction and selection of important and relevant features from raw data or input [39]. The Common Spatial Pattern (CSP) technique is a popular method used in the domains of brain-computer interfaces and EEG signal processing for extracting features [42]. For the diagnosis of PD, the feature extraction method called CSP has been utilized to extract valuable information from EEG signals. On the other hand, when diagnosing AD, the EEG dataset was given as input to the 2D-CNN model without any future extraction, resulting in a 128*19 matrix after dividing the dataset into epochs.

2.1.4 Shallow Learning Classification

The AB, CatBoost, XGBoost, DT, GNB, LDA, and KNN algorithms were applied in this study for diagnosing PD and AD. The following subsections provide an overview of the fundamental principles of each classification algorithm.

2.1.4.1 AdaBoost

AdaBoost is a machine learning (ML) technique that combines numerous weak classifiers to produce a powerful classifier in the field of machine learning. Its technique, called boosting, involves progressively improving the performance of weak classifiers by giving higher importance to instances that were misclassified by previous weak classifiers. In each round of iteration, AdaBoost trains a weak classifier on a weighted version of the training data. Once the weak classifier is trained, it provides predictions for each instance in the training data. The weights of misclassified instances are then amplified, whereas the weights of correctly classified instances are reduced. This process is repeated for a specified number of iterations or until all instances are correctly classified. Ultimately, the final strong classifier is produced by merging the weighted predictions made by each weak classifier [43, 44]. In this study, boosting for AD and PD classification is stopped when it reaches a maximum of 150 estimators.

2.1.4.2 CatBoost

CatBoost is a type of ML algorithm that falls under the gradient boosting algorithm category. Its effectiveness lies in its ability to utilize ordered boosting and a technique for handling categorical features. By employing ordered boosting and incorporating ordered target statistics, based on the principle of ordering, CatBoost effectively manages to address target leakage and prediction shift issues that arise due to disparities in the distribution of training and testing datasets [45]. In this study, the default settings of the CatBoost algorithm were used.

2.1.4.3 XGBoost

XGBoost, also referred to as eXtreme Gradient Boosting, is a highly favored algorithm in the field of ML due to its remarkable performance and accuracy when it comes to predicting outcomes. It operates on the gradient boosting framework, which is a powerful method in ensemble learning. Through a sequential process, XGBoost combines multiple weak prediction models, predominantly decision trees, creating a robust predictive model. In essence, XGBoost aims to optimize both the overall prediction error and a regularization term that aids in preventing overfitting. Additionally, XGBoost incorporates numerous important enhancements and optimizations that include parallel processing, tree pruning, regularization techniques, and handling of missing values [46]. In this study, boosting for AD and PD classification is stopped when it reaches a maximum of 150 estimators.

2.1.4.4 Decision Tree

The Decision Tree acts as a classification algorithm that utilizes a tree-shaped structure, enabling the representation of decisions and their outcomes. It starts with a single node, called the root node, which represents the entire dataset. This dataset is then partitioned into subsets at each internal node of the tree, utilizing one or more attributes. This process of partitioning is repeated recursively until certain criteria are met, such as the attainment of a maximum depth or a minimum number of instances within a node. The leaf nodes at the end of the tree indicate the class labels or outcomes [47]. In this study, the maximum depth of a decision tree classification is 50.

2.1.4.5 Gaussian Naïve Bayes

The Gaussian Naive Bayes (GNB) algorithm falls under the category of supervised algorithms in machine learning. Its primary purpose is for classification tasks. It is most commonly used and performs exceptionally well in text classification. Examples of its application include sentiment analysis and spam filtration. The algorithm is known for its speed, ease of implementation, ability to work with limited training data, and scalability. Its performance is superior to other classification techniques when the data showcases conditional independence. On the other hand, the Nearest Neighbor classifier is a supervised lazy algorithm that is not suitable for real-time prediction. The Random Forest algorithm is designed to handle complex data behavior. However, if not properly constructed, it can lead to overfitting [48]. In this study, the default settings of the GNB algorithm were used.

2.1.4.6 Linear Discriminant Analysis

Linear Discriminant Analysis (LDA), a supervised ML technique, is commonly employed for performing classification tasks. It is particularly beneficial when dealing with datasets where the number of features is larger than the number of instances. The main goal of LDA is to find a mix of attributes that enhances the differentiation between classes while reducing the variability within each category. This is achieved by mapping the data onto a lower-dimensional space where the classes are as distinct as possible. In order to classify a new instance, LDA projects it onto the lower-dimensional space and assigns it to the class with the highest probability of occurrence. The probabilities are estimated based on the Gaussian distributions of each class and the discriminant functions [49, 50].

2.1.4.7 K-Nearest Neighbors

The K-Nearest Neighbors (KNN) algorithm is a technique used in supervised ML to perform classification tasks. KNN exhibits excellent performance in pattern recognition and making predictions. When dealing with a new data point, the KNN algorithm initially gathers nearby data points. Attributes that have the potential to vary significantly can exert a profound influence on the distances between the data points. The algorithm then sorts these nearest data points based on their distance from the incoming data point. This distance can be measured in various ways, with the Euclidean distance being the recommended metric. The next step involves selecting a specific number of data points that have the smallest distances and categorizing them accordingly. In KNN, it is customary to select an odd number of closest data points when dealing with two classes. The group with the highest number of data points determines the category to which the new data point belongs [51]. In this study, the default settings of the KNN algorithm were used.

2.1.5 Deep Learning Classification

The DNN algorithm was applied for diagnosing PD, and 2D-CCN algorithm was applied for diagnosing AD in this study. The following subsections provide an overview of the fundamental principles of each classification algorithm.

2.1.5.1 Deep Neural Network

DNN algorithms are a type of ML algorithms that take inspiration from how the human brain is structured and functions. They consist of multiple layers of artificial neurons known as nodes or units. Within a DNN, information passes through these layers of nodes. Each node receives inputs from nodes in the preceding layer, conducts a mathematical operation on these inputs, and generates an output. The outputs of the nodes in each layer are then passed as inputs to the nodes in the subsequent layer. This cycle continues until the final layer, which generates the output of the network. The nodes in a DNN are structured hierarchically, with each layer progressively learning more abstract features based on the previous layer's outputs. This hierarchical arrangement allows the network to acquire intricate representations of the input data [52]. Figure 4 presents the block diagram of the Deep Neural Network classification of PD. In this study, DNN algorithm has 4 hidden layers, and each hidden layer has 100 neurons.

2.1.5.2 2-Dimensional Convolutional Neural Network

The 2D Convolutional Neural Network (2D-CNN) is a popular deep learning technique used in computer vision tasks, particularly for signal classification. This

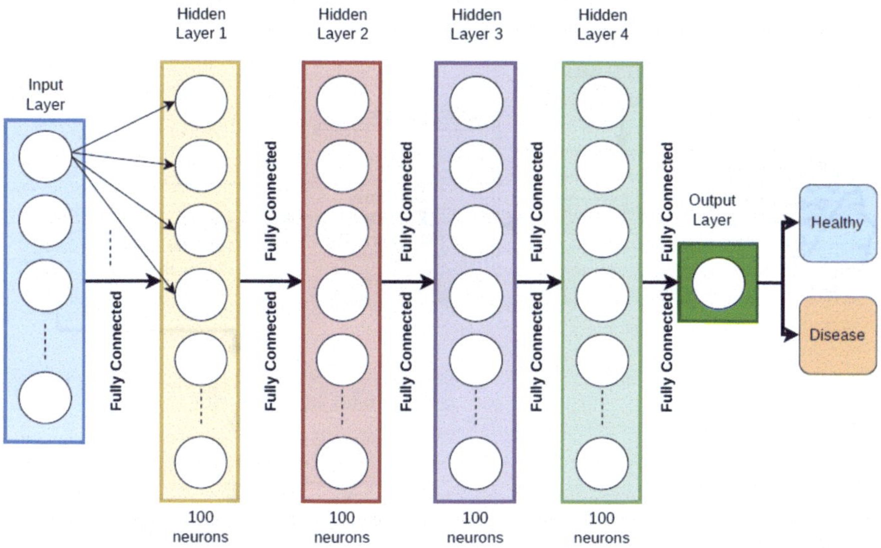

Fig. 4 The block diagram of the Deep Neural Network classification of Parkinson's disease

algorithm is designed to work with 2-dimensional input data, which is commonly in the form of images. By employing convolutional layers, the algorithm is able to extract essential features from the input data. Additionally, pooling layers are used to reduce the size of the feature maps. Following the convolutional and pooling layers, fully connected layers are commonly included for classification or other subsequent tasks. The key steps in a 2D-CNN algorithm involve convolution, non-linearity, pooling, fully connected layers, and the output layer. The 2D-CNN method used to classify EEG signals relies on the capabilities of deep learning and convolutional layers. These components aid in extracting spatial and temporal characteristics from the EEG data, thus enabling accurate and automated identification of different brain states or conditions [53, 54]. Figure 5 presents the block diagram of the 2-Dimensional Convolutional Neural Network classification of AD. In this study, randomly, a set of 5 different models for 2D-CNN classifiers were chosen. Next, the input data undergoes the application of 256 filters (with a 3×3 kernel) and 128 filters (with a 3×3 kernel) in the convolution layer. Then, the Flatten layer is employed to bridge the convolutional layers to the fully connected layers. Finally, the last layer consists of dense layers, known for their effectiveness in grasping intricate patterns and correlations within the data. These layers are comprised of 512 dense units and 216 dense units.

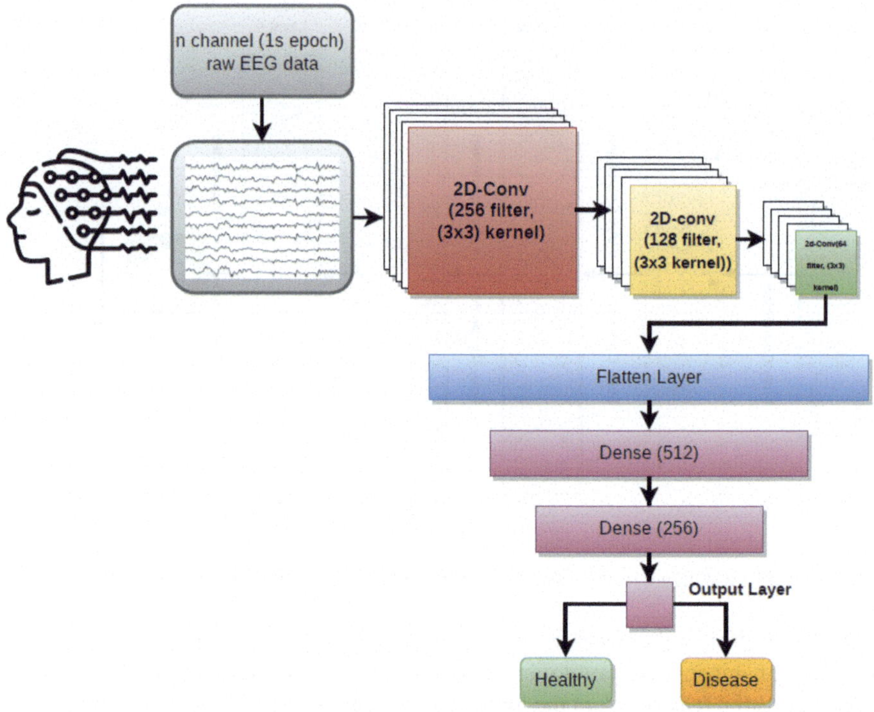

Fig. 5 The block diagram of the 2-Dimensional Convolutional Neural Network classification of Alzheimer's disease

3 The Simulation Results and Discussion

In this part, we examined the diagnosis of Parkinson's disease (PD) and Alzheimer's disease (AD) by analyzing electroencephalography (EEG) signals. We utilized two different approaches: shallow learning (SL) and deep learning (DL). The SL algorithms, such as AdaBoost (AB), CatBoost, XGBoost, Decision Tree (DT), Gaussian Naïve Bayes (GNB), Linear Discriminant Analysis (LDA), and K-Nearest Neighbors (KNN), were used in this study. On the other hand, the Deep Learning (DL) algorithms focused on Deep Neural Network (DNN) and 2-Dimensional Convolutional Neural Network (2D-CNN). Various metrics, including classification accuracy, f-measure score, kappa score, Jaccard score, ROC score, recall, precision, and elapsed time, were considered to evaluate the performance of these methods. A comparative analysis was conducted to determine the effectiveness of the SL algorithms in comparison to each other and to the DL algorithms.

3.1 Performance Measures

The evaluation of classification performance is done through the use of the confusion matrix. Table 1 illustrates the components of the confusion matrix, namely, True Positive (TP), True Negative (TN), False Positive (FP), and False Negative (FN).

The classification model's performance can be evaluated by utilizing the confusion matrix technique. To assess the classification performance, the subsequent equations were implemented:

$$\text{Classification accuracy }(\%) = \frac{\mathbf{TP + TN}}{\mathbf{TP + FP + TN + FN}} \times \mathbf{100} \tag{1}$$

$$\text{Kappa} = \frac{\mathbf{2 \times (TP \times TN - FN \times FP)}}{\mathbf{(TP + FP) \times (FP + TN) \times (TP + FN) \times (FN + TN)}} \tag{2}$$

$$\text{Precision} = \frac{\mathbf{TP}}{\mathbf{TP + FP}} \tag{3}$$

$$\text{Jaccard} = \frac{\mathbf{TP}}{\mathbf{TP + FP + FN}} \tag{4}$$

$$\text{Recall} = \frac{\mathbf{TP}}{\mathbf{TP + FN}} \tag{5}$$

$$\text{True positive rate (TPR)} = \frac{\mathbf{TP}}{\mathbf{TP + FN}} \tag{6}$$

$$\text{f} - \text{measure} = \frac{\mathbf{2 \times Precision \times Recall}}{\mathbf{Precision + Recall}} \tag{7}$$

$$\text{False positive rate (FPR)} = \frac{\mathbf{FP}}{\mathbf{TN + FP}} \tag{8}$$

$$\text{Area under curve (AUC)} = \int \mathbf{TPR\ d(FPR)} \tag{9}$$

3.2 Results and Discussion

The study involved the utilization of shallow learning and deep learning algorithms to identify and diagnose both PD and AD. During the preliminary stage of processing for both shallow learning and deep learning, an FIR filter referred to as

Table 1 Confusion matrix

	Predicted	
Confusion matrix	Negative	Positive
Negative	TN	FP
Positive	FN	TP

Table 2 Confusion matrix of the SL (AB, CatBoost, XGBoost, DT, GNB, LDA, and KNN) algorithms for the diagnosis of PD and AD

Confusion matrix		Parkinson's disease		Alzheimer's disease	
		0	1	0	1
AdaBoost	0	593	33	28	2
	1	34	510	14	33
CatBoost	0	619	7	33	0
	1	6	538	6	41
XGBoost	0	617	9	33	0
	1	6	538	11	36
Decision tree	0	578	48	27	3
	1	51	493	15	32
Gaussian Naïve Bayes	0	369	257	23	7
	1	89	455	17	30
Linear discriminant analysis	0	555	71	29	1
	1	28	516	15	32
K-nearest neighbors	0	624	2	30	0
	1	6	538	17	30

Table 3 Confusion matrix of the DL (DNN and 2D-CNN) algorithms for the diagnosis of PD and AD

Confusion matrix		Parkinson's disease	
		0	1
Deep neural network	0	615	11
	1	7	537
		Alzheimer's disease	
		0	1
2-dimensional convolutional neural network	0	29	1
	1	3	44

Finite Impulse Response is utilized to transmit EEG signals through a range of 7–30 Hz, successfully eliminating unwanted noise and artifacts. To further enhance the signals, we then implemented ICA, which stands for Independent Component Analysis, to the output produced by the filter. The Parkinson's EEG signal dataset contains a total of 32 channels and is divided into 1-s epochs. Conversely, the Alzheimer's EEG signal dataset is composed of 19 channels and is also split into 1-s epochs. Afterward, the dataset underwent fivefold cross-validation (FCW) for the SL modeling process. During the process of DL modeling, the dataset underwent a hold-out method with 80% allocated for training and 20% for testing. Table 2 presents confusion matrix of the SL (AB, CatBoost, XGBoost, DT, GNB, LDA, and KNN) algorithms for the diagnosis of PD and AD. Table 3 presents confusion matrix of the DL (DNN and 2D-CNN) algorithms for the diagnosis of PD and AD.

The RoC-AUC is a performance metric for binary classification models that quantifies its ability to differentiate between positive and negative instances. It is a valuable tool for comparing and selecting different models, as well as for determining an optimal classification threshold [55]. Figure 6 presents the obtained RoC-AUC curve using SL (CatBoost, XGBoost, and KNN) algorithms for the diagnosis of PD. Figure 7 presents the obtained RoC-AUC curve using SL (CatBoost, XGBoost, and KNN) algorithms for the diagnosis of AD. The three best models for SL are given RoC-AUC curve in Figs. 6 and 7. Figure 8 presents the obtained RoC-AUC curve using DL (DNN and 2D-CNN) algorithms for the diagnosis of PD and AD. Among SL algorithms, the KNN algorithm exhibited the highest RoC-AUC curve for detecting of PD, while the CatBoost algorithm proved to be the most effective in diagnosing AD, based on its RoC-AUC curve. After examining the RoC-AUC curves for shallow learning and deep learning algorithms, it was determined that the 2D-CNN algorithm performed superiorly to SL algorithms when detecting AD. Additionally, findings suggested that the DNN algorithm exhibited similar accuracy to KNN in diagnosing PD.

Table 4. presents the comparison of the SL (AB, CatBoost, XGBoost, DT, GNB, LDA, and KNN) algorithms performance for the diagnosis of PD. According to Table 4, the KNN algorithm shows the highest accuracy rate for diagnosing PS. The GNB algorithm, however, has a relatively low accuracy rate. Meanwhile, other shallow learning algorithms display acceptable rates of accuracy. The performance of different machine learning (ML) algorithms (AB, CatBoost, XGBoost, DT, GNB, LDA, and KNN) in diagnosing AD is compared in Table 5. The results indicate that the CatBoost algorithm achieves the highest accuracy rate for AD diagnosis. On the other hand, the GNB algorithm has a relatively lower accuracy rate. Table 6 illustrates the performance comparison of DL algorithms (DNN and 2D-CNN) in the diagnosis of AD and PD. For diagnosing PD, the DNN algorithm yields similar results to the KNN algorithm and generally surpasses shallow learning algorithms in accuracy. However, when assessing Elapsed Time, the KNN algorithm proves to be significantly faster than the DNN algorithm. In the case of diagnosing AD, the 2D-CNN algorithm delivers higher accuracy compared to shallow learning algorithms. However, considering Elapsed Time, the 2D-CNN algorithm is notably slower than the shallow learning algorithm. Figure 6 presents the obtained classification accuracy of the PD using AB, CatBoost, XGBoost, DT, GNB, LDA, KNN, and DNN algorithms. Figure 9 presents the obtained classification accuracy of the AD using AB, CatBoost, XGBoost, DT, GNB, LDA, KNN, and 2D-CNN algorithms. According to Fig. 6, KNN is the algorithm that achieves the highest accuracy when it comes to diagnosing of PD. On the other hand, Fig. 10 demonstrates that 2D-CNN is the algorithm with the highest accuracy when it comes to the diagnosis of AD.

<table>
<tr><th></th><th>Parkinson Disease</th></tr>
<tr><td>CatBoost</td><td>zzzz
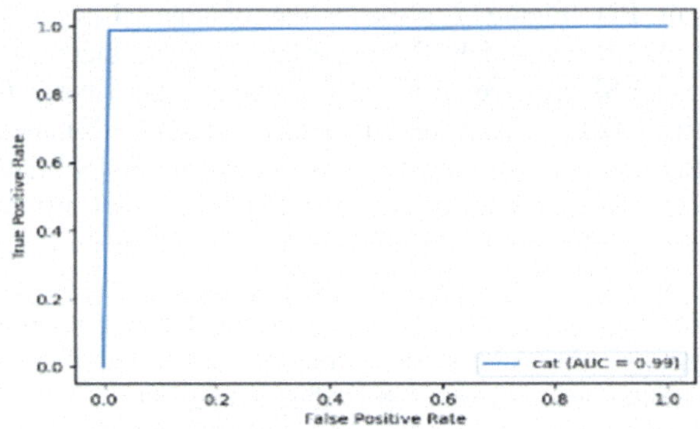
</td></tr>
<tr><td>XGBoost</td><td>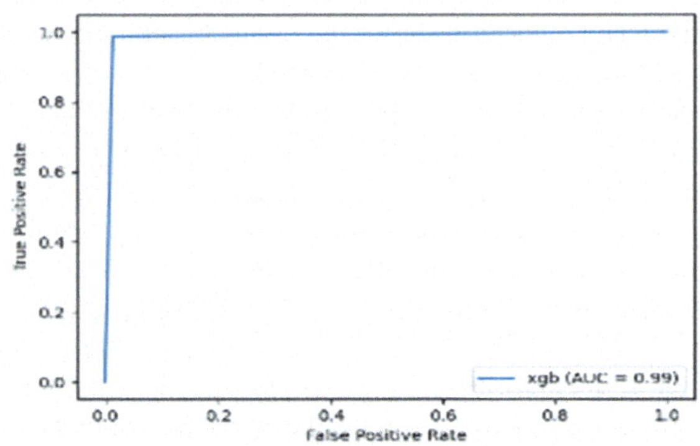
</td></tr>
<tr><td>K-Nearest Neighbors</td><td>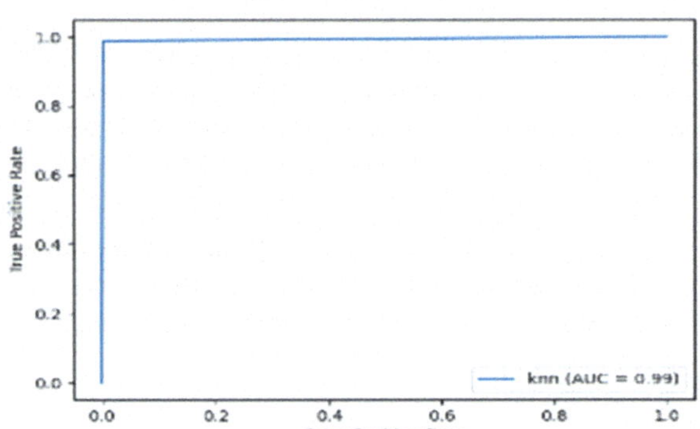
</td></tr>
</table>

Fig. 6 The obtained RoC-AUC curve using SL (CatBoost, XGBoost, and KNN) algorithms for the diagnosis of PD

<table>
<tr><th></th><th>Alzheimer Disease</th></tr>
<tr><td>CatBoost</td><td>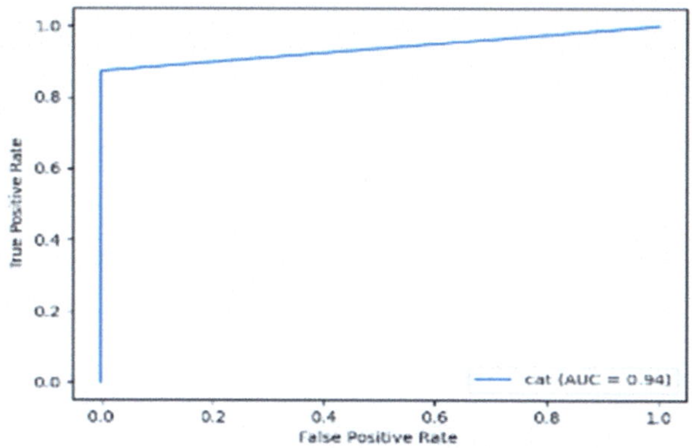
</td></tr>
<tr><td>XGBoost</td><td>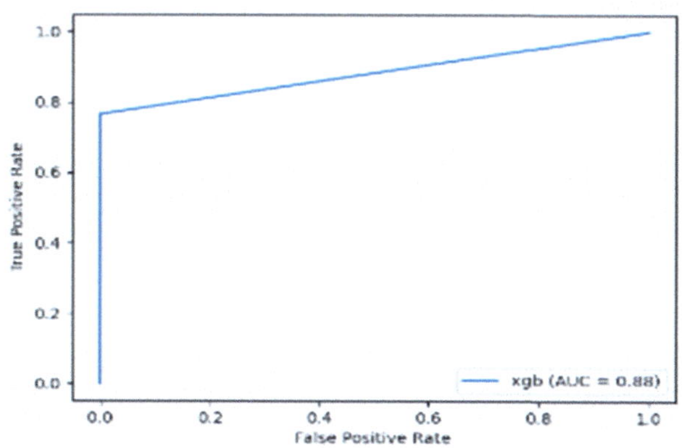
</td></tr>
<tr><td>K-Nearest Neighbors</td><td>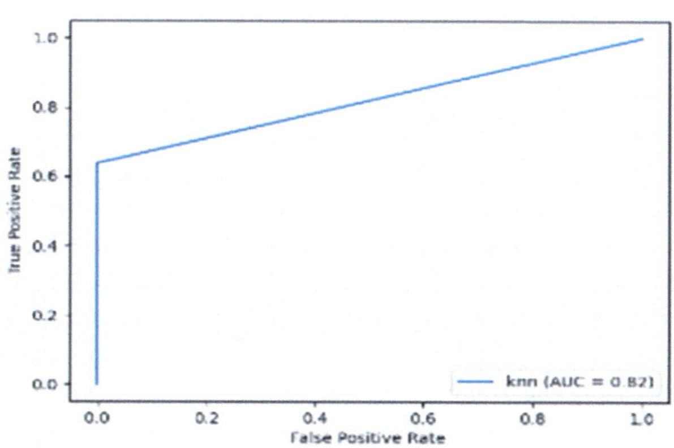
</td></tr>
</table>

Fig. 7 The obtained RoC-AUC curve using SL (CatBoost, XGBoost, and KNN) algorithms for the diagnosis of AD

<table>
<tr><th></th><th>Parkinson Disease</th></tr>
<tr><td>Deep Neural Network</td><td>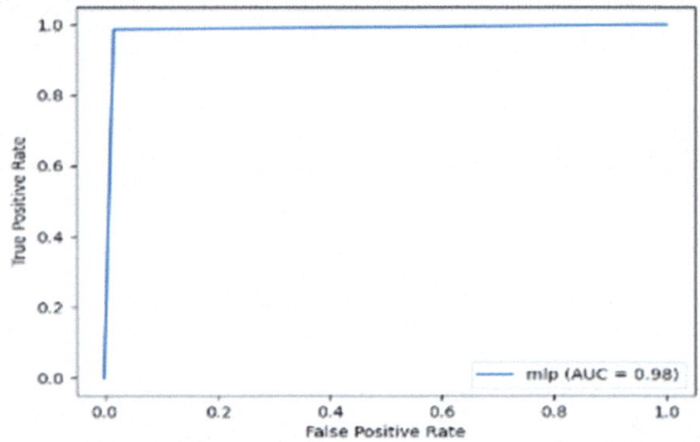
</td></tr>
<tr><th></th><th>Alzheimer Disease</th></tr>
<tr><td>Dimensional Convolutional Neural Network</td><td>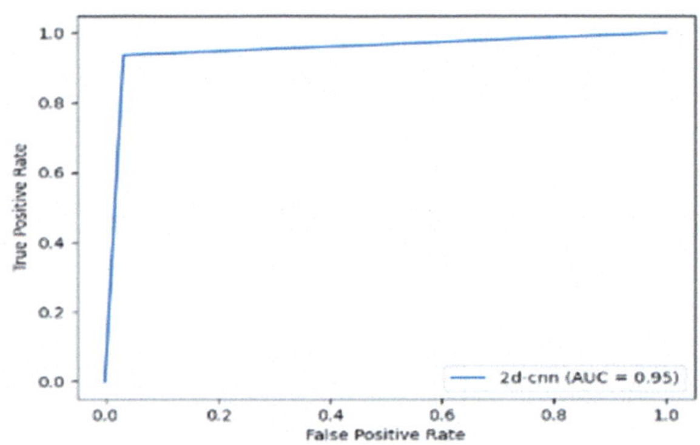
</td></tr>
</table>

Fig. 8 The obtained RoC-AUC curve using DL (DNN and 2D-CNN) algorithms for the diagnosis of PD and AD

4 Conclusions

The main objective of this study is to examine the various techniques employed in diagnosing Parkinson's disease (PD) and Alzheimer's disease (AD). These techniques encompass both shallow learning (SL) and deep learning (DL) methods. Electroencephalography (EEG) signals play a crucial role in disease diagnosis, but they need to undergo preprocessing. In this study, the preprocessing step involved utilizing the Butterworth band-pass filter. The same categorization was performed in this paper, using the Butterworth band-pass filter. For both AD and PD, Common Spatial Patterns (CSP) were used, and features were extracted. The CSP played a role in the entire classification process. However, in the case of AD, only raw EEG signals were processed (2-Dimensional Convolutional Neural Network (2D-CNN)). In the AD dataset, EEG signal recordings lasted 8 s, while in the PD dataset, they

Table 4 Comparison of the SL (AB, CatBoost, XGBoost, DT, GNB, LDA, and KNN) algorithms performance for the diagnosis of PD

	Accuracy (%)	f-measure score	Kappa score	Jaccard score	ROC score	Recall	Precision	Elapsed time (s)
AdaBoost	94.3	0.942	0.885	0.884	0.942	0.942	0.943	0.943
CatBoost	98.9	0.989	0.978	0.976	0.989	0.989	0.989	0.989
XGBoost	98.7	0.987	0.974	0.973	0.987	0.987	0.987	0.987
Decision tree	91.5	0.915	0.830	0.833	0.915	0.915	0.915	0.915
Gaussian Naïve Bayes	70.4	0.703	0.417	0.568	0.713	0.713	0.722	0.704
Linear discriminant analysis	91.5	0.915	0.831	0.839	0.918	0.918	0.916	0.915
K-nearest neighbors	**99.3**	**0.993**	**0.986**	**0.985**	**0.993**	**0.993**	**0.993**	**0.993**

Table 5 Comparison of the SL (AB, CatBoost, XGBoost, DT, GNB, LDA, and KNN) algorithms performance for the diagnosis of AD

	Accuracy (%)	f-measure score	Kappa score	Jaccard score	ROC score	Recall	Precision	Elapsed time (s)
AdaBoost	79.2	0.791	0.593	0.673	0.818	0.818	0.805	0.887
CatBoost	**92.2**	**0.920**	**0.842**	**0.872**	**0.936**	**0.936**	**0.917**	**19.369**
XGBoost	85.7	0.856	0.718	0.766	0.883	0.883	0.866	1.049
Decision tree	76.6	0.765	0.542	0.640	0.790	0.790	0.778	0.034
Gaussian Naïve Bayes	68.8	0.686	0.382	0.556	0.702	0.702	0.693	0.002
Linear discriminant analysis	79.2	0.792	0.597	0.667	0.824	0.824	0.814	0.007
K-nearest neighbors	77.9	0.779	0.579	0.638	0.819	0.819	0.819	0.001

Table 6 Comparison of the DL (DNN and 2D-CNN) algorithms performance for the diagnosis of AD and PD

	Parkinson's disease Deep Neural Network	Alzheimer's disease 2-Dimensional Convolutional Neural Network
Accuracy (%)	98.5	94.8
f-measure score	0.985	0.946
Kappa score	0.969	0.892
Jaccard score	0.968	0.917
ROC score	0.985	0.951
Recall	0.985	0.951
Precision	0.984	0.942
Elapsed time (s)	0.985	60.826

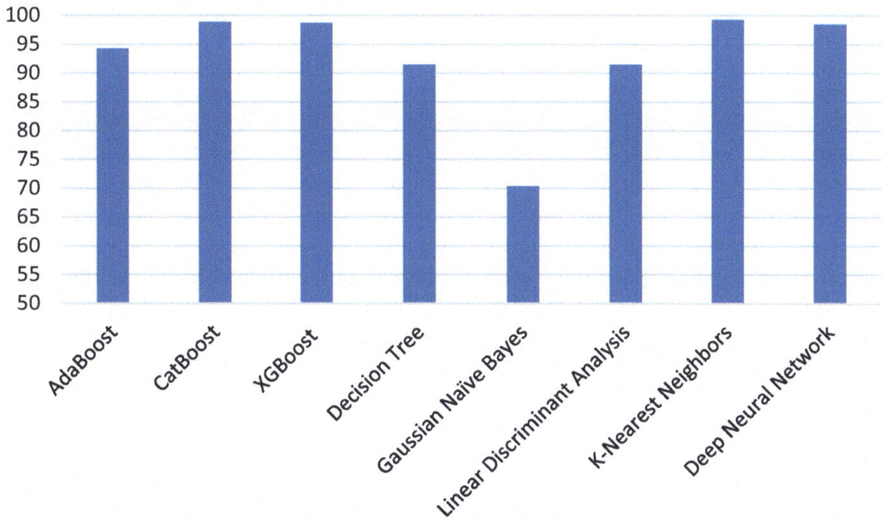

Fig. 9 The obtained classification accuracy of the PD using AB, CatBoost, XGBoost, DT, GNB, LDA, KNN, and DNN algorithms

lasted significantly longer. Both datasets were divided into 1-s epochs and classified accordingly. In diagnosing both PD and AD, the same shallow learning algorithms, including AdaBoost (AB), CatBoost, XGBoost, Decision Tree (DT), Gaussian Naïve Bayes (GNB), Linear Discriminant Analysis (LDA), and K-Nearest Neighbors (KNN), have been utilized. In shallow learning, the KNN algorithm demonstrates a remarkable accuracy of 99.3% in diagnosing PD, while the CatBoost algorithm achieves the best result of 92.2% in diagnosing AD. The DL utilizes the

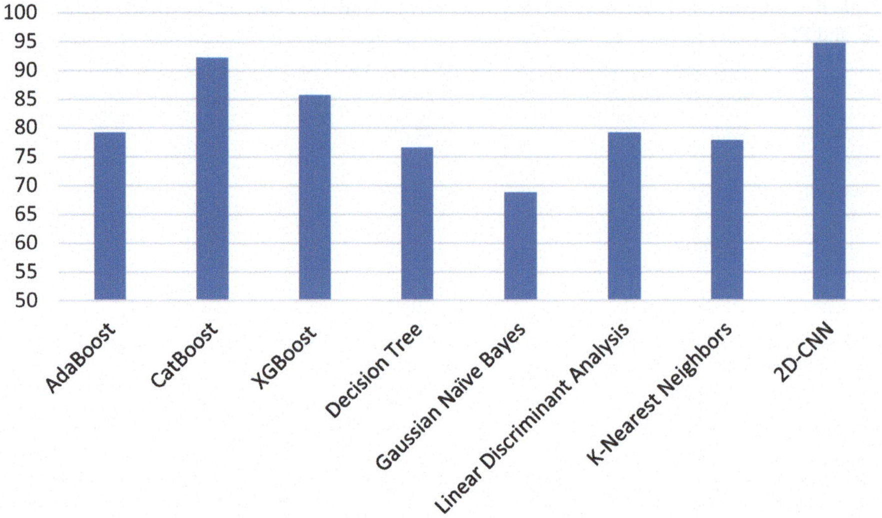

Fig. 10 The obtained classification accuracy of the AD using AB, CatBoost, XGBoost, DT, GNB, LDA, KNN, and 2D-CNN algorithms

DNN algorithm to achieve a remarkable accuracy of 98.5% in diagnosing PD. Moreover, the 2D-DNN algorithm, also employed in deep learning, attains an accuracy of 94.8% when diagnosing AD. In the diagnosis of AD, the DL method gives much better results than the SL method, but the elapsed time is quite high in DL. In the diagnosis of PD, SL, and DL are almost equal, but elapsed time is quite high in deep learning. For future studies, clinical data will be collected to diagnose diseases using the created model, along with proposing new models, instead of relying on a ready-made dataset like in this study.

References

1. Liu, F., Zhou, L., Shen, C., & Yin, J. (2014). Multiple kernel learning in the primal for multimodal Alzheimer's disease classification. *IEEE Journal of Biomedical and Health Informatics, 18*(3), 984–990. https://doi.org/10.1109/JBHI.2013.2285378
2. Özseven, T., & Polat, G. (2023). A *survey of machine learning methods for diagnosing parkinson's disease with handwriting*. In 2023 7th international symposium on multidisciplinary studies and innovative technologies (ISMSIT) (pp. 1–9). Ankara, Turkiye, https://doi.org/10.1109/ISMSIT58785.2023.10304996.
3. Prashanth, R., & Roy, S. D. (2018). Early detection of Parkinson's disease through patient questionnaire and predictive modelling. *International Journal of Medical Informatics., 119*, 75–87. https://doi.org/10.1016/j.ijmedinf.2018.09.008
4. Leifer, B. P. (2003). Early diagnosis of Alzheimer's disease: Clinical and economic benefits. *Journal of the American Geriatrics Society, 51*, S281–S288. https://doi.org/10.1046/j.1532-5415.5153.x

5. Pasluosta, C. F., Gassner, H., Winkler, J., Klucken, J., & Eskofier, B. M. (2015). An emerging era in the management of Parkinson's disease: Wearable technologies and the internet of things. *IEEE Journal of Biomedical and Health Informatics, 19*(6), 1873–1881. https://doi.org/10.1109/JBHI.2015.2461555
6. Khan, P., et al. (2021). Machine learning and deep learning approaches for brain disease diagnosis: Principles and recent advances. *IEEE Access, 9*, 37622–37655. https://doi.org/10.1109/ACCESS.2021.3062484
7. Pahuja, G., & Nagabhushan, T. N. (2021). A comparative study of existing machine learning approaches for Parkinson's disease detection. *IETE Journal of Research, 67*(1), 4–14. https://doi.org/10.1080/03772063.2018.1531730
8. Nissar, I., Raza Rizvi, D., Masood, S., & Nazir Mir, A. (2019). Voice based detection of Parkinson's disease through ensemble machine learning approach: A performance study. *EAI Endorsed Transactions on Pervasive Health and Technology, 5*(19), e2. https://doi.org/10.4108/eai.13-7-2018.162806
9. Sajal, M. S. R., Ehsan, M. T., Vaidyanathan, R., et al. (2020). Telemonitoring Parkinson's disease using machine learning by combining tremor and voice analysis. *Brain Informatics, 7*, 12. https://doi.org/10.1186/s40708-020-00113-1
10. Hussain, M. M., Weslin, D., Kumari, S., Umamaheswari, S., & Kamalakannan, K. (2023). Enhancing Parkinson's disease identification using ensemble classifier and data augmentation techniques in machine learning. *Clinical eHealth, 6*, 1–9. https://doi.org/10.1016/j.ceh.2023.11.002
11. Önder, M., Şentürk, Ü., Polat, K., & Paulraj, D. (2023). *Classification of freezing of Gait in Parkinson's disease using machine learning algorithms.* In 2023 international conference on research methodologies in knowledge management, artificial intelligence and telecommunication engineering (RMKMATE) (pp. 1–5). Chennai, India, 2023, https://doi.org/10.1109/RMKMATE59243.2023.10368706.
12. Aljalal, M., Aldosari, S. A., Molinas, M., et al. (2022). Detection of Parkinson's disease from EEG signals using discrete wavelet transform, different entropy measures, and machine learning techniques. *Scientific Reports, 12*, 22547. https://doi.org/10.1038/s41598-022-26644-7
13. Yang, C. Y., & Huang, Y. Z. (2022). Parkinson's disease classification using machine learning approaches and resting-state EEG. *Journal of Medical and Biological Engineering, 42*, 263–270. https://doi.org/10.1007/s40846-022-00695-7
14. Li, K., Ao, B., Xin, W., Wen, Q., Haq, E. U., & Yin, J. (2023). Parkinson's disease detection and classification using EEG based on deep CNN-LSTM model. *Biotechnology and Genetic Engineering Reviews*, 1. https://doi.org/10.1080/02648725.2023.2200333
15. Chawla, P., Rana, S. B., Kaur, H., Singh, K., Yuvaraj, R., & Murugappan, M. (2023). A decision support system for automated diagnosis of Parkinson's disease from EEG using FAWT and entropy features. *Biomed. Signal Process., 79*, 104116. https://doi.org/10.1016/j.bspc.2022.104116
16. Kara Gulay, B., Demirel, N., Vahaplar, A., & Guducu, C. (2023). A novel feature extraction method using chemosensory EEG for Parkinson's disease classification. *Biomedical Signal Processing and Control, 79*, 104147.
17. Kumar, K., & Ghosh, R. (2023). Parkinson's disease diagnosis using recurrent neural network based deep learning model by analyzing online handwriting. *Multimedia Tools and Applications, 83*, 11687. https://doi.org/10.1007/s11042-023-15811-1
18. Sarin, K., Bardamova, M., Svetlakov, M., Koryshev, N., Ostapenko, R., Hodashinskaya, A., & Hodashinsky, I. (2023). PA three-stage fuzzy classifier method for Parkinson's disease diagnosis using dynamic handwriting analysis. *Decision Analytics Journal, 8*, 1–23. https://doi.org/10.1016/j.dajour.2023.100274
19. Rao, K. N., Gandhi, B. R., Rao, M. V., Javvadi, S., Vellela, S. S., & Khader Basha, S. (2023). *Prediction and classification of Alzheimer's disease using machine learning techniques in 3D MR images.* In 2023 international conference on sustainable computing and smart systems

(ICSCSS) (pp. 85–90). Coimbatore, India, https://doi.org/10.1109/ICSCSS57650.2023.10169550.
20. Bahadori, R., Kiaei, A. A., Malek Zadeh, H., Boush, M., Abadijou, S., Safaei, D., & Mehdikhani, A. H. (2023). Diagnosing Alzheimer's disease levels using machine learning and MRI: A novel approach. *Preprint*, 2023061184. https://doi.org/10.20944/preprints202306.1184.v1
21. EL-Geneedy, M., Moustafa, H. E.-D., Khalifa, F., Khater, H., & AbdElhalim, E. (2023). An MRI-based deep learning approach for accurate detection of Alzheimer's disease. *Alexandria Engineering Journal, 63*, 211–221. https://doi.org/10.1016/j.aej.2022.07.062
22. Lee, G., Nho, K., Kang, B., et al. (2019). Predicting Alzheimer's disease progression using multi-modal deep learning approach. *Scientific Reports, 9*, 1952. https://doi.org/10.1038/s41598-018-37769-z
23. Xiao, Z., Ding, Y., Lan, T., Zhang, C., Luo, C., & Qin, Z. (2017). Brain MR image classification for Alzheimer's disease diagnosis based on multifeature fusion. *Computational and Mathematical Methods in Medicine, 2017*, 1. https://doi.org/10.1155/2017/1952373
24. Sudharsan, M., & Thailambal, G. (2021). Alzheimer's disease prediction using machine learning techniques and principal component analysis (PCA). *Materials Today: Proceedings, 81*, 182. https://doi.org/10.1016/j.matpr.2021.03.061
25. Miah, Y., Prima, C. N. E., Seema, S. J., Mahmud, M., & Kaiser, M. S. (2021). Performance comparison of machine learning techniques in identifying dementia from open access clinical datasets. In *Advances on smart and soft computing* (pp. 79–89). Springer. https://doi.org/10.1007/978-981-15-6048-4_8
26. Fouad, I. A., El-Zahraa, F., & Labib, M. (2023). Identification of Alzheimer's disease from central lobe EEG signals utilizing machine learning and residual neural network. *Biomedical Signal Processing and Control, 86*, 105266. https://doi.org/10.1016/J.BSPC.2023.105266
27. Kulkarni, N. (2018). Use of complexity based features in diagnosis of mild Alzheimer disease using EEG signals. *International Journal of Information Technology, 10*, 59–64. https://doi.org/10.1007/s41870-017-0057-0
28. Khare, S. K., & Acharya, U. R. (2023). Adazd-net: Automated adaptive and explainable Alzheimer's disease detection system using EEG signals. *Knowledge-Based Systems, 278*, 110858. https://doi.org/10.1016/j.knosys.2023.110858
29. Pirrone, D., Weitschek, E., Di Paolo, P., De Salvo, S., & De Cola, M. C. (2022). EEG signal processing and supervised machine learning to early diagnose Alzheimer's disease. *Applied Sciences, 12*(11). https://doi.org/10.3390/app12115413
30. Fouladi, S., Safaei, A. A., Mammone, N., et al. (2022). Efficient deep neural networks for classification of Alzheimer's disease and mild cognitive impairment from scalp EEG recordings. *Cognitive Computation, 14*, 1247–1268. https://doi.org/10.1007/s12559-02210033-3
31. OpenNeuro. (2023). *UC San Diego resting state EEG data from patients with Parkinson's disease*. https://openneuro.org/datasets/ds002778/versions/1.0.5. Accessed 24 Dec 2023.
32. Jackson, N., et al. (2019). Characteristics of waveform shape in Parkinson's disease detected with scalp electroencephalography. *Eneuro, 6*(3), ENEURO.0151.
33. Imani, M. (2023). Alzheimer's diseases diagnosis using fusion of high informative BiLSTM and CNN features of EEG signal. *Biomedical Signal Processing and Control, 86*, 105298. https://doi.org/10.1016/J.BSPC.2023.105298
34. Nobukawa, S., Yamanishi, T., Kasakawa, S., Nishimura, H., Kikuchi, M., & Takahashi, T. (2020). Classification methods based on complexity and synchronization of electroencephalography signals in Alzheimer's disease. *Frontiers in Psychiatry, 11*, 511787. https://doi.org/10.3389/FPSYT.2020.00255/BIBTEX
35. Dash, S. (2023). Chapter 22 – A systematic review of adaptive machine learning techniques for early detection of Parkinson's disease. In A. Abraham, S. Dash, S. K. Pani, & L. García-Hernández (Eds.), *Artificial Intelligence for Neurological Disorders* (pp. 361–385). Academic. https://doi.org/10.1016/B978-0-323-90277-9.00018-3

36. Alawad, M., & Lin, M. (2018). Memory-efficient probabilistic 2-D finite impulse response (FIR) filter. *IEEE Transactions on Multi-Scale Computing Systems, 4*(1), 69–82. https://doi.org/10.1109/TMSCS.2017.2695588
37. Daud, S. S., & Sudirman, R. (2015). *Butterworth bandpass and stationary wavelet transform filter comparison for electroencephalography signal.* In 2015 6th international conference on intelligent systems, modelling and simulation (pp. 123–126). Kuala Lumpur, Malaysia, 2015, https://doi.org/10.1109/ISMS.2015.29.
38. Zhang, S., Wang, S., & Zhao, C. (2024). Vertices packaging-based interval independent component analysis (VP-I2CA) for fault detection with process uncertainty. *IEEE Transactions on Industrial Informatics, 20*(1), 919–930. https://doi.org/10.1109/TII.2023.3271737
39. Procházka, A., Mudrová, M., Vyšata, O., Háva, R., & Suárez Araujo, C. P. (2010). *Multi-channel EEG signal segmentation and feature extraction.* In 2010 IEEE 14th international conference on intelligent engineering systems (pp. 317–320). Las Palmas, Spain, 2010, https://doi.org/10.1109/INES.2010.5483824.
40. Malaekah, E., & Cvetkovic, D. (2013). *Automatic detection of the wake and stage 1 sleep stages using the EEG sub-epoch approach.* In 2013 35th annual international conference of the ieee engineering in medicine and biology society (EMBC) (pp. 6401–6404), Osaka, Japan, 2013, https://doi.org/10.1109/EMBC.2013.6611019.
41. Haartsen, R., van der Velde, B., Jones, E. J. H., et al. (2020). Using multiple short epochs optimises the stability of infant EEG connectivity parameters. *Scientific Reports, 10*, 12703. https://doi.org/10.1038/s41598-020-68981-5
42. Ang, K. K., Chin, Z. Y., Zhang, H., & Guan, C. (2008, June). *Filter bank common spatial pattern (FBCSP) in brain-computer interface.* In 2008 IEEE international joint conference on neural networks (IEEE world congress on computational intelligence) (pp. 2390–2397). IEEE. https://doi.org/10.1109/IJCNN.2008.4634130.
43. Schapire, R. E. (2013). Explaining AdaBoost. In B. Schölkopf, Z. Luo, & V. Vovk (Eds.), *Empirical inference*. Springer. https://doi.org/10.1007/978-3-642-41136-6_5
44. Önder, M., Şentürk, Ü., Polat, K., & Paulraj, D. (2023, November). *Diagnosis of Alzheimer's disease using boosting classification algorithms.* In 2023 international conference on research methodologies in knowledge management, artificial intelligence and telecommunication engineering (RMKMATE) (pp. 1–5). IEEE. https://doi.org/10.1109/RMKMATE59243.2023.10369418.
45. Wang, Y., Huang, X., Ren, X., et al. (2022). In-process belt-image based material removal rate monitoring for abrasive belt grinding using CatBoost algorithm. *International Journal of Advanced Manufacturing Technology, 123*, 2575–2591. https://doi.org/10.1007/s00170-022-10341-w
46. Osman, A. I. A., Ahmed, A. N., Chow, M. F., Huang, Y. F., & El-Shafie, A. (2021). Extreme gradient boosting (Xgboost) model to predict the groundwater levels in Selangor Malaysia. *Ain Shams Engineering Journal, 12*, 1545–1556. https://doi.org/10.1016/j.asej.2020.11.011
47. Gong, Z., Ding, Y., Chen, Y., & Cree, M. J. (2023). Wearable microwave medical sensing for stroke classification and localization: A space-division-based decision-tree learning method. *IEEE Transactions on Antennas and Propagation, 71*(8), 6906–6917. https://doi.org/10.1109/TAP.2023.3283131
48. Athish, V. P., Rajeswari, D., & SS, S.N. (2023, March). *Football prediction system using gaussian naïve bayes algorithm.* In 2023 second international conference on electronics and renewable systems (ICEARS) (pp. 1640–1643). IEEE.
49. Park, C. H., & Park, H. (2008). A comparison of generalized linear discriminant analysis algorithms. *Pattern Recognition, 41*(3), 1083–1097. https://doi.org/10.1109/TNNLS.2014.2306844
50. Huang, D., Quan, Y., He, M., et al. (2009). Comparison of linear discriminant analysis methods for the classification of cancer based on gene expression data. *Journal of Experimental & Clinical Cancer Research, 28*, 149. https://doi.org/10.1186/1756-9966-28-149

51. Islam, M. M., Iqbal, H., Haque, M. R., & Hasan, M. K. (2017). *Prediction of breast cancer using support vector machine and K-Nearest neighbors*. In 2017 IEEE region 10 humanitarian technology conference (R10-HTC) (pp. 226–229), Dhaka, Bangladesh, 2017, https://doi.org/10.1109/R10-HTC.2017.8288944.
52. Deng, L., Hinton, G., & Kingsbury, B. (2013). *New types of deep neural network learning for speech recognition and related applications: An overview*. In 2013 IEEE international conference on acoustics, speech and signal processing (pp. 8599–8603), Vancouver, BC, Canada, 2013, https://doi.org/10.1109/ICASSP.2013.6639344.
53. Trivizakis, E., et al. (2019). Extending 2-D convolutional neural networks to 3-D for advancing deep learning cancer classification with application to MRI liver tumor differentiation. *IEEE Journal of Biomedical and Health Informatics, 23*(3), 923–930. https://doi.org/10.1109/JBHI.2018.2886276
54. Ghaderizadeh, S., Abbasi-Moghadam, D., Sharifi, A., Zhao, N., & Tariq, A. (2021). Hyperspectral image classification using a hybrid 3D-2D convolutional neural networks. *IEEE Journal of Selected Topics in Applied Earth Observations and Remote Sensing, 14*, 7570–7588. https://doi.org/10.1109/JSTARS.2021.3099118
55. Önder, M., Dogan, M. U., & Polat, K. (2023). Classification of smart grid stability prediction using cascade machine learning methods and the internet of things in smart grid. *Neural Computing and Applications, 35*, 17851–17869. https://doi.org/10.1007/s00521-023-08605-x

Shallow Learning vs. Deep Learning in Anomaly Detection Applications

Ismail A. Mageed, Ashiq H. Bhat, and Hafeez Ur Rehman

1 Background

1.1 Anomaly Detection and Its Application Domains

The increasing use of the Internet in communication and commerce, along with developments in networking technology [1], has increased the threat posed by spammers, attackers, and illicit businesses. Like cyberspace burglar alarms, intrusion detection systems have become essential for network security. Unfortunately, to detect novel attacks, most currently available commercial solutions rely on existing attack signatures and necessitate regular updates. Network anomaly detection systems (NADSs), on the other hand, model normal system behaviour and are effective in detecting both known and unknown attacks.

A major issue that has been researched in several application domains and research fields is anomaly detection (AD). Various methods have been devised to identify irregularities; some are customized for certain fields, while others are more all-encompassing. The work in [2] has provided a structured overview of anomaly detection research by categorizing existing techniques based on their underlying approach and identifying key assumptions used to distinguish normal from

I. A. Mageed (✉)
Faculty of Engineering and Digital Technologies, School of Computer Science, AI, and Electronics, University of Bradford, Bradford, UK
e-mail: iamoham@bradford.ac.uk

A. H. Bhat
Department of Biostatistics, Rajendra Memorial Research, Institute of Medical Sciences, Patna, India

H. U. Rehman
School of Computing and Data Science, OUC with Liverpool J. Mores University, Doha, Qatar
e-mail: h.urrehman@ljmu.ac.uk

Ö. F. Ertuğrul et al. (eds.), *Shallow Learning vs. Deep Learning*, The Springer Series in Applied Machine Learning, https://doi.org/10.1007/978-3-031-69499-8_7

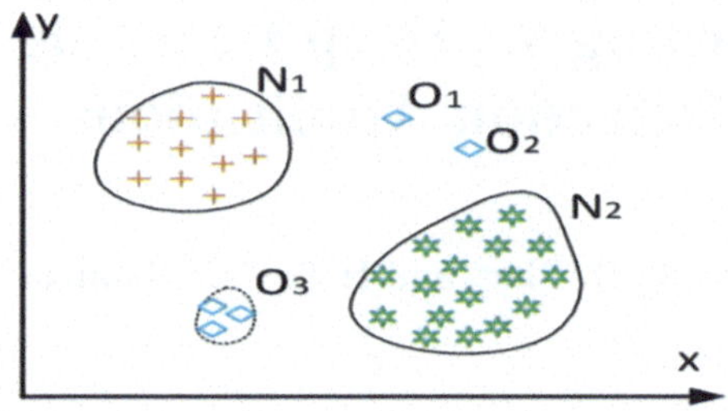

Fig. 1 A display of data patterns that diverge from estimated normal behaviour is referred to as anomalies. There are two normal zones in the data in the example provided; points that deviate significantly from these regions are regarded as anomalies

anomalous behaviour. These assumptions can serve as guidelines when applying a technique to assess its effectiveness in a particular domain.

Finding patterns [2] in data that differ from expected behaviour is referred to as AD. These patterns, also known as anomalies, outliers, or exceptions, do not conform to the normal patterns observed in the data. Anomaly detection techniques aim to categorize and understand these unusual patterns, which can be useful in various domains such as data mining and database applications.

AD is extensively employed in many different domains like fraud detection, intrusion detection, defect detection, and military surveillance, words like anomalies [2] and outliers are frequently used. Finding anomalies in data is crucial because they can reveal useful and frequently crucial information, such as the presence of malicious tumours in MRI pictures, the presence of credit cards or identity theft, or the presence of flaws in spacecraft parts. It plays a crucial role in uncovering actionable insights in diverse application areas. Figure 1 (c.f., [2]) displays a two-dimensional data set anomaly.

A popular approach of locating anomalies is to develop a zone that corresponds to usual behaviour and label any observation outside of it as an abnormality [2]. However, several challenges arise, such as the difficulty in precisely defining the normal region, the adaptability of malicious actors to appear normal, the evolving nature of normal behaviour, the variation in the notion of anomalies across different domains, the scarcity of labelled data for training, and the presence of noise that resembles anomalies. These challenges make the general problem of anomaly detection complex, leading to the development of specific techniques tailored to different application domains.

To create methods for anomaly detection, researchers in AD have taken motivation from ideas in a variety of fields, including information theory, statistics, ML, data mining, and spectral analysis. The methodology utilized for identifying anomalies in the context of AD is frequently dictated by the features of the application domain whereby the anomalies ought to be identified [2]. These interdisciplinary approaches help address the challenges associated with detecting anomalies in diverse datasets. Researchers have applied AD techniques [2] to specific problem

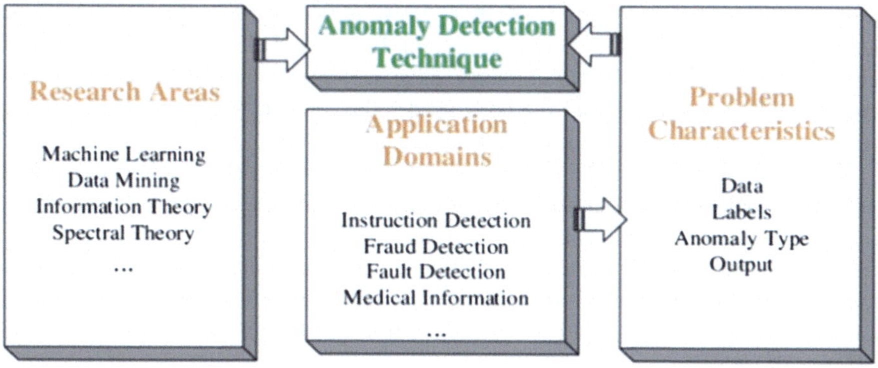

Fig. 2 AD essential components that form the basis of such techniques and help in detecting anomalies effectively

formulations. Figure 2 (c.f., [2]) illustrates the main components associated with any AD technique.

Contextual anomalies [2] are determined by both contextual attributes (e.g. location or time) and behavioural attributes (e.g. rainfall amount). These concepts are commonly explored in time-series and spatial data analysis, such as detecting credit card fraud or identifying unusual temperature patterns.

AD has several application domains, namely intrusion detection, fault detection, fraud/damage detection, medical and public health AD, and AD for textual data.

The process of discovering and detecting malicious activity in a computer system, such as unauthorized access or computer abuse, is known as intrusion detection [2]. Computer security is interested in anomalies, or deviations from typical system behaviour. In this domain, anomaly detection techniques are used; however, because tagged intrusion data is scarce, managing false alarms, processing vast amounts of data, and utilizing semi-supervised or unsupervised algorithms provide challenges.

Unexpected events or wear and tear can cause damage to industrial units [2], such as motors and turbines. Like this, structural defect detection approaches use methods like statistical profiling, parametric statistical modelling, mixture of models, and neural networks to find structures' anomalies, including internal beams cracking or airframe straining. These methods support taking preventive actions and aid in the early discovery of damage.

Fraud detection [2] involves identifying criminal activities within commercial organizations like banks, credit card companies, insurance agencies, and stock markets. These activities can include identity theft, where malicious users either pose as customers or are actual customers and consume the organization's resources unlawfully. To safeguard the organization and its clients, fraud detection aims to identify and stop such fraudulent activity.

AD in the medical and public health domains [2] involves identifying abnormal patterns in patient records, which can arise from factors like abnormal patient conditions, errors in instrumentation, or recording errors. Diverse methods are utilized to identify irregularities, such as those aimed at identifying disease epidemics in certain regions. Patient age, weight, blood type, and other characteristics are frequently included in this domain's data, which may also have temporal and spatial components. Most AD methods used today, which frequently take a semi-supervised approach, are designed to find unusual records.

The main objective in the context of AD in text data [2] is to find novel and uncommon subjects, happenings, or news items within a set of papers or articles. These irregularities result from the presence of fascinating or unusual subjects. Because of the gradual collection of high-dimensional data documents, sparsely populated, with a temporal component. A challenge in this field is effectively handling the significant variations present within documents belonging to a single category or topic.

1.2 Foundational Background on Shallow Learning and Deep Learning

In artificial intelligence systems, ML is becoming more and more widespread. It describes a system's ability to automatically create analytical models and solve related tasks by learning from training data [3]. By using this method, the system can learn from the training data and forecast future events by identifying patterns and relationships. Artificial neural networks are used in the ML concept known as deep learning (DL) [3]. In several situations, it has demonstrated better performance than both shallow learning (SL) models and conventional data analysis techniques.

It is critical to distinguish between several words and concepts to comprehend the topic of artificial intelligence [3]. Artificial intelligence (AI) comprises methods that let computers behave like humans and figure out difficult tasks on their own. Hierarchical AI components such as deep neural networks, artificial neural networks, and ML algorithms are essential for automating processes and deriving insights from data.

ML is a field of study [3] where computer programs improve their performance in specific tasks by learning from experience. It automates the process of building analytical models and can be applied to tasks like object detection, natural language translation, classification, regression, and clustering. ML algorithms learn from training data, extract patterns from large databases, and enable computers to make reliable and repeatable decisions. This has led to successful applications in various domains, including fraud detection, and credit scoring.

Three types of ML can be distinguished in the context of electronic markets [3]: supervised, unsupervised, and reinforcement learning. In electronic markets, supervised learning is frequently utilized for tasks including product search, customer perception analysis, stock market forecasting, and understanding client wants. However, there are also implementations of unsupervised learning, such as market segmentation using customer reviews, and reinforcement learning, like market-making strategies.

Distinct classes of algorithms are available for distinct learning tasks in ML [3]. Artificial neural networks (ANNs), which draw inspiration from the information-processing mechanisms of biological systems, are one type of algorithm. An ANN is made up of interconnected artificial neurons that exchange signals, with the signals' strength being modulated by weights. These networks can have multiple layers, including hidden layers, and can be modified to suit different contexts. Deep neural networks, a type of ANN, have more than one hidden layer and often employ advanced operations or multiple activations in neurons, allowing them to handle complex tasks.

DL refers to the core capability of neural networks [3] to perform complex tasks by learning from large and high-dimensional data. On the other end of the spectrum, SL techniques do not have this ability, such as decision trees and basic artificial neural networks like shallow autoencoders. Although the lines separating DL from SL are not perfectly drawn, they can be identified by the features they offer.

It is stated that [3] most sophisticated ML algorithms are untraceable in their decision-making process (black boxes) unless specifically stated; however, certain shallow ML systems are thought to be interpretable by humans (white boxes). On the other hand, processing vast and high-dimensional data, such as text, images, videos, audio, and speech, is especially well-suited for deep learning. Shallow machine learning can still yield better and easier-to-understand results, nonetheless, when dealing with low-dimensional data and sparse training set availability. It is also mentioned that deep learning has its limitations in solving problems that require strong AI capabilities, as highlighted in Searle's Chinese room argument. There is no precise distinction between the two concepts in the literature, and as a result, a dashed line is used in Fig. 4 (c.f., [4]) to represent this lack of demarcation. This suggests that there is ambiguity or overlap between the concepts being discussed. Figure 3, inspired by [4] visually represents different approaches of ML approaches and their relationships.

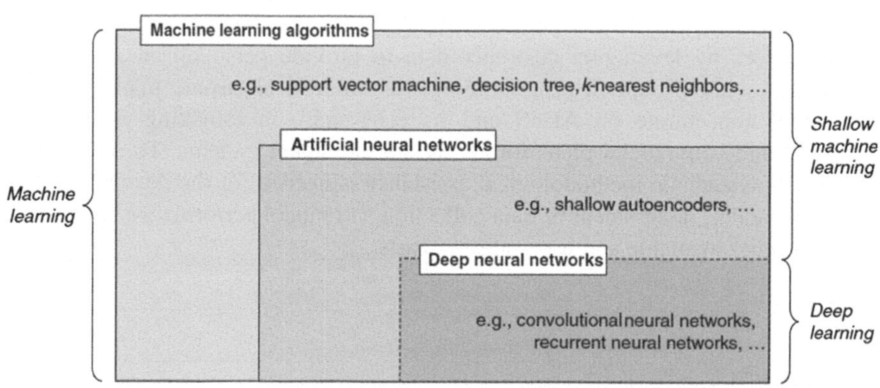

Fig. 3 The Venn diagram illustrates the different concepts and classes within ML

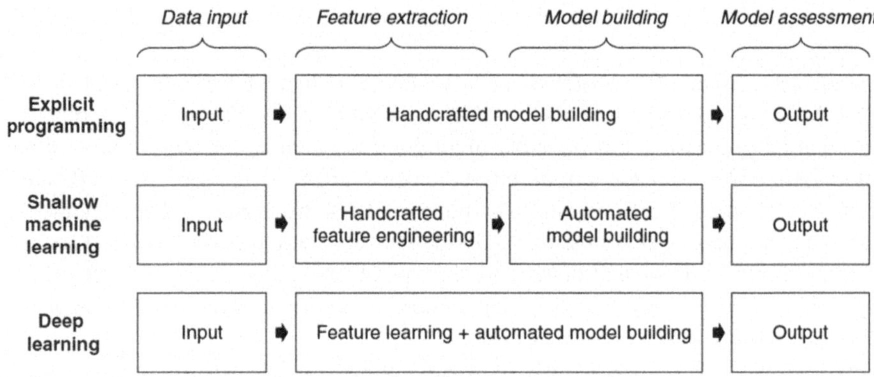

Fig. 4 Schematic of analytical model building

Deep neural networks [3] are advantageous as they do not require manual feature engineering because they use automated feature learning. Their effectiveness stems on their ability to handle enormous volumes of noisy, unstructured data. DL algorithms [3] can operate directly on high-dimensional raw input data to automatically extract features throughout the model-building process. Feature representations [3] may be employed as end-to-end systems or in conjunction with other ML techniques, such as support vector machines (SVMs) or decision trees once they are extracted for additional analysis. Convolutional neural networks (CNNs) and recurrent neural networks (RNNs) are two more architectural modifications that have been developed for a range of data and applications [3].

Analytical model building [3] is an approach displayed in Figure 4 (c.f., [4]) and involves using learning algorithms to find patterns and relationships in the input data that are pertinent to the learning objective. While DL can operate directly on extremely complex raw input data and automatically build discriminative feature representations, SL necessitates well-designed features for such a form of assignment.

AI-as-a-Service (AIaaS) [3] is an emerging electronic market that has the potential to significantly impact existing service platforms. It enables the enhancement of smart services by leveraging customer data to provide personalized advice and instructions without explicit programming. Research in electronic markets in the future will concentrate on AIaaS and its ecosystem, investigating novel uses, functions, and commercial plans for DL-based intelligent systems. To solve these challenges, systematic methodological assistance is needed for the development of analytical models, assessment of data collection and model performance, and assurance of readily available and user-safe forecasts.

2 Performance Comparison of Shallow and Deep Learning Algorithms in Anomaly Detection Applications

In recent years, the exploration of AD methods has become pivotal, particularly in the context of leveraging both shallow and deep learning algorithms. AD is the task of identifying data points that deviate significantly from normal patterns. It is an important task in a variety of applications, including fraud detection, network intrusion detection, medical diagnosis, etc. Shallow learning algorithms, such as k-nearest neighbours (KNN) [5, 6] and local outlier factor (LOF) [7], have been traditionally used for anomaly detection. However, DL algorithms, such as autoencoders and CNNs, have recently gained popularity due to their ability to learn complex patterns from large datasets. The overall classification scheme for AD is shown in Fig. 5. The classification is arranged on the plane defined by two key components, namely, the model and feature map, which categorizes various AD approaches. Grouping them based on shared principles, we discern three main categories: one-class classification, probabilistic models, and reconstruction models [8]. All these categories articulate both shallow and deep models, as detailed in the nomenclature section for a list of abbreviations. In addition, these three categories can be improved by simply distance-based strategies. We highlight loss, regularization, and inference mode as additional critical modelling elements in tackling the AD problem, alongside the model and feature map.

2.1 *Shallow Learning Algorithms for Anomaly Detection Applications*

Traditional strategies known as "shallow anomaly detection approaches" make use of statistical and machine learning methods to find data points that exhibit a

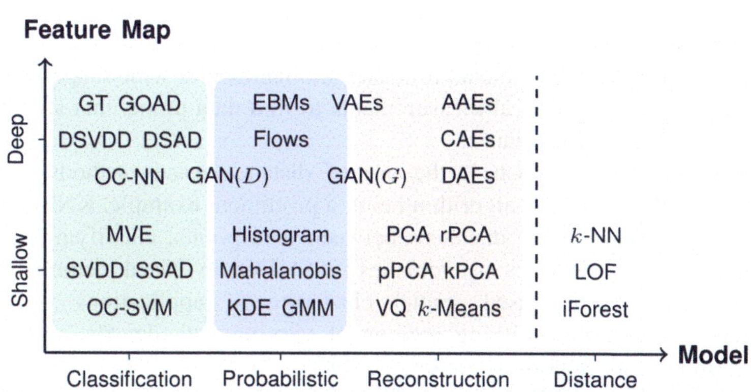

Fig. 5 The classification of deep and shallow learning approaches [8]

considerable departure from usual patterns [5–11]. These methods are typically less complex and computationally efficient than deep learning approaches, making them suitable for real-time applications.

The most common shallow anomaly detection techniques include:

- *K-Nearest Neighbours (KNN)*: KNN compares a new data point to its k-nearest neighbours in the training data. If the new data point is significantly different from its neighbours, it is considered an anomaly [5, 6].
- *Local Outlier Factor (LOF)*: LOF assigns a local outlier factor score to each data point, based on the density of its surrounding data points. Data points with low LOF scores are considered outliers [7].
- *Support Vector Data Description (SVDD)*: SVDD constructs a hypersphere that tightly encloses the training data. Data points that fall outside the hypersphere are considered anomalies [9].
- *Isolation Forest*: Isolation Forest randomly partitions the training data into subsets until each subset contains only a single data point. Outliers tend to be isolated earlier than normal data points [10].
- *One-Class Support Vector Machine (OC-SVM)*: OC-SVM learns a decision boundary that separates the training data from a hypothetical origin. Anomalies are defined as data points that lie outside of the decision boundary [11].

The common characteristics among the aforementioned methods lie in their shared type of being unsupervised, representing the prevailing approach in AD. This prevalence arises due to the frequent absence of labelled anomalous data in standard AD settings. Even when such labelled data is accessible, it often proves inadequate to comprehensively characterize all facets of anomalousness [8]. Consequently, a supervised approach tends to be ineffective. Instead, a fundamental concept in AD involves unsupervised learning of a model of normality from normal data, enabling the detection of anomalies through discernible deviations from the established model.

The literature on classical anomaly detection techniques encompasses a diverse array of methods developed to identify irregularities or outliers within datasets [5–12]. Among these techniques, one of the earliest and widely used approaches is statistical methods. These techniques, including the z-score, modified z-score, and Grubbs' test, offer key instruments to locate anomalies in a wide range of regions because they employ statistical measurements to find data points that significantly move from the mean or median.

Another classical approach is the use of distance-based methods, with the k-nearest neighbours (KNN) algorithm being a prominent example. KNN identifies anomalies by measuring the distances between data points, classifying instances with unusually large distances as anomalies [5, 6]. Density-based methods, such as the well-known density-based spatial clustering of applications with noise (DBSCAN), focus on identifying regions of varying data density, categorizing points in low-density regions as anomalies [13].

Within the realm of ML, one-lass SVM has gained prominence based on its ability to create a boundary around normal instances, flagging instances outside this

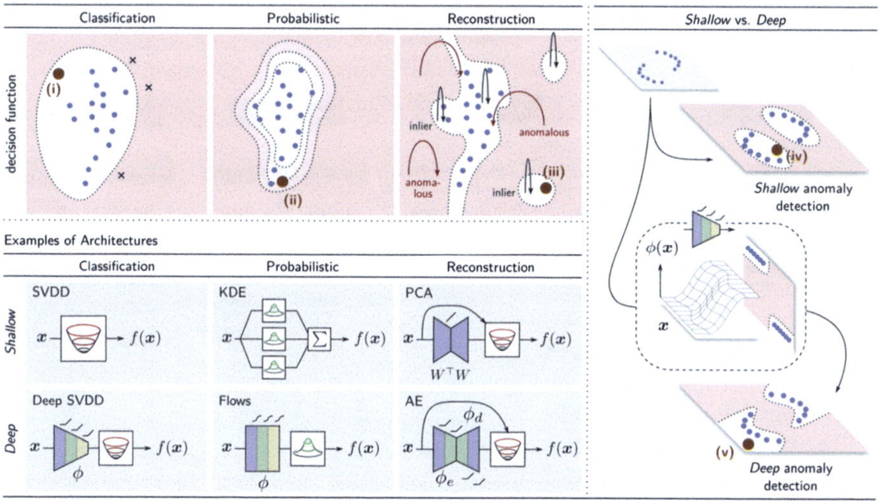

Fig. 6 Various anomaly detection techniques: diagrams depict architectures with deep and shallow feature maps, highlighting model-specific phenomena at selected input space locations (i)–(v) [8].

boundary as anomalies [11]. Decision tree-based methods, exemplified by the Isolation Forest algorithm, leverage the concept of isolation to identify anomalies efficiently, particularly in high-dimensional datasets [9–11] (Fig. 6).

Ensemble techniques, such as the combination of multiple classifiers or models, have also been explored. Bagging and boosting methods, applied to anomaly detection, aim to enhance overall performance and robustness by aggregating the outputs of multiple base detectors applied on datasets of network intrusion detection to building energy consumption related anomaly detection [13, 14].

While classical anomaly detection techniques have shown effectiveness in various applications, their performance can be contingent on factors, such as dataset characteristics, the nature of anomalies, and the chosen algorithm parameters. As the field evolves, researchers continue to explore hybrid models and advanced techniques to address the challenges posed by complex and dynamic datasets, reflecting the ongoing quest for improved anomaly detection solutions.

2.2 Deep Learning Algorithms for Anomaly Detection Applications

Recently, DL approaches have emerged as powerful tools for anomaly detection, offering the ability to capture intricate patterns and relationships in complex datasets [15]. One of the key advantages of DL is its capacity to automatically learn hierarchical representations from data, making it well-suited for tasks where the

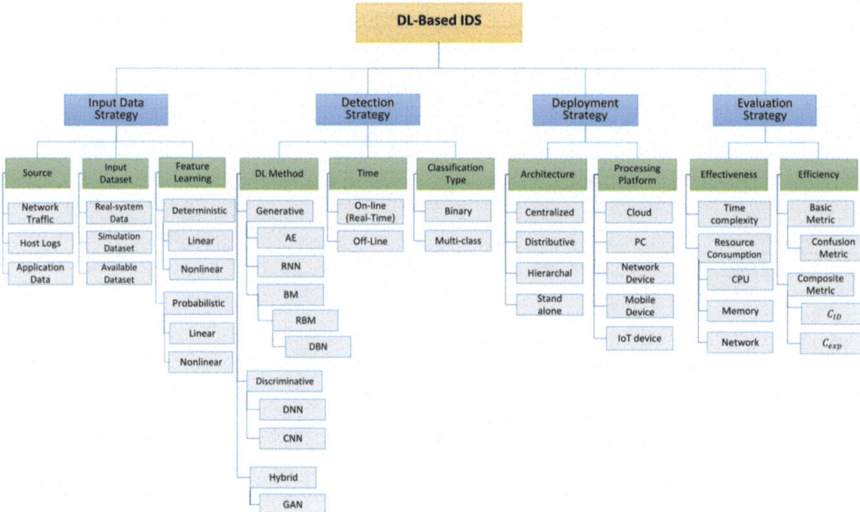

Fig. 7 Hierarchy of deep learning models for anomaly detection [17]

underlying structures may be highly intricate or abstract [16]. The general hierarchy of DL models for anomaly detection is shown in Fig. 7, as proposed in [17].

Autoencoders, a type of neural network, have gained prominence in deep learning-based anomaly detection. These networks consist of an encoder and a decoder, trained to reconstruct input data. Anomalies are identified by observing significant discrepancies between the input and the reconstructed output. Variational autoencoders (VAEs) introduce probabilistic elements into this architecture, allowing for the modelling of data distribution and enhancing the detection of subtle anomalies [18, 19]. The intrinsic ability of autoencoders to capture and encode the salient features of normal patterns within a dataset makes them particularly well-suited for anomaly detection tasks. In the realm of finance, autoencoders have been harnessed to scrutinize transactional data, identifying deviations from established behavioural norms and thereby fortifying fraud detection mechanisms [18]. Similarly, in network security, these neural networks prove invaluable in discerning unusual patterns within communication flows, enabling the swift identification of potential cyber threats and enhancing the overall resilience of computer systems.

Moreover, the versatility of autoencoders extends to critical sectors such as manufacturing, healthcare, and environmental monitoring. In manufacturing, these models play a pivotal role in quality control by detecting anomalies in production processes, safeguarding against defects, and ensuring product integrity [19]. Within healthcare, autoencoders analyse patient data to identify irregular physiological readings or aberrant patterns in medical imagery, offering a proactive approach to healthcare monitoring.

Recurrent neural networks (RNNs) and long short-term memory networks (LSTMs) are particularly effective for sequential data, such as time series data

[20]. These architectures learn temporal dependencies and can capture patterns evolving over time, making them valuable for applications like fraud detection in financial transactions or monitoring industrial processes. One prominent application lies in the field of finance, where RNNs excel at capturing temporal dependencies in time-series data. Financial transactions often exhibit intricate patterns, and RNNs, with their ability to retain memory of past information, prove invaluable in discerning normal transaction sequences from potentially fraudulent ones. The utilization of long short-term memory (LSTM) networks, a variant of RNNs designed to overcome the vanishing gradient problem, enhances the model's capacity to identify anomalies by preserving contextual information over extended time periods [21]. The amalgamation of RNNs and LSTM architectures contributes significantly to the efficacy of anomaly detection systems, fostering advancements in financial security.

By training a discriminator network to distinguish between real-life generated data and a generator network to produce realistic data, generative adversarial networks (GANs) have demonstrated potential in AD [22]. Anomalies are then identified as instances where the discriminator struggles to distinguish between the two, highlighting deviations from the learned data distribution. Transfer learning has also been employed by a few authors in DL-based anomaly detection, where they utilize pre-trained models on big datasets and improve them to specific AD tasks [22, 23]. This approach has demonstrated effectiveness within labelled anomaly scarce data-scenarios.

Research on hybrid models, typically combining conventional and DL techniques, is ongoing to utilize the advantages of both approaches for more reliable and flexible anomaly detection in a range of applications. On one hand, the DL methods have demonstrated remarkable success in various domains, including anomaly detection; however, they are not without their limitations. One significant challenge lies in the requirement for large amounts of labelled training data. DL models, particularly neural networks, thrive on extensive datasets for effective learning and generalization. In the context of anomaly detection, obtaining a sufficiently diverse set of labelled anomalies can be inherently challenging. Anomalies are often rare occurrences, and the limited availability of labelled instances hinders the model's ability to accurately identify and generalize to novel types of anomalies [16–21]. This limitation can lead to suboptimal performance, especially in scenarios where anomalies exhibit diverse and evolving patterns.

Another notable constraint is the lack of interpretability in DL architecture for AD. Neural networks, especially complex architectures like deep autoencoders and generative models, are often treated as "black box" systems [15]. Understanding the inner workings of these models and interpreting the learned representations can be challenging, making it difficult to provide meaningful explanations for the detection decisions. In applications where interpretability is crucial, such as in healthcare or finance, the inscrutable nature of DL methods may pose a barrier to adoption [16, 17]. Striking a balance between the high-dimensional representation power of DL and the interpretability requirements of practical applications remains an ongoing challenge in the deployment of these methods for anomaly detection.

In conclusion, the extensive exploration of deep learning networks in the realm of anomaly detection has unfolded a promising avenue for addressing complex challenges across diverse domains. The efficacy of deep learning models, particularly autoencoders, in capturing intricate patterns within data has been demonstrated through a multitude of applications. From enhancing cybersecurity measures to fortifying industrial processes against unexpected faults, the adaptability and robustness of deep learning networks have proven instrumental [13–18]. As we navigate the landscape of anomaly detection, it is evident that these sophisticated models hold the potential to revolutionize our approach to identifying irregularities in a wide range of datasets.

2.3 *Comparative Evaluation of Anomaly Detection Algorithms*

This section aims to analyse and compare the performance of different algorithms in AD applications, shedding light on their strengths, weaknesses, and applicability. AD is a critical aspect of data analytics, machine learning, and various application domains, requiring robust algorithms to identify patterns deviating from expected behaviours [15]. This comparative evaluation aims to assess the performance of several prominent anomaly detection algorithms in diverse scenarios, shedding light on their strengths, limitations, and applicability.

Several algorithms have emerged as key players in anomaly detection, including Isolation Forests [10], One-Class SVM [11], and autoencoders [19, 20]. The Isolation Forest algorithm excels in isolating anomalies by leveraging decision trees, demonstrating effectiveness in high-dimensional datasets and scenarios with varying data distributions. One-Class SVM, on the other hand, is renowned for its ability to define a hyperplane that encapsulates normal instances, making it particularly useful when dealing with skewed datasets. Autoencoders, a class of neural networks, exhibit versatility by learning intrinsic representations of data and detecting anomalies through reconstruction errors, proving effective in applications with complex, nonlinear patterns.

To conduct a comprehensive evaluation, we consider multiple performance metrics such as precision, recall, F1 score, and execution time across diverse datasets [18]. The analysis encompasses real-world scenarios like cybersecurity, industrial systems, and healthcare, where anomaly detection is crucial. By comparing these algorithms in various contexts, this evaluation seeks to guide practitioners and researchers in selecting the most suitable anomaly detection algorithm based on the specific requirements of their applications. The insights gained from this comparative study contribute to advancing the field and facilitating informed decisions for implementing AD solutions.

Numerous studies have focused on the efficacy of SL algorithms, such as k-nearest neighbours (k-NN) [5, 6], support vector machines (SVM) [8, 9] and

Isolation Forests [10], in identifying anomalies within datasets. These algorithms, known for their simplicity and interpretability, have found success in various domains, ranging from finance to cybersecurity. The review synthesizes findings from recent research, highlighting benchmark datasets, evaluation metrics, and scenarios where shallow learning excels.

Concurrently, DL algorithms, including CNNs and RNNs, long short-term memory (LSTM) have gained prominence due to their capacity to capture intricate patterns and data- dependencies [11, 15–18]. A list of key features of all recently proposed deep learning models for anomaly detection problems is listed in Table 1. The columns DA, DP, PT, and Archit. are short for data augmentation, dropout, pre-training, and architecture, respectively. The "# layers" account for all layers except the input layer. The term "lReLU" represents leaky ReLU. This review critically assesses the performance of DL approaches in AD, considering factors such as model complexity, training data requirements, and computational efficiency. Noteworthy applications and case studies illustrating the effectiveness of DL in specific domains are scrutinized (Table 2).

In a recent paper [15], the authors reviewed 12 different deep learning approaches to anomaly detection [15]. They identified several key characteristics of these methods, including their underlying assumptions, their ability to handle different types of data, and their interpretability. Additionally, they [15] addressed the difficulties and prospects for further AD research as well as providing an empirical evaluation of several of key techniques.

In another paper [16], some authors compared the performance of 30 anomaly detection algorithms, including both shallow and deep learning methods, on 57 benchmark datasets. They found that DL algorithms generally outperformed shallow learning algorithms, especially on datasets with high-dimensional or complex data. However, shallow learning algorithms were still competitive in some cases, and they were often more interpretable than DL algorithms.

Furthermore, the literature review investigates recent efforts to bridge the gap between shallow and deep learning paradigms in anomaly detection. Hybrid models that leverage the interpretability of shallow algorithms while harnessing the feature representation capabilities of deep models are explored. Comparative studies that evaluate the trade-offs and synergies between these two approaches provide valuable insights for researchers and practitioners.

2.4 Areas of Improvement for Anomaly Detection

Enhancing anomaly detection methodologies involves addressing several key areas for improvement. Firstly, algorithmic robustness and adaptability to diverse datasets need attention. Anomaly detection systems should exhibit resilience to variations in data patterns and distributions to ensure consistent performance across different domains [11]. Additionally, interpretability remains a crucial facet; developing models that offer clear explanations for anomaly identifications can enhance user

Table 1 Key features of standardized recent DL-based anomaly detection algorithms [17]

Method	Sup.	Objective	DA	DP	PT	Archit.	Activation	# Layers	Loss	Data
OADA	Semi	Reconstruction	Yes	No	No	AE	ReLU	3	MSE	Video
Replicator	Unsup.	Reconstruction	No	No	No	AE	Tanh	2	MSE	Tabular
RandNet	Unsup.	Reconstruction	No	Yes	Yes	AE	ReLU	3	MSE	Tabular
RDA	Semi	Reconstruction	No	No	No	AE	Sigmoid	2	MSE	Tabular
UODA	Semi	Reconstruction	No	No	Yes	AE & RNN	Sigmoid	4	MSE	Sequence
AnoGAN	Semi	Generative	No	No	No	Conv.	ReLU	4	MAE	Image
EBGAN	Semi	Generative	No	No	No	Conv. & MLP	ReLU/lReLU	3–4	GAN	Image & Tabular
LSA	Semi	Predictive	No	No	No	Conv.	lReLU	4–7	MSE & KL	video
GT	Semi	Classification	Yes	Yes	No	Conv.	ReLU	10–16	CE	Image
E^3Outlier	Semi	Classification	Yes	Yes	No	Conv.	ReLU	10	CE	Image
REPEN	Unsup.	Distance	No	No	No	MLP	ReLU	1	Hinge	Tabular
RDP	Unsup.	Distance	No	No	No	MLP	lReLU	1	MSE	Tabular
AE-1SVM	Unsup.	One-class	No	No	No	AE & Conv.	Sigmoid	2-5	Hinge	Tabular & image
DeepOC	Semi	One-class	No	No	No	3D Conv.	ReLU	5	Hinge	Video
Deep SVDD	Semi	One-class	No	No	Yes	Conv.	lReLU	3–4	Hinge	Image
Deep SAD	Semi	One-class	No	No	Yes	Conv. & MLP	lReLU	3–4	Hinge	Image & tabular
DEC	Unsup.	Clustering	No	Yes	Yes	MLP	ReLU	4	KL	Image & tabular
DAGMM	Unsup.	Clustering	No	Yes	No	AE & MLP	Tanh	4–6	Likelihood	Tabular
SDOR	Unsup.	Anomaly scores	No	No	Yes	ResNet & MLP	ReLU	50 + 2	MAE	Video
PReNet	Weak	Anomaly scores	Yes	No	No	MLP	ReLU	2-4	MAE	Tabular
MIL	Weak	Anomaly scores	No	Yes	Yes	3DConv. & MLP	ReLU	18/34 + 3	Hinge	Video
PUP	Unsup.	Anomaly scores	No	No	No	MLP	ReLU	3	Likelihood	Sequence

DevNet	Weak	Anomaly scores	No	No	No	MLP	ReLU	2–4	Deviation	Tabular
APE	Unsup.	Anomaly scores	No	No	No	MLP	Sigmoid	3	Softmax	Tabular
AEHE	Unsup.	Anomaly scores	No	No	No	AE & MLP	ReLU	4	Softmax	Graph
ALOCC	Semi	Anomaly scores	Yes	No	No	AE & CNN	lReLU	5	GANs	Image
OCAN	Semi	Anomaly scores	No	No	Yes	LSTM-AE & MLP	ReLU	4	GANs	Sequence
Fence GAN	Semi	Anomaly scores	No	Yes	No	Conv. & MLP	lReLU/Sigmoid	4–5	GANs	Image & tabular
OCGAN	Semi	Anomaly scores	No	No	No	Conv.	ReLU/Tanh	3	GANs	Image

Table 2 Standardized datasets for anomaly detection [17]

Domain	Data	Size	Dimension	Anomaly (%)	Type
Intrusion detection	KDD Cup 99 [13]	4091–567,497	41	0.30–7.70	Tabular
Intrusion detection	UNSW-NB15 [100]	257,673	49	≤9.71	Streaming
Excitement prediction	KDD Cup 14	619,326	10	6.00	Tabular
Dropout prediction	KDD Cup 15	35,091	27	0.10–0.40%	Sequence
Malicious URLs detection	URL [93]	2.4 m	3.2m	33.04	Streaming
Spam detection	Webspam [160]	350,000	16.6m	39.61	Tabular/text
Fraud detection	Credit-card-fraud [34]	284,807	30	0.17	Streaming
Vandal detection	UMD Wikipedia [76]	34,210	N/A	50.00	Sequence
Mutant activity detection	p53 Mutants [13]	16,772	5,408	0.48	Tabular
Internet ads detection	AD [13]	3279	1,555	14.00	Tabular
Disease detection	Thyroid [13]	7200	21	7.40	Tabular
Disease detection	Arrhythmia [13]	452	279	14.60	Tabular
Defect detection	MVTec AD	5354	N/A	35.26	Image
Video surveillance	UCSD Ped 1 [81]	14,000 frames	N/A	28.6	Video
Video surveillance	UCSD Ped 2 [81]	4560 frames	N/A	35.9	Video
Video surveillance	UMN [106]	7739 frames	N/A	15.5–18.1	Video
Video surveillance	Avenue [90]	30,652 frames	N/A	12.46	Video
Video surveillance	ShanghaiTech Campus	317,398 frames	N/A	5.38	Video
Video surveillance	UCF-Crime	Videos (13.8 m frames)	N/A	13 crimes	Video
System log analysis	HDFS Log [164]	11.2 m	N/A	2.90	Sequence
System log analysis	OpenStack log	1.3 m	N/A	7.00	Sequence

trust and facilitate decision-making. Scalability is another critical concern, as anomaly detection algorithms should efficiently handle large datasets without compromising on accuracy [15]. Furthermore, incorporating advanced techniques, such as transfer learning and ensemble methods, can augment the generalization capabilities of models, making them more effective in real-world scenarios [16, 17]. Addressing these areas of improvement will contribute to the evolution of anomaly detection, fostering more robust, interpretable, and scalable solutions tailored to diverse application domains.

Shallow anomaly detection approaches offer several advantages [5–11], over deep learning approaches, including:

- *Interpretability*: Shallow models are generally more interpretable than deep models, making it easier to understand why a particular data point was identified as an anomaly.
- *Computational efficiency*: Shallow models are appropriate for real-time applications since they are usually less computationally expensive than deep models.
- *Data requirements*: Shallow models can often be trained on relatively small datasets, compared to deep models.

However, shallow anomaly detection approaches also have some limitations [8], the most prominent of which are outlined below:

- *Performance*: Shallow models may not perform as well as deep models on complex or high-dimensional data.
- *Sensitivity to noise*: Shallow models can be sensitive to noise in the training data, which can lead to false positives.
- *Lack of adaptability*: Shallow models may not be able to adapt to changes in the underlying data distribution over time.

While deep learning has shown impressive results in anomaly detection, it also comes with several limitations as reported by several recent works [15–23]. The main limitations of deep learning models are:

1. *Data requirements*: DL algorithms often require large amounts of labelled data for training. This can be a significant challenge in many applications where labelled data is scarce or expensive to obtain.
2. *Interpretability*: Deep learning models can be complex and difficult to understand, making it challenging to interpret their results and explain why a particular data point was identified as an anomaly. This lack of interpretability can hinder trust in the model and limit its application in critical domains.
3. *Computational cost*: Training and deploying DL models can be computationally expensive, requiring powerful hardware and specialized software. This can be a barrier to entry for smaller organizations or applications with limited resources.
4. *Sensitivity to noise*: Deep learning models can be sensitive to noise and outliers in the training data, which can lead to false positives. This limitation requires careful data preprocessing and cleaning to ensure accurate results.
5. *Overfitting*: DL models can overfit the training data and fail to generalize to unseen data. This can result in poor performance when the model encounters anomalies that were not present in the training data.
6. *Vulnerability to adversarial attacks*: DL models can be susceptible to adversarial attacks, where attackers manipulate the data to cause the model to make incorrect

predictions. This vulnerability can be a serious concern for security-critical applications.

7. *Lack of robustness*: Deep learning models may not be robust to changes in the underlying data distribution over time. This can lead to performance degradation if the model is not constantly updated with new data.
8. *Bias*: Deep learning models can inherit biases from the data they are trained on. This can lead to discriminatory or unfair outcomes, particularly in sensitive applications.
9. *Ethical considerations*: The use of DL for anomaly detection raises ethical concerns about privacy and discrimination. Carefully considering these implications is crucial before deploying such models in real-world applications.

Despite these limitations, DL remains a powerful tool for anomaly detection. Researchers continuously come up with novel strategies that get beyond these limitations that improve the interpretability and performance of DL models. With ongoing advancements, DL can potentially revolutionize AD in the upcoming years.

3 Some Challenging Open Problems

A MDS (Malware Detection System) [23] is still frequently employed as a defensive solution in commercial goods, even though it is incapable of detecting future assaults or variants of current attack types. In contrast, a NADS has difficulties in its efficient design yet can discover major dangers. These difficulties, which are investigated using an anomaly-based methodology, include the difficulty of creating a thorough profile that correctly discriminates between typical and aberrant behaviour.

- Creating a complete profile [23] that encompasses all potential legitimate behaviours is challenging due to the difficulty in accurately distinguishing between normal and abnormal patterns using network features. This can lead to false positive and false negative errors, where normal behaviour is classified as an attack and vice versa.
- Adaptive and scalable NADS architecture [23] requires the use of autonomous approaches capable of handling the enormous throughput and massive sizes of contemporary network systems. These autonomous techniques can automatically adjust their threshold, which determines the boundary between normal and malicious events, ensuring effective detection and adaptation in dynamic network environments.
- Real-time detection of network anomalies is challenging due to the presence of noisy or irrelevant features in network traffic data [23]. This can lead to increased processing time and false alarm rates. Overcoming those challenges requires the development of feature reduction techniques to eliminate irrelevant attributes and lightweight detection methods that can accurately distinguish between normal and abnormal patterns with low variations.

- In the field of network security [23], having access to reliable and diverse datasets for Intrusion Detection Systems (IDS), is crucial for learning and validating models of NADSs. On the other hand, inadequate network information, a lack of attack diversity, and incorrect labelling are common problems with current IDS datasets. To overcome this, realistic settings covering a range of normal and attack circumstances must be designed. Additionally, ground truth attack events must be included in the dataset to ensure its credibility while testing new IDSs.
- Organizations are in greater danger of not being able to defend their assets due to the rising incidence of new ransomware strains [23]. Computer data is encrypted by malware known as ransomware, preventing accessibility until a ransom payment is made. Since the payload of packets is encrypted, it is essential to build flow features to identify and mitigate ransomware attacks. For handling large volumes of network traffic, effective approaches for flow aggregation and feature selection must also be established.
- Within various sectors, including industrial control systems, cloud/fog computing, software-defined networks, and the Internet of Things, it is imperative to design efficient anomaly detection systems (ADSs) [23] to detect potential cyber threats. To safeguard various nodes in wide-area networks, these ADSs must be able to scale and adapt while continuously monitoring high-speed networks.
- Traditional NADSs [23] need to build profiles for common occurrences in network traffic and telemetry data from IoT devices to detect harmful behaviours. DL methods and mixture models can help NADS perform better, but they come with challenges when it comes to scalability, interoperability, and processing enormous amounts of data that need to be detected quickly and effectively.
- It is important to note that the Tennessee Eastman process (TEP) data [24] used in the experiments are synthetic, meaning they are artificially generated. While the quality of the simulation is high, it is crucial to consider that real chemical processes are complex and may involve additional parameters that cannot be accurately simulated. Despite this limitation, all the methods evaluated in the study have achieved high scores in their performance. In principle, while the simulation used in the experiments is of good quality, real chemical processes are more complex, and there may be additional factors that cannot be accurately simulated. Despite this, all the methods evaluated in the study have achieved high scores, indicating their effectiveness in the given context. So, by examining synthetic data with intentionally introduced faults in a fault-free scenario, while this serves as a preliminary step, the task becomes more difficult when dealing with real chemical process data. The main challenge lies in identifying and accurately labelling anomalies within the data, and it is important to consider additional metrics for a comprehensive analysis.
- Research in intrusion detection faces several challenges [25, 26], including the lack of datasets that showcase recent attacks, the trade-off between increasing accuracy and the complexity and computational resources required for machine learning models, and the issue of overfitting, this results in testing data performance that is low yet training data performance that is good. These challenges highlight the need for careful consideration and optimization when developing intrusion detection systems based on ML.

4 Concluding Remarks and Future Research Directions

This chapter discusses the utilization of SL and DL techniques for anomaly detection applications. It explores the benefits and drawbacks of employing these methods in such applications, offering insights into their effectiveness and limitations. An examination of the advantages and drawbacks associated with implementing these methods in anomaly detection applications is presented, providing a clear understanding of their effectiveness and limitations. Additionally, the chapter discusses strategies for evaluating the efficacy of various approaches in anomaly detection and guides the reader in selecting the most suitable methods. Emphasizing the incorporation of observations, unaddressed challenges, and avenues for future research, this chapter highlights the dynamic landscape of anomaly detection, fostering a comprehensive perspective on its current state and future directions.

References

1. Patcha, A., & Park, J. M. (2007). An overview of anomaly detection techniques: Existing solutions and latest technological trends. *Computer Networks, 51*(12), 3448–3470.
2. Chandola, V., Banerjee, A., & Kumar, V. (2009). Anomaly detection: A survey. *ACM Computing Surveys (CSUR), 41*(3), 1–58.
3. Janisch, C., Zschech, P., & Heinrich, K. (2021). Machine learning and deep learning. *Electronic Markets, 31*(3), 685–695.
4. Goodfellow, I., Bengio, Y., & Courville, A. (2016). *Deep learning*. MIT Press.
5. Sarmadi, H., & Karamodin, A. (2020). A novel anomaly detection method based on adaptive Mahalanobis-squared distance and one-class kNN rule for structural health monitoring under environmental effects. *Mechanical Systems and Signal Processing, 140*, 106495.
6. Su, M. Y. (2011). Real-time anomaly detection systems for Denial-of-Service attacks by weighted k-nearest-neighbour classifiers. *Expert Systems with Applications, 38*(4), 3492–3498.
7. Alghushairy, O., Alsini, R., Soule, T., & Ma, X. (2020). A review of local outlier factor algorithms for outlier detection in big data streams. *Big Data and Cognitive Computing, 5*(1), 1.
8. Ruff, L., Kauffmann, J. R., Vandermeulen, R. A., Montavon, G., Samek, W., Kloft, M., et al. (2021). A unifying review of deep and shallow anomaly detection. *Proceedings of the IEEE, 109*(5), 756–795.
9. Turkoz, M., Kim, S., Son, Y., Jeong, M. K., & Elsayed, E. A. (2020). Generalized support vector data description for anomaly detection. *Pattern Recognition, 100*, 107119.
10. Xu, D., Wang, Y., Meng, Y., & Zhang, Z. (2017, December). *An improved data anomaly detection method based on isolation forest*. In 2017 10th international symposium on computational intelligence and design (ISCID) (Vol. 2, pp. 287–291). IEEE.
11. Miao, X., Liu, Y., Zhao, H., & Li, C. (2018). Distributed online one-class support vector machine for anomaly detection over networks. *IEEE Transactions on Cybernetics, 49*(4), 1475–1488.
12. Çelik, M., Dadaşer-Çelik, F., & Dokuz, A. Ş. (2011, June). Anomaly detection in temperature data using DBSCAN algorithm. In *2011 international symposium on innovations in intelligent systems and applications* (pp. 91–95). IEEE.
13. Zhong, Y., Chen, W., Wang, Z., Chen, Y., Wang, K., Li, Y., et al. (2020). HELAD: A novel network anomaly detection model based on heterogeneous ensemble learning. *Computer Networks, 169*, 107049.

14. Araya, D. B., Grolinger, K., ElYamany, H. F., Capretz, M. A., & Bitsuamlak, G. (2017). An ensemble learning framework for anomaly detection in building energy consumption. *Energy and Buildings, 144*, 191–206.
15. Pang, G., Shen, C., Cao, L., & Hengel, A. V. D. (2021). Deep learning for anomaly detection: A review. *ACM Computing Surveys (CSUR), 54*(2), 1–38.
16. Chalapathy, R., & Chawla, S. (2019). *Deep learning for anomaly detection: A survey*. arXiv preprint arXiv:1901.03407.
17. Aldweesh, A., Derhab, A., & Emam, A. Z. (2020). Deep learning approaches for anomaly-based intrusion detection systems: A survey, taxonomy, and open issues. *Knowledge-Based Systems, 189*, 105124.
18. An, J., & Cho, S. (2015). Variational autoencoder based anomaly detection using reconstruction probability. *Special Lecture on IE, 2*(1), 1–18.
19. Xu, H., Chen, W., Zhao, N., Li, Z., Bu, J., Li, Z., ... & Qiao, H. (2018, April). *Unsupervised anomaly detection via variational auto-encoder for seasonal kpis in web applications*. In Proceedings of the 2018 world wide web conference (pp. 187–196).
20. Al Mamun, S. A., & Beyaz, M. (2019). *LSTM recurrent neural network (RNN) for anomaly detection in cellular mobile networks*. In Machine learning for networking: First international conference, MLN 2018, Paris, France, November 27–29, 2018, Revised Selected Papers 1 (pp. 222–237). Springer.
21. Ergen, T., & Kozat, S. S. (2019). Unsupervised anomaly detection with LSTM neural networks. *IEEE Transactions on Neural Networks and Learning Systems, 31*(8), 3127–3141.
22. Schlegl, T., Seeböck, P., Waldstein, S. M., Schmidt-Erfurth, U., & Langs, G. (2017, May). *Unsupervised anomaly detection with generative adversarial networks to guide marker discovery*. In International conference on information processing in medical imaging (pp. 146–157).-Cham: Springer.
23. Sabuhi, M., Zhou, M., Bezemer, C. P., & Musilek, P. (2021). Applications of generative adversarial networks in anomaly detection: A systematic literature review. *IEEE Access, 9*, 161003–161029.
24. Moustafa, N., Hu, J., & Slay, J. (2019). A holistic review of network anomaly detection systems: A comprehensive survey. *Journal of Network and Computer Applications, 128*, 33–55.
25. Hartung, F., Franks, B. J., Michels, T., Wagner, D., Liznerski, P., Reithermann, S., ... & Kloft, M. (2023). Deep anomaly detection on tennessee Eastman process data. *Chemie Ingenieur Technik*.
26. Suthishni, D. N. P., & Kumar, K. S. (2022, March). A review on machine learning based security approaches in intrusion detection system. In *2022 9th International conference on computing for sustainable global development (INDIACom)* (pp. 341–348). IEEE.

Shallow Learning Versus Deep Learning in Natural Language Processing Applications

Lina Sawalha and Tahir Cetin Akinci

1 Introduction

Natural language processing (NLP) [1] has *significantly* developed overtime. The evolution of NLP has led to the emergence of a myriad of new applications that have transformed the way we interact with technology. The continuous development of NLP has resulted in higher precision and thus a much better understanding of human natural language. The developments in machine learning algorithms, along with the vast amounts of linguistic data today, have played a pivotal role in the development of more sophisticated NLP models. This has allowed for new venues of innovative applications such as effective personal assistants and chatbots.

With all the advancements in NLP, many challenges still exist. However, it is anticipated that NLP will undergo further breakthroughs, leading to a sea change in the way we interact with the digital world. This chapter reviews NLP and the different artificial intelligence (AI) and machine learning (ML) methods that have been used over time. Because of the vast amount of advancements in NLP in recent years and the huge amount of publications, we consider recent surveys [2–13]. Previous work described the history of NLP [6], reviewed NLP applications and their use in various domains [5, 7–10, 12–16], or evaluated different deep learning models [2, 3, 6] and pretrained models [17].

L. Sawalha
Department of Electrical and Computer Engineering, Western Michigan University, Kalamazoo, MI, USA
e-mail: lina.sawalha@wmich.edu

T. C. Akinci (✉)
Bourns College of Engineering - Center for Environmental Research and Technology (Ce-Cert), University of California Riverside, Riverside, CA, USA
e-mail: tahircetin.akinci@ucr.edu

Ö. F. Ertuğrul et al. (eds.), *Shallow Learning vs. Deep Learning*, The Springer Series in Applied Machine Learning, https://doi.org/10.1007/978-3-031-69499-8_8

This chapter compares shallow learning (SL) methods with deep learning (DL) methods for NLP and their effects on performance and accuracy. It also classifies NLP models based on shallow and deep learning techniques, and discusses the strengths and challenges of each model. Given the extensive literature on NLP [18], it is nearly impossible to consider all previous works in this study. This chapter discusses efforts using SL, DL and pretrained models for some NLP applications and classifies recent work. Additionally, it discusses the challenges of NLP and its models and paves the way for future directions.

The rest of the chapter is organized as follows. Section 2 briefly summarizes NLP, its history, the levels of NLP, and the core NLP tasks. The chapter also classifies the existing NLP models based on the learning algorithms used (Section 3). Section 4 describes shallow learning techniques and their challenges, while Section 5 describes the deep learning methods and their challenges. Then, Section 6 compares shallow and deep learning models and classifies recent NLP task processing efforts, and Section 7 provides further discussion of the ML models. Section 8 describes the current challenges of NLP and its applications and discusses future directions. Finally, Section 9 concludes the chapter.

2 Overview of Natural Language Processing

Natural language processing is the field dedicated to enabling computers to understand human language. It is considered an interdisciplinary field of computer science and linguistics [19, 20]. It is also considered a subset of artificial intelligence (AI) [21, 22]. NLP involves the analysis, understanding, and representation of natural language, enabling natural interaction between humans and machines.

NLP can be divided into natural language understanding (NLU) and natural language generation (NLG) [20]. NLP consists of algorithms, models, and systems designed to interpret various forms of human communication, including text, speech, and expressions, in a meaningful and contextually relevant manner for computer software. Moreover, NLP facilitates human-computer interaction by allowing computers to generate human language. Developed over the past century, NLP has been applied to core tasks such as machine translation and text classification. It has also spurred emerging applications like personal assistants, chatbots, customer review handling, and fake news detection. In addition, NLP has been recently used in multiple other domains of applications, such as medicine [8, 10, 13, 15], digital forensics [7], and management [5].

Throughout the evolution of NLP, various algorithms have been used, including those rooted in AI and ML (Fig. 1)—both shallow and deep learning. While some NLP methods are AI methods, others are based on ML approaches, and some of the ML approaches used by NLP depend on deep learning. This section describes a brief history of NLP and discusses the different NLP levels and core tasks.

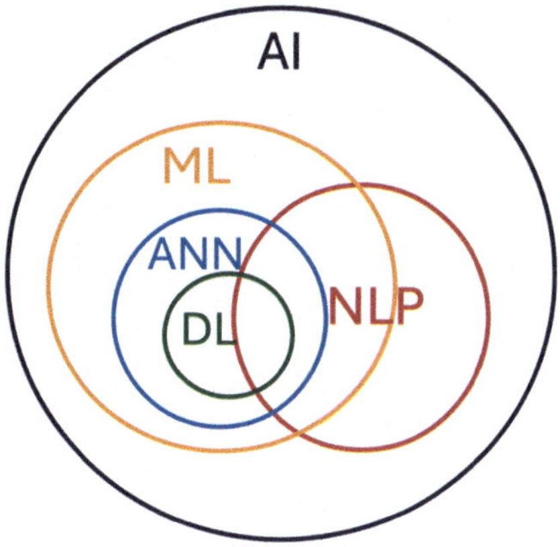

Fig. 1 NLP and AI subfields

2.1 *Brief History of NLP*

The development of NLP began in the late 1940s with research on machine translation (MT) [1, 21]. Weaver and Booth worked on one of the earliest MT projects in 1946 after World War II [23]. Weaver discussed in his Memorandum of 1949 [24] the importance and difficulty of MT, which inspired many people to work on it [21]. In 1950, Alan Turing envisioned machines engaging in human conversations [25]. NLP applications such as speech recognition started to emerge after that. In the following decades, NLP relied heavily on rule-based systems, while statistical information theory emerged.

The 1960s and 1970s saw significant developments in machine translation and the development of linguistic theories that influenced NLP methodologies. The 1980s saw a significant use of statistical and probabilistic theories. The trend continued later with the use of different ML techniques for NLP, as ML has established itself in the field of computer science. Different ML techniques were used for different types of applications, including several NLP tasks [26]. During the past decade, deep ML techniques have been investigated and used for various NLP levels and applications [3]. Recently, pretrained deep learning and transformer models have been utilized and showed much better performance for NLP applications [27].

2.2 *NLP Levels*

NLP traditionally has consisted of methods for text understanding and generation, such as parsing, text and word segmentation, language modeling, word-sense disambiguation, part-of-speech tagging, classification, information extraction, and machine translation. These methods depend on features extracted at different NLP levels, which represent the "levels of a language" [21]. NLP levels are also known as the "synchronous model of language" [28]. There are seven main levels of NLP: phonological, morphological, lexical, syntactic, semantic, discourse, and pragmatic [20, 21, 28].

- Phonological level: Phonology is the study of the sound of words and word segments and their pronunciations. The phonological level features and analysis are used for speech-related applications. It also includes the semantics of a sound to interpret the meaning of human language [20].
- Morphological level: Morphemes are the smallest parts of a word that have meaning, such as prefixes, suffixes, and root words [20]. Morphological analysis is concerned with finding the morphemes of words and their meanings. It deals with the internal structure of words and how words are formed using morphemes.
- Lexical level: The lexical level is focused on single words and their meanings. Lexical analysis can use features of each word at the lexical, morphological, and/or phonological levels [29]. The features often depend on manually extracted language forms. Rule-based algorithms, finite-state machines, and statistical algorithms have been used at this level. Rule-based methods rely on an expertly crafted set of rules that determine how to identify and analyze words based on their morphology, spelling patterns, etc. Lexical analysis algorithms include N-grams [30], word tokenization [31], lemmatization [32], and classification [33].
- Syntactic Level: The syntactic level involves parsing sentences and analyzing their grammar and structure. It helps to understand the relationships between words and their roles in the formation of sentences, phrases, and clauses. There are different types of parsing including top-down [34], bottom-up, probabilistic parsing, constituency parsing, and dependency parsing.
- Semantic Level: The semantic level focuses on the meaning of words, phrases, and sentences. It aims to understand the intended meaning and context of the text. The set of features extracted at this level depends on the application. For instance, features extracted for machine translation can be different from those extracted for text-speech synthesis. The first requires the meaning of the words and morphemes, and the second requires their pronunciation. Semantic analysis includes algorithms such as word sense disambiguation [35, 36], sentiment analysis [37, 38], and semantic parsing [39].
- Discourse Level: While the semantic and pragmatic levels involve sentences, the discourse level deals with the structure and organization of larger chunks of text, such as paragraphs, articles, or conversations. Discourse analysis considers the properties of the text as a whole and how sentences connect and contribute to the overall meaning [21].
- Pragmatic Level: The pragmatic level goes beyond the literal meaning of words. It involves understanding the social and cultural context of sentences, language use, situational context, and the relationship between speakers. Pragmatic analysis aims to interpret the implied meanings of phrases, sentences, and silence and to grasp the speaker's intentions. Examples of pragmatic analysis include sarcasm detection [40] and joke understanding [41].

2.3 NLP Tasks

Each NLP level described above consists of different tasks. Some tasks are performed at more than one NLP level. This section describes the most common NLP core tasks such as tokenization, part of speech tagging, named entity recognition, parsing, semantic role labeling, information extraction, and text classification. Below is a brief description of the core NLP tasks.

- Tokenization: It is the process of breaking text into smaller parts or "tokens." Tokenization is an essential task in NLP that can be applied to break text into sentences, phrases, and words. It can also be applied at the morphological level to break words into subwords, such as prefixes, roots, and suffixes.
- Part of Speech Tagging: POS tagging refers to the process of assigning tags for each token in a text. It involves labeling grammatical features using a standard label set for each token [42]. It is a cruical task in NLP that helps in understanding the syntactic structure of sentences.
- Named Entity Recognition: NER is a process of extracting specific information from a text by identifying named entities. The named entity is a phrase or a word that identifies a certain text [43], for example, names of people and organizations, locations, dates, and other distinct names that helps identifying and classifying entities.
- Parsing: It is the process of breaking down text into smaller parts to understand its structure and meaning. It involves determining parts of speech, word relationships, and the hierarchical structure of sentences. Parsing is used at different levels, mainly the syntactic and semantic levels. Syntactic parsing focuses on the grammatical structure of text, and it includes dependency parsing and constituency parsing. Dependency parsing [44] finds the relationships between words in a sentence, while constituency parsing [45] divides sentences into phrases based on formal grammar. Syntactic parsing is useful for tasks such as sentiment analysis and understanding. Semantic parsing aims to understand the meaning of sentences by converting sentences into their logical form [39]. In addition, discourse parsing aims to parse the discourse structure of a text [46]. Semantic and discourse parsing are important for applications such as question answering and text categorization.
- Word Sense Disambiguation: WSD aims to determine the correct meaning of a word in context [35, 36]. This is important when a word has multiple possible meanings (or senses). WSD is required for other NLP tasks such as machine translation and information retrieval.
- Text Classification: TC is a fundamental task of NLP that is involves assigning text to a predefined class based on its context [46]. It is used to structure and organize text, documents, emails, customer reviews, etc., and assign them a category. This is useful for automating the process of sorting large volumes of text data, spam detection, and sentiment analysis.
- Information Extraction: IE is the process of extracting useful structured information from large and unstructured texts [47]. IE is used to automatically extract

relations, entities, events, emotions, or facts depending on the use or application in a systematic and structured way. Examples of applications that use IE include question answering, text summarization, event detection, sentiment analysis, and knowledge base construction.

- Semantic Role Labeling: SRL is the process of identifying and labeling the semantic roles of words and phrases. The semantic role represents a conceptual relation of words and phrases that conveys a meaning. SRL aims at understanding the meaning of a sentence by finding the relationships between the different roles. It can be used for information extraction, sentiment analysis, question answering, text summarization, and machine translation.
- Sentiment Analysis: SA is used for text mining to find opinions, sentiments, and text subjectivity [37]. It requires sentiment identification and classification. SA is particularly useful for customer service and customer feedback analysis.

3 Classification of NLP Models

In recent years, a vast amount of NLP models and algorithms have been developed, prompting numerous surveys and review articles. This section focuses on the literature from recent studies published within the past 4 years. Previous work classified NLP models from various perspectives. Singh and Mahmood [27] classified NLP based on models for natural language understanding (NLU) and natural language generation (NLG), model size reduction, information retrieval, long sequence models, and computationally efficient architectures. Another study classified NLP levels according to language understanding and generation [48]. Furthermore, a taxonomy of NLP has been created according to the fields of study of NLP [49]. Hupkes et al. [50] created a new taxonomy by categorizing existing literature based on their motivation, the specific type of generalization they target, the nature of data shifts they address, the origin of these shifts, and wherein the NLP modeling process these shifts occur. In this section, we classify NLP models and algorithms based on their architectures (shallow or deep) and their data/training (Fig. 2).

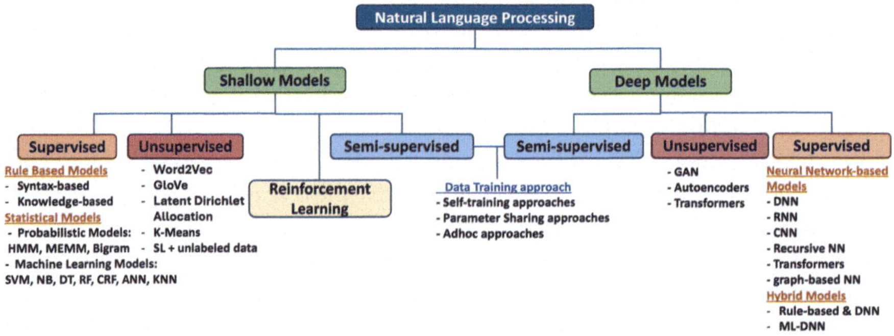

Fig. 2 Classification of NLP models

3.1 Shallow NLP Models

Shallow NLP models have a long history of use in various NLP tasks and applications. Shallow techniques include algorithms that have a shallow architecture and do not require multiple levels of calculations. They can be further classified into supervised, unsupervised, semi-supervised, and reinforcement learning methods. Supervised shallow methods include rule-based and statistical-based models. Rule-based models represent one of the oldest approaches in NLP. They rely on a set of predefined language rules, often designed by linguistic experts, to process and analyze text. They consist of syntax-based and knowledge-based methods.

Syntax-based models utilize the syntax structure of a language to comprehend its text and develop rules. Examples of syntax-based models include part-of-speech tagging [51], constituency parsing, dependency parsing, and transformation-based syntax. Various algorithms have been applied to these models to increase their accuracy over the years including machine learning. On the other hand, knowledge-based models take into account information about words beyond the syntax. They include named entity recognition, ontology-based information extraction, semantic role labeling, and lexical-semantic resources.

Furthermore, statistical-based supervised shallow models have been used in NLP. They include probabilistic-based algorithms such as the Hidden Markov Model (HMM) and ML models. Supervised ML algorithms use labeled data for training and are divided into classification and regression algorithms [52, 53]. However, NLP mostly uses classification algorithms such as support vector machine (SVM), naive Bayes (NB), decision trees (DT), K-nearest neighbors (KNN), and artificial neural networks (ANN).

Unsupervised learning relies on unlabeled data where algorithms cannot know the correct answer, so unsupervised learning algorithms do not work for regression but work for classification [53]. Unsupervised learning aims to find relationships and patterns among data, resulting in unsupervised data clustering. Examples of shallow unsupervised ML include K-means clustering. ANN and other ML algorithms can also be used for classification in unsupervised mode, using unlabeled data. Additionally, other common unsupervised NLP models include *Word2Vec* [54, 55], *GloVe* [56], and latent Dirichlet allocation (LDA) [57, 58].

Semi-supervised learning techniques combine aspects of both supervised and unsupervised learning. Collobert et al. [59] categorized semi-supervised learning into self-training [59, 60] approaches, ad hoc approaches, and parameter sharing approaches. Self-training is used to leverage a large number of unlabeled data to improve the accuracy of a model. It involves an iterative self-training paradigm that uses a small amount of labeled data to generate pseduo-labels for a larger dataset to refine the model iteratively. Ad hoc models rely on techniques to exploit unlabeled data in conjuntion with the labeled data. Parameter sharing techniques leverage both labeled and unlabeled data to improve the performance of a model. On the other hand, reinforcement learning (RL) is a machine learning algorithm that is neither supervised nor unsupervised. In reinforcement learning, the information about the correctness of the answer is provided but not how to improve it as the learner has to find that out. RL is different from supervised and unsupervised methods as it utilizes feedback to maximize a reward. Thus, RL is often considered a separate type of learning.

3.2 Deep NLP Models

Deep learning consists of models with several layers and more complex structures. It often requires more complex data and longer training. Similar to the shallow models, DL models can be classified into supervised, unsupervised, and semi-supervised. The supervised methods include deep neural network models such as deep ANN (DNN), recurrent neural network (RNN) [61, 62], convolution neural network (CNN) [63, 64], and transformer networks [65–67]. In addition, hybrid rule-based and DL models have been used for NLP. Supervised DL models have been used for many NLP applications, including text classification [68, 69] and understanding [70], machine translation [71–73], and question answering [72].

Unsupervised deep learning models used for NLP include autoencoders [74, 75] and generative adversarial networks (GANs) [67, 72]. Autoencoders used for NLP, such as text clustering [75] and summarization [76, 77], typically use unlabeled data with a large number of text documents for training. Although GANs are not as common as other deep learning methods for NLP, they have been increasingly used [67]. Semi- supervised deep learning methods are similar to those of shallow learning methods but with deep layers. On the other hand, deep reinforcement learning integrates RL and deep learning for feedback. This method has been used in many NLP applications [78, 79].

4 Shallow Methods for NLP

Shallow methods for NLP consist of rule-based and statistical methods as mentioned in Sect. 3. Rule-based approaches rely on creating a set of rules, applying them, processing them, and finally refining the rules if necessary. They can be further classified into syntax-based or knowledge-based approaches. The rule-based methods use simple, complex, and special rules [80]. Simple rules rely on basic syntactic and lexical features such as POS tags, orders of tokens, or n-grams [81–83]. Complex rules include a combination of multiple simple rules together [84] and special rules. In addition, extraction of relationships using multiple patterns and simple rules [80, 82, 84] can be used for knowledge-based approaches. Semantic features are also essential for knowledge-based NLP systems. Rule-based methods have shown success in different NLP tasks; however, finding rules depends on expertise, which is a tedious and time-consuming method [85]. Rule-based approaches have been recently used for information extraction [83, 86], text interpretation [87], and text classification [88].

Statistical methods, on the other hand, provide a more systematic approach to solving NLP tasks, but they require training data. They consist of probabilistic methods [89] and machine learning methods and do not require predefined rules. Probabilistic methods include HMM [90–92], bigram and trigram hidden Markov model [93–95], and maximum entropy Markov model (MEMM) [96, 97]. ML models require the collection of data sets to be used for training. Supervised learning requires a labeled data set, while unlabeled data set can be used for unsupervised learning. Shallow ML algorithms include support vector machine, naive Bayes, random forest, decision trees, K-nearest neighbors, and conditional random fields.

4.1 ML-Based Shallow Learning

Below is a description of common shallow ML algorithms used for NLP tasks.

- Support Vector Machine: SVM [98] is a supervised ML classification algorithm that has been applied to different NLP levels, tasks, and applications. SVM has been successfully used for POS tagging [99–101], word sense disambiguation [102, 103], information extraction [104], relation extraction [105], semantic role labeling [106], dependency analysis [107, 108], and chunking [109].
- Naive Bayes: NB [110] is a supervised ML classification algorithm that learns the probabilities of features for each class. NB has been used for different NLP applications, including text classification [111–113], information retrieval [114], and spam filtering [115, 116].
- Decision Trees: DT [117] is a supervised ML algorithm that is structured like a tree with nodes, where each node represents a decision based on a feature. DT has been used for POS tagging [118].
- Random Forest: RF [119] is an ensemble-based supervised classifier that uses multiple DTs. Each DT uses a subset of the training data and a subset of the features. RF finds the importance of features and thus does not require an explicit feature extraction mechanism. It has been applied to text classification [120] and sentiment analysis [121].
- Conditional Random Fields: CRF [122] is a discriminative probabilistic algorithm that models dependencies between labels. CRF can model more complex relationships and dependencies compared to HMM. It has been used for different NLP tasks separately, such as entity recognition [123], POS [124], and chunking, or as part of a hybrid ML algorithm [125, 126].
- K-Nearest Neighbors: KNN is a supervised ML algorithm that is used to classify text based on its K-th nearest neighbor. KNN has been used for many NLP tasks such as word sense disambiguation [127], text classification [128], sentiment analysis [129], and speech analysis [130]. KNN finds the nearest neighbors of a data point by measuring the distance between that point and all other points in the

dataset. Common distance metrics include the Euclidean distance, the Manhattan distance, and cosine similarity. Then, the *K* neighbors with the shortest distance are found. In classification tasks, KNN labels the cluster based on the majority vote among the *K* neighbors. In text classification tasks, for example, KNN classifies a new text by finding the *K* most similar texts from the training set and assigning the majority class among them to the new document.

- K-Means: K-means [131] is an unsupervised classification algorithm used for various Tasks, such as document clustering [132], topic modeling [133], word clustering [134], and text summarization [135]. K-means can group similar items together using unlabeled data by dividing the dataset into *k* clusters. The algorithm iterates through finding the centroid of clusters and assigning data points to the nearest centroid by finding their distance to the centroid. For example, K-means clusters documents together based on the similarity of their word vectors or bag-of-words representations. It can also be applied to cluster words based on their semantic similarity, which can be useful for tasks like word sense disambiguation or creating word embeddings.
- Logistic Regression: LR [136] is a linear classification algorithm that can be applied to tasks like text classification [137], sentiment analysis [138, 139], and spam detection [116]. Each feature in LR is assigned a weight, and the weighted sum of features is passed through a logistic (sigmoid) function to produce the probability of a given class. LR is a simple, interpretable, and efficient algorithm, making it a suitable choice for tasks where linear decision boundaries are effective; however, it cannot explore the non-linear relationships of features and complex linguistic structures.
- Artificial Neural Network: Shallow ANN [140] typically consists of a single hidden layer between the input and output layers. Because of their limited depth, shallow ANN may struggle with capturing complex linguistic patterns and tasks. However, it offers a balance between computational efficiency and accuracy and have been used in tasks like sentiment analysis [141], text classification [142], and named entity recognition [143]. Shallow ANN can be trained using techniques like backpropagation.
- Reinforcement Learning: RL [144] has been applied to different NLP tasks, including dialogue systems [145], machine translation [146], and text summarization [147]. RL is used to perform NLP tasks by optimizing the model to maximize a reward based on the feedback it receives considering predefined metrics. For example, RL has been applied to text summarization tasks to generate concise summaries by maximizing a reward signal based on the relevance and informativeness of the summaries. However, RL in NLP requires careful design of the reward function as it often faces challenges such as sparse rewards and high-dimensional action spaces.

Today, the data deluge resulting from technological advances has enabled the collection of more features, which provide an opportunity for higher performance of NLP models. However, having a large number of features can result in poor performance for SL algorithms, because they depend on careful feature selection to achieve good performance. While many NLP tasks use manual feature selection, the reliance on feature selection and extraction algorithms is essential in the current big data era.

Furthermore, the data deluge problem facilitated the emergence of new and innovative applications such as chatbots and personal assistants. Such complex applications require complex processing. While shallow NLP with advanced feature extraction methods have been successfully applied for different NLP tasks, they do not provide the required performance for such complex applications. However, they can be used as part of hybrid ML models for complex applications.

In addition, hybrid shallow NLP algorithms have been applied by using two or more shallow algorithms. Hybrid algorithms combine different rule-rule-based, rule-statistical, and statistical-statistical algorithms. For example, Shien and Nyunt proposed a hybrid algorithm based on ontology and SVM for sentimental classification [148]. Another hybrid algorithm combined syntax-based and knowledge-based techniques for interpreting textual regulations [87]. Finaly, shallow ML algorithms can be combined with DL algorithms for NLP tasks.

5 Deep Learning Methods for NLP

Nowadays, the use of deep models is an important part of NLP [149]. DL techniques are effectively used to process large datasets and understand complex language structures [150]. DL techniques offer several advantages over traditional approaches in NLP. First, DL models can utilize large datasets to better understand complex language structures. This allows for more accurate and comprehensive results. DL models can provide high accuracy in tasks, such as translation, text classification, and sentiment analysis. Moreover, DL models are often better at generalization, thanks to features such as transfer learning and recursive structures. A deep model trained on one task can perform better on similar tasks, improving overall efficiency [151, 152].

Deep learning models for NLP are fundamentally based on artificial neural network architectures that can process and understand human language data. These models learn hierarchical representations of text data by capturing semantic and syntactic relationships, enabling them to perform various NLP tasks such as sentiment analysis, named entity recognition, machine translation, text generation, and text classification [153]. Common neural network models used in the literature in this area are RNNs, the first deep learning models for sequential data. RNNs process input data one token at a time while maintaining a hidden state that captures context. Long short-term memory (LSTM) and gated recurrent unit (GRU) variants address the problem of vanishing gradients [154].

CNNs, originally designed for image processing, have been adapted for NLP tasks, such as text classification and sentiment analysis [155]. This adaptation involves treating words or embeddings as "images" and applying convolutional layers to extract features. On the other hand, transformative models, exemplified by the seminal paper "Attention is All You Need," utilize self-attention mechanisms to capture global dependencies within input sequences [156]. Models like BERT (bidirectional encoder representations from transformers) and GPT (generative pretrained transformer) have demonstrated state-of-the-art performance across various NLP benchmarks. BERT, in particular, revolutionized NLP by introducing bidirectional context understanding through pretraining on extensive text corpora, resulting in significant advancements in tasks such as question answering and text classification. Variants of BERT, including RoBERTa, ALBERT, and DistilBERT, further optimize its architecture for efficiency and task-specific performance enhancements [157]. Among these, the most widely used algorithms in this domain include the following: (a) recurrent neural networks (LSTM, GRU), (b) convolutional neural networks (TextCNN, CharCNN), (c) transformer models (BERT, GPT, RoBERTa, T5, Transformer-XL), and (d) sequence-to-sequence models (Seq2Seq, transformer-based Seq2Seq models) for machine translation and summarization [158, 159].

With the development of highly computational devices, deep neural networks have recently started to play an important role in many artificial intelligence applications. DNNs have been used in many ML applications, from image classification to object detection and even malware detection. In addition, DNNs can be used in speech recognition, voice analysis and synthesis, language translation, and dictionary applications [20, 160, 161]. Challenges in this area include data efficiency, computational resource issues, interpretability, fine-tuning, generalization, and ethical concerns. Despite these challenges, ongoing research and advances in DL are enabling significant advances in NLP. However, the use of DL models in NLP also faces some challenges. One of the main ones is the large datasets required for training DL models. The preparation of these datasets and ethical issues can be a serious concern. Also, training and tuning DL models can be quite time-consuming and computationally demanding [162, 163]. Another challenge is the explainability and interpretability of deep models' decisions. Deep learning can be more opaque than traditional methods for understanding decision processes. This raises issues of reliability and ethics [163, 164]. A controversial issue is the sensitivity of deep models. In particular, they can learn social biases in language, which can lead to injustices. Therefore, the model's training data and especially the data preprocessing stages are of great importance.

6 Comparison

Choosing the best model for NLP applications depends on the problem being addressed, the available data, and the need to interpret the results. Both shallow and deep learning can be suited to different types of NLP tasks. Table 1 summarizes the differences between shallow and deep learning NLP models based on feature extraction, interpretability, dataset size, architecture, performance, and computational power.

Table 1 Shallow vs deep learning for NLP

Aspect	Shallow learning	Deep learning
Feature engineering	Needs feature extraction/selection	Does not require feature extraction
Interpretability	Easier to interpret	Harder to interpret
Dataset size	Smaller	Larger
Architecture	Simple/few hidden layers	Complex/several hidden layers
Performance	Works well for certain applications	Higher performance for complex tasks
Computational power	Less computing power	Higher computing power

Shallow learning techniques often require feature engineering, a process that requires feature selection and extraction algorithms to reduce the curse of the dimensionality problem and result in higher performance. Sometimes SL relies on manual extraction for domain-specific tasks and for higher interpretability. Additionally, SL techniques work with a smaller data set compared to DL techniques. On the other hand, DL methods extract features as part of their hidden layer processing and often require larger datasets to learn with high accuracy due to the complexity of the architecture [165, 166].

Additionally, DL models employ intricate architectures comprising multiple hidden layers. In contrast, SL models have simpler architectures and may only incorporate one or a few hidden layers, making them more effective with limited datasets. Due to their straightforward architecture, the results of SL techniques are easier to interpret compared to DL models. This is because DL models often function as black boxes, which makes it challenging to comprehend the rationale behind their decisions.

DL models require higher computational power and take longer latency to train and run due to the complexity of their architectures, compared to SL models. However, DL models can provide higher accuracy for many different NLP tasks and applications because of their ability to find complex relationships and hidden features in the data set. For that reason, DL models are suitable for personal assistants and chatbots. On the contrary, SL requires less computational power and takes less latency, so they better fit smaller tasks and tasks that cannot afford higher computational power such as energy-efficient and real-time systems.

Table 2 shows recent work for the different NLP tasks described in Sect. 2.3 using SL, DL, and hybrid learning models. The hybrid models include two or more models that combine different SL and/or DL models together. Due to the huge number of publications on NLP, we include a representative sample of recent publications (2020–2024).

Recently, pretrained models have been increasingly used for many complex NLP tasks [211]. This surge in popularity stems from their ability to capture rich linguistic representations from massive amounts of text data [212]. However, training these models often requires significant computational resources and large datasets, which

Table 2 Shallow, deep, and hybrid approaches for NLP tasks

NLP task	Shallow models	Deep models	Hybrid models
POS	HMM [167, 168], CRF [124, 168–170]	Bi-LSTM [170, 171]	[172]
NER	HMM [173]	CNN [174]	[175]
Parsing	SVM, LR, RF [176]	CNN [176, 177], RNN [178], LSTM [176]	[179–183]
TC	KNN [184], SVM [185]	CNN [186, 187]	[188–190]
IE	NLP [191], LDA [192]	CNN [193]	[191]
SRL	NLP [191]	CNN [194]	[195, 196]
SA	LR [139], SVM [139], ANN [141], KNN [197]	SiEBERT [198], CNN [199, 200], RNN [199], DNN [199, 201]	[202–205]
SR	KNN [206]	CNN [207]	[208–210]

can be a limitation for certain applications. Hybrid approaches that combine shallow learning techniques with pretrained models have emerged as an effective solution.

In addition, LSTM models can capture long-term dependencies in text sequences, maintain context over varying lengths of input, and mitigate the vanishing gradient problem. $Pre-trained$ LSTM and other pretrained models such as BERT or GPT can be integrated with SL models and fine-tuned for specific tasks with a relatively small dataset required for tuning. This combination proves particularly advantageous for training on small datasets or when there is limited computational capacity such as edge computing applications.

7 Discussion

When looking at NLP from a rationalist perspective, we can conclude that a significant portion of knowledge in the human mind is not acquired through the senses but is instead genetically predetermined [20]. Many researchers assume that some fundamental knowledge and reasoning mechanisms in linguistics are directly encoded with rules or other forms of representation. This assumption is also compatible with statistics and mathematics. In other words, a specific algorithm uses an iterative process to characterize the basic algorithm optimized with numerical parameters and a numerical measure characterizing the learning phase [213, 214]. Machine learning models can primarily be categorized as generative or discriminative. Generative methods can produce synthetic data as they construct rich probability distribution models. Discriminative methods, on the other hand, are more functional, accurately predicting final probabilities based on observations [215]. Discriminative methods rely on a less information-intensive approach and making distinctions between languages, while generative models may become problematic when many features are utilized. In addition, discriminative models allow for the use of more features. This necessitates the utilization of methods such as Bayes, Markov, neural networks, and BERT, aimed at revealing evaluation metrics used to assess the model's performance and relevant challenges [216, 217].

Recent technological research in NLP indicates that most applications focus on text classification. While some studies discuss the application of text analytics in industry, the task of monitoring and collecting text data is not thoroughly addressed, and the scope of the proposed models remains limited to specific domains [217, 218]. Moreover, research often neglects feature selection or extraction methods based on information extraction, instead typically focusing on text classification for search engines. These studies also fail to reveal the latest techniques used in NLP. However, they do include details concerning subtasks related to text classification, such as data collection, data preprocessing, and semi-supervised or unsupervised data labeling for training machine learning models [219].

8 Challenges and Future Work

With all of the advancements in models for NLP applications and datasets, there exist some challenges and limitations. The challenges include understanding the ambiguity and context of languages, some difficulties with multilingual processing, translation tasks from/to multiple languages, data quality and availability for accurate NLP, computational demands, and scalability especially in the current big data era. In addition, NLP struggles with ethical considerations and biases that need to be addressed.

One of the main challenges that NLP faces today is the ambiguity associated with the meaning of words, text, or speech coming from several aspects including language use, culture, intention, expressions, tone, and silence. Some languages can have different meanings of words based on the context and how words are used. Also, for speech recognition, some words have the same pronunciations but different spellings and meanings. In addition, NLP systems lack common sense reasoning and struggle to understand the context of a text-based on intention, culture, and personal styles. It often struggles with understanding humor and sarcasm as humans might understand them differently [20]. While new contextual embeddings and models such as BERT, transformer-based architectures, and GPT allow for capturing deeper linguistic contexts and nuances, there is still a need for more improvements. For example, future models can dynamically expand the context window based on the relevance of the surrounding text to increase the accuracy of NLP models and reduce context ambiguity.

Another major challenge facing NLP today is interpretability and explainability. Although deep learning models for NLP are complex and act as black boxes, ensuring explainability and interpretability is crucial to understanding results and decisions and ensuring compliance with regulations for some applications like medical and financial applications. Explainable AI models for NLP aim to have explainable, interpretable, and rational models [220, 221]. Existing models include

LIME (Local Interpretable Model-Agnostic Explanations) [222], shapely [223], visualization, and feature importance models. However, these models require special expertise to understand the results [221]. New explainable AI techniques are required to include deeper understanding and reasoning models, expand the scope of explanations, reduce expert dependency, and detect any malicious factors.

In addition, ethical considerations and biases are major concerns surrounding NLP models, stemming from the vast amount of data they rely on. Training NLP models on a large amount of data often involves personal data, raising ethical and privacy concerns. Techniques like anonymization help to mitigate the risk, but systematic and strict privacy measures are still crucial to protect user data. While NLP and ML have been used to detect privacy issues [224, 225] and ethical guidelines for NLP and AI have been established, enforcing guidelines remains a challenge. Examples, of ongoing ethical guideline efforts, include the IEEE Global Initiative on Ethics of Autonomous and Intelligent Systems [226] and the AI Ethics Guidelines Global Inventory [227]. In addition, because of the amount of regulations that exist, there need to be a consensus on global regulations [228]. Finally, an interdisciplinary effort among researchers, policymakers, and industry professionals to outline ethical guideline allows for forming informative global regulations [229].

Furthermore, bias in training data and NLP models poses a significant challenge, resulting in unfair decisions and discrimination. Diversifying training data and ensuring representativeness can help reduce bias and improve fairness. This issue, however, is exacerbated by political and social biases. While there have been several attempts to address biases in NLP, a major challenge lies in the lack of clear definitions for what types of bias are harmful, why, and for whom [230].

Finally, a wealth of new and emerging applications for NLP can be further explored: for example, the use of NLP for the social good to investigate global challenges such as misinformation, multilingual global communication, and climate change. Additionally, NLP could power next-generation virtual assistants that can not only understand our words but also recognize emotions and offer natural personalized assistance to enhance productivity, accessibility, and user experience in many different domains.

9 Conclusion

As an in-depth review of the field of NLP, this chapter provides an overview of NLP, its levels, and core tasks and examines in detail shallow learning and deep learning methodologies. Shallow methods include approaches based on predetermined rules and statistical patterns derived from labeled/unlabeled datasets. Models such as HMM and SVM offer simplicity and transparency but struggle to capture the complex nuances of language. On the other hand, deep learning architectures, especially CNNs and RNNs, work on multiple layers of abstraction, enabling the automatic extraction of hierarchical features from raw text data. In this chapter, we discuss the specific advantages and challenges of each of the shallow learning and deep learning methods and provide information for comprehensively evaluating these two paradigms in the field of NLP, in order to understand how both approaches are suitable for different tasks and applications. Finally, the chapter discusses the current challenges that NLP encounters and provides future research directions.

References

1. Jones, K. S. (1994). Natural language processing: A historical review. In *Current issues in computational linguistics: In honour of Don Walker* (pp. 3–16).
2. Jing, K., & Xu, J. (2019). A survey on neural network language models. arXiv preprint arXiv:1906.03591.
3. Otter, D. W., Medina, J. R., & Kalita, J. K. (2020). A survey of the usages of deep learning for natural language processing. *IEEE Transactions on Neural Networks and Learning Systems, 32*(2), 604–624.
4. Zhou, M., Duan, N., Liu, S., & Shum, H.-Y. (2020). Progress in neural NLP: Modeling, learning, and reasoning. *Engineering, 6*(3), 275–290.
5. Kang, Y., Cai, Z., Tan, C.-W., Huang, Q., & Liu, H. (2020). Natural language processing (NLP) in management research: A literature review. *Journal of Management Analytics, 7*(2), 139–172.
6. Johri, P., Khatri, S. K., Al-Taani, A. T., Sabharwal, M., Suvanov, S., & Kumar, A. (2021). Natural language processing: History, evolution, application, and future work. In A. Abraham, O. Castillo, & D. Virmani (Eds.), *Proceedings of 3rd international conference on computing informatics and networks* (pp. 365–375). Springer.
7. Ukwen, D. O., & Karabatak, M. (2021). *Review of NLP-based systems in digital forensics and cybersecurity*. In 2021 9th international symposium on digital forensics and security (ISDFS) (pp. 1–9).
8. Locke, S., Bashall, A., Al-Adely, S., Moore, J., Wilson, A., & Kitchen, G. B. (2021). Natural language processing in medicine: a review. *Trends in Anaesthesia and Critical Care, 38*, 4–9.
9. Hall, K., Chang, V., & Jayne, C. (2022). A review on natural language processing models for covid-19 research. *Healthcare Analytics, 2*, 100078.
10. Lareyre, F., Nasr, B., Chaudhuri, A., Di Lorenzo, G., Carlier, M., & Raffort, J. (2023). Comprehensive review of natural language processing (NLP) in vascular surgery. In *EJVES Vascular Forum*.
11. Min, B., Ross, H., Sulem, E., Veyseh, A. P. B., Nguyen, T. H., Sainz, O., Agirre, E., Heintz, I., & Roth, D. (2023). Recent advances in natural language processing via large pre-trained language models: A survey. *ACM Computing Surveys, 56*(2), 1–40.
12. Tounsi, A., & Temimi, M. (2023). A systematic review of natural language processing applications for hydrometeorological hazards assessment. *Natural Hazards, 116*(3), 2819–2870.
13. Gorenstein, L., Konen, E., Green, M., & Klang, E. (2024). Bert in radiology: A systematic review of natural language processing applications. *Journal of the American College of Radiology*.
14. Oshikawa, R., Qian, J., & Wang, W. Y. (2018). A survey on natural language processing for fake news detection. arXiv preprint arXiv:1811.00770.
15. Esteva, A., Kale, A., Paulus, R., Hashimoto, K., Yin, W., Radev, D., & Socher, R. (2021). Covid-19 information retrieval with deep-learning based semantic search, question answering, and abstractive summarization. *NPJ Digital Medicine, 4*(1), 68.
16. Joshi, A., Dabre, R., Kanojia, D., Li, Z., Zhan, H., Haffari, G., & Dippold, D. (2024). Natural language processing for dialects of a language: A survey. arXiv preprint arXiv:2401.05632.

17. Rizinski, M., Jankov, A., Sankaradas, V., Pinsky, E., Mishkovski, I., & Trajanov, D. (2024). Comparative analysis of NLP-based models for company classification. *Information, 15*, 1–32.
18. Mohammad, S. M. (2020). *NLP scholar: A dataset for examining the state of NLP research*. In: Proceedings of the twelfth language resources and evaluation conference (pp. 868–877). European Language Resources Association, Marseille.
19. Jurafsky, D., & Martin, J. H. (2014). *Speech and language processing* (2nd ed.).
20. Khurana, D., Koli, A., Khatter, K., & Singh, S. (2023). Natural language processing: State of the art, current trends and challenges. *Multimedia Tools and Applications, 82*(3), 3713–3744.
21. Liddy, E.D.: Natural language processing (2001).
22. Chowdhary, K. R. (2020). *Natural language processing* (pp. 603–649). Springer.
23. Locke, W. N., & Booth, A. D. (1955). *Machine translation of languages: Fourteen essays*. Wiley.
24. Weaver, W. (1952). *Translation*. In: Proceedings of the conference on mechanical translation.
25. Alan, M. (1950). Turing. Computing machinery and intelligence. *Mind, 59*(236), 433–460.
26. Nagarhalli, T. P., Vaze, V., & Rana, N. (2021). *Impact of machine learning in natural language processing: A review*. In Third international conference on intelligent communication technologies and virtual mobile networks (pp. 1529–1534).
27. Singh, S., & Mahmood, A. (2021). The NLP cookbook: Modern recipes for transformer based deep learning architectures. *IEEE Access, 9*, 68675–68702.
28. Reshamwala, A., Mishra, D., & Pawar, P. (2013). Review on natural language processing. *IRACST Engineering Science and Technology: An International Journal, 3*(1), 113–116.
29. Sproat, R. (2000). Lexical analysis. In *Handbook of natural language processing* (2nd ed., pp. 37–57). Marcel Dekker Inc.
30. Yannakoudakis, E. J., Tsomokos, I., & Hutton, P. J. (1990). N-grams and their implication to natural language understanding. *Pattern Recognition, 23*(5), 509–528.
31. Song, X., Salcianu, A., Song, Y., Dopson, D., & Zhou, D. (2020). Fast wordpiece tokenization. arXiv preprint arXiv:2012.15524.
32. Müller, T., Cotterell, R., Fraser, A., & Schütze, H. (2015). *Joint lemmatization and morphological tagging with lemming*. In Proceedings of the conference on empirical methods in natural language processing (pp. 2268–2274).
33. Hasanovna, D. M. (2022). Aspectual and lexical-semantic classification of verbs. *Open Access Repository, 8*(2), 116–121.
34. Roark, B., Bachrach, A., Cardenas, C., & Pallier, C. (2009). *Deriving lexical and syntactic expectation-based measures for psycholinguistic modeling via incremental top-down parsing*. In Proceedings of the conference on empirical methods in natural language processing (pp. 324–333).
35. Navigli, R. (2009). Word sense disambiguation: A survey. *ACM Computing Surveys, 41*(2), 1–69.
36. Bevilacqua, M., Pasini, T., Raganato, A., & Navigli, R. (2021). *Recent trends in word sense disambiguation: A survey*. In International joint conference on artificial intelligence (pp. 4330–4338).
37. Medhat, W., Hassan, A., & Korashy, H. (2014). Sentiment analysis algorithms and applications: A survey. *Ain Shams Engineering Journal, 5*(4), 1093–1113.
38. Wankhade, M., Rao, A. C. S., & Kulkarni, C. (2022). A survey on sentiment analysis methods, applications, and challenges. *Artificial Intelligence Review, 55*(7), 5731–5780.
39. Kamath, A., & Das, R. (2018). A survey on semantic parsing. arXiv preprint arXiv:1812.00978.
40. Joshi, A., Bhattacharyya, P., & Carman, M. J. (2017). Automatic sarcasm detection: A survey. *ACM Computing Surveys, 50*(5), 1–22.
41. Hasan, M. K., Rahman, W., Zadeh, A., Zhong, J., Tanveer, M. I., Morency, L.-P., & Hoque, M. (2019). UR-FUNNY: A multimodal language dataset for understanding humor. arXiv preprint arXiv:1904.06618.

42. Yogish, D., Manjunath, T., & Hegadi, R. S. (2019). Review on natural language processing trends and techniques using nltk. *Communications in Computer and Information Science, 1037*, 589–606.
43. Sharnagat, R. (2014). Named entity recognition: A literature survey. *Center For Indian Language Technology*, 1–27.
44. Nivre, J. (2010). Dependency parsing. Language and linguistics. *Compass, 4*(3), 138–152.
45. Zhu, M., Zhang, Y., Chen, W., Zhang, M., & Zhu, J. (2013). *Fast and accurate shift- reduce constituent parsing*. In Proceedings of the 51st annual meeting of the association for computational linguistics (pp. 434–443).
46. Li, J., Liu, M., Qin, B., & Liu, T. (2022). A survey of discourse parsing. *Frontiers of Computer Science, 16*(5), 165329.
47. Sarawagi, S. (2008). Information extraction. *Foundations and Trends® in Databases, 1*(3), 261–377.
48. Olujimi, P. A., & Ade-Ibijola, A. (2023). NLP techniques for automating responses to customer queries: A systematic review. *Discover Artificial Intelligence, 3*(1), 20.
49. Schopf, T. (2023). *A taxonomy of natural language processing*. https://towardsdatascience.com/a-taxonomy-of-natural-language-processing-dfc790cb4c01. Last accessed Apr 2024.
50. Hupkes, D., Giulianelli, M., Dankers, V., Artetxe, M., Elazar, Y., Pimentel, T., Christodoulopoulos, C., Lasri, K., Saphra, N., Sinclair, A., & Ulmer, D. (2023). A taxonomy and review of generalization research in NLP. *Nature Machine Intelligence, 5*(10), 1161–1174.
51. Schmid, H. (1999). Improvements in part-of-speech tagging with an application to German. In *Natural language processing using very large corpora* (pp. 13–25). Springer.
52. Nafea, A. A., Alameri, S. A., Majeed, R. R., Khalaf, M. A., & AL-Ani, M. M. (2024). A short review on supervised machine learning and deep learning techniques in computer vision. *Babylonian Journal of Machine Learning, 2024*, 48–55.
53. Hapke, H., Howard, C., & Lane, H. (2019). *Natural language processing in action: Understanding, analyzing, and generating text with python* (p. 15). Simon and Schuster.
54. Church, K. W. (2017). Word2vec. *Natural Language Engineering, 23*(1), 155–162.
55. Di Gennaro, G., Buonanno, A., & Palmieri, F. A. (2021). Considerations about learning word2vec. *The Journal of Supercomputing*, 1–16.
56. Pennington, J., Socher, R., & Manning, C. D. (2014). *Glove: Global vectors for word representation*. In Proceedings of the 2014 conference on empirical methods in natural language processing (pp. 1532–1543).
57. Blei, D. M., Ng, A. Y., & Jordan, M. I. (2003). Latent dirichlet allocation. *Journal of Machine Learning Research, 3*(Jan), 993–1022.
58. Chauhan, U., & Shah, A. (2021). Topic modeling using latent dirichlet allocation: A survey. *ACM Computing Surveys, 54*(7), 1–35.
59. Collobert, R., Weston, J., Bottou, L., Karlen, M., Kavukcuoglu, K., & Kuksa, P. (2011). Natural language processing (almost) from scratch. *Journal of Machine Learning Research, 12*, 2493–2537.
60. Joachims, T. (1999). Transductive inference for text classification using support vector machines. *Icml, 99*, 200–209.
61. Xiao, J., & Zhou, Z. (2020). *Research progress of rnn language model*. In IEEE inter-national conference on artificial intelligence and computer applications (pp. 1285–1288).
62. Dhyani, M., & Kumar, R. (2021). An intelligent chatbot using deep learning with bidirectional rnn and attention model. *Materials today: proceedings, 34*, 817–824.
63. Khan, A., Sohail, A., Zahoora, U., & Qureshi, A. S. (2020). A survey of the recent architectures of deep convolutional neural networks. *Artificial Intelligence Review, 53*, 5455–5516.
64. Hema, C., & Marquez, F. P. G. (2023). Emotional speech recognition using CNN and deep learning techniques. *Applied Acoustics, 211*, 109492.

65. Gillioz, A., Casas, J., Mugellini, E., & Abou Khaled, O. (2020). *Overview of the transformer-based models for NLP tasks*. In 15th conference on computer science and information systems (pp. 179–183).
66. Rothman, D. (2021). *Transformers for natural language processing: Build innovative deep neural network architectures for NLP with python, PyTorch, TensorFlow, BERT, RoBERTa, and more*. Packt Publishing Ltd.
67. Jain, P. K., Quamer, W., & Pamula, R. (2023). Consumer sentiment analysis with aspect fusion and GAN-BERT aided adversarial learning. *Expert Systems, 40*, 13247.
68. Li, Q., Peng, H., Li, J., Xia, C., Yang, R., Sun, L., Yu, P. S., & He, L. (2020). A survey on text classification: From shallow to deep learning. arXiv preprint arXiv:2008.00364.
69. Minaee, S., Kalchbrenner, N., Cambria, E., Nikzad, N., Chenaghlu, M., & Gao, J. (2021). Deep learning–based text classification: A comprehensive review. *ACM Computing Surveys, 54*(3), 1–40.
70. Zhan, J., & Dahal, B. (2017). Using deep learning for short text understanding. *Journal of Big Data, 4*, 1–15.
71. Popel, M., Tomkova, M., Tomek, J., Kaiser, L., Uszkoreit, J., Bojar, O., & Žabokrtskŷ, Z. (2020). Transforming machine translation: A deep learning system reaches news translation quality comparable to human professionals. *Nature Communications, 11*(1), 1–15.
72. Li, X., Mao, K., Lin, F., & Feng, Z. (2023). Feature-aware conditional GAN for category text generation. *Neurocomputing, 547*(4), 126358.
73. Nivaashini, M., Priyanka, G., & Aarthi, S. (2024). Deep neural machine translation (DNMT) hybrid deep learning architecture-based english-to-indian language translation. In *Automatic speech recognition and translation for low resource languages* (pp. 331–373).
74. Asudani, D. S., Nagwani, N. K., & Singh, P. (2023). Impact of word embedding models on text analytics in deep learning environment: A review. *Artificial Intelligence Review, 56*, 10345–10425.
75. Hosseini, S., & Varzaneh, Z. A. (2022). Deep text clustering using stacked autoencoder. *Multimedia Tools and Applications, 81*(8), 10861–10881.
76. Zou, Y., Lin, J., Zhao, L., Kang, Y., Jiang, Z., Sun, C., Zhang, Q., Huang, X., & Liu, X. (2021). *Unsupervised summarization for chat logs with topic-oriented ranking and context-aware auto-encoders*. In Proceedings of the AAAI conference on artificial intelligence (vol. 35, pp. 14674–14682).
77. Joshi, A., Fidalgo, E., Alegre, E., & Fern'andez-Robles, L. (2019). Summcoder: An unsupervised framework for extractive text summarization based on deep auto-encoders. *Expert Systems with Applications, 129*, 200–215.
78. Uc-Cetina, V., Navarro-Guerrero, N., Martin-Gonzalez, A., Weber, C., & Wermter, S. (2023). Survey on reinforcement learning for language processing. *Artificial Intelligence Review, 56*(2), 1543–1575.
79. Latif, S., Cuayahuitl, H., Pervez, F., Shamshad, F., Ali, H. S., & Cambria, E. (2023). A survey on deep reinforcement learning for audio-based applications. *Artificial Intelligence Review, 56*(3), 2193–2240.
80. Wu, C., Li, X., Guo, Y., Wang, J., Ren, Z., Wang, M., & Yang, Z. (2022). Natural language processing for smart construction: Current status and future directions. *Automation in Construction, 134*, 104059.
81. Lin, J.-R., Hu, Z.-Z., Zhang, J.-P., & Yu, F.-Q. (2016). A natural-language-based approach to intelligent data retrieval and representation for cloud bim. *Computer-Aided Civil and Infrastructure Engineering, 31*(1), 18–33.
82. Zhang, J., & El-Gohary, N. M. (2016). Semantic NLP-based information extraction from construction regulatory documents for automated compliance checking. *Journal of Computing in Civil Engineering, 30*(2), 04015014.
83. Xu, N., Ma, L., Wang, L., Deng, Y., & Ni, G. (2021). Extracting domain knowledge elements of construction safety management: Rule-based approach using chinese natural language processing. *Journal of Management in Engineering, 37*(2), 04021001.

84. Liu, B. Q. F. (2023). Hierarchical model rule based NLP for semantic training representation using multi level structures. *Informatica, 48*(7), 29–38.
85. Chiche, A., & Yitagesu, B. (2022). Part of speech tagging: A systematic review of deep learning and machine learning approaches. *Journal of Big Data, 9*(1), 10.
86. Ren, R., & Zhang, J. (2021). Semantic rule-based construction procedural information extraction to guide jobsite sensing and monitoring. *Journal of Computing in Civil Engineering, 35*(6), 04021026.
87. Xu, X., & Cai, H. (2021). Ontology and rule-based natural language processing approach for interpreting textual regulations on underground utility infrastructure. *Advanced Engineering Informatics, 48*, 101288.
88. Karystianis, G., Thayer, K., Wolfe, M., & Tsafnat, G. (2017). Evaluation of a rule-based method for epidemiological document classification towards the automation of systematic reviews. *Journal of Biomedical Informatics, 70*, 27–34.
89. Pisceldo, F., Adriani, M., & Manurung, R. (2009). *Probabilistic part of speech tagging for bahasa indonesia*. In Third international MALINDO workshop (pp. 1–6).
90. Jiang, J. (2012). In C. C. Aggarwal & C. Zhai (Eds.), *Information extraction from text* (pp. 11–41). Springer.
91. Morwal, S., Jahan, N., & Chopra, D. (2012). Named entity recognition using hidden markov model (HMM). *International Journal on Natural Language Computing, 1*(4).
92. Anandika, A., Mishra, S. P., & Das, M. (2021). *Review on usage of hidden markov model in natural language processing*. In Intelligent and cloud computing: Proceedings of ICICC 2019 (Vol. 1, pp. 415–423).
93. Ojokoh, B., Zhang, M., & Tang, J. (2011). A trigram hidden markov model for metadata extraction from heterogeneous references. *Information Sciences, 181*(9), 1538–1551.
94. Sarkar, K., & Gayen, V. (2013). *A trigram HMM-based POS tagger for indian languages*. In: Proceedings of the international conference on frontiers of intelligent computing: theory and applications (pp. 205–212).
95. Huda, A. F., Al-Hamro, I. Z., Awalluddin, A. S., & Pamungkas, M. I. (2021). *Arabic part of speech (POS) tagging analysis using HMM trigram method on al-qur'an ayah sentences*. In 7th international conference on wireless and telematics (pp. 1–4).
96. McCallum, A., Freitag, D., & Pereira, F. C. (2000). Maximum entropy markov models for information extraction and segmentation. *Icml, 17*, 591–598.
97. Alam, F., & Islam, M. A. (2020). *A proposed model for bengali named entity recognition using maximum entropy markov model incorporated with rich linguistic feature set*. In Proceedings of the international conference on computing advancements (pp. 1–6).
98. Cortes, C., & Vapnik, V. (1995). Support-vector networks. *Machine Learning, 20*, 273–297.
99. Nakagawa, T., Kudo, T., & Matsumoto, Y. (2001). Unknown word guessing and part-of-speech tagging using support vector machines. In *NLPRS* (pp. 325–331).
100. Gimenez, J., & Marquez, L. (2004). Fast and accurate part-of-speech tagging: The SVM approach revisited. In *Recent advances in natural language processing III* (pp. 153–163).
101. Binulal, G. S., Goud, P. A., & Soman, K. (2009). A SVM based approach to telugu parts of speech tagging using svmtool. *International Journal of Recent Trends in Engineering, 1*(2), 183.
102. Lee, Y. K., Ng, H. T., & Chia, T. K. (2004). *Supervised word sense disambiguation with support vector machines and multiple knowledge sources*. In Proceedings of the third international workshop on the evaluation of systems for the semantic analysis of text (pp. 137–140).
103. Zhong, L., & Wang, T. (2020). Towards word sense disambiguation using multiple kernel support vector machine. *International Journal of Innovative Computing, Information and Control, 16*(2), 555–570.
104. Li, Y., Bontcheva, K., & Cunningham, H. (2009). Adapting SVM for natural language learning: A case study involving information extraction. *Natural Language Engineering, 15*(2), 241–271.

105. Zhou, G., Su, J., Zhang, J., & Zhang, M. (2005). *Exploring various knowledge in relation extraction*. In Proceedings of the 43rd annual meeting of the association for computational linguistics (pp. 427–434).
106. Hacioglu, K. (2004). *Semantic role labeling using dependency trees*. In Proceedings of the 20th international conference on computational linguistics (pp. 1273–1276).
107. Kudo, T. (2001). Japanese dependency structure analysis based on support vector machines.
108. Yamada, H., & Matsumoto, Y. (2003). *Statistical dependency analysis with support vector machines*. In Proceedings of the eighth international conference on parsing technologies (pp. 195–206).
109. Kudo, T., & Matsumoto, Y. (2001). *Chunking with support vector machines*. In Second meeting of the North American chapter of the association for computational linguistics.
110. Frank, E., Trigg, L., Holmes, G., & Witten, I. H. (2000). Naive bayes for regression. *Machine Learning, 41*, 5–25.
111. McCallum, A., & Nigam, K. (1998). *A comparison of event models for naive bayes text classification*. In AAAI-98 workshop on learning for text categorization (vol. 752, pp. 41–48).
112. Phuc, D., & Phung, N. T. K. (2007). *Using na¨ıve bayes model and natural language processing for classifying messages on online forum*. In IEEE international conference on research, innovation and vision for the future (pp. 247–252).
113. Xu, S. (2018). Bayesian na¨ıve bayes classifiers to text classification. *Journal of Information Science, 44*(1), 48–59.
114. Lewis, D. D. (1998). *Naive (Bayes) at forty: The independence assumption in information retrieval.* In European conference on machine learning (pp. 4–15).
115. Metsis, V., Androutsopoulos, I., & Paliouras, G. (2006). *Spam filtering with naive bayes – Which naive bayes?* In Third Conference on email and anti-spam (vol. 17, pp. 28–69).
116. Othman, N. F., & Din, W. (2019). YouTube spam detection framework using na¨ıve bayes and logistic regression. *Indonesian Journal of Electrical Engineering and Computer Science, 14*(3), 1508–1517.
117. Myles, A. J., Feudale, R. N., Liu, Y., Woody, N. A., & Brown, S. D. (2004). An introduction to decision tree modeling. *Journal of Chemometrics: A Journal of the Chemometrics Society, 18*(6), 275–285.
118. Giorgos, O., Dimitris, K., Thanasis, P., & Dimitris, C. (1999). *Decision trees and NLP: A case study in POS tagging*. In Proceedings of annual conference on artificial intelligence (ACAI).
119. Breiman, L. (2001). Random forests. *Machine Learning, 45*, 5–32.
120. Agrawal, R., Gupta, A., Prabhu, Y., & Varma, M. (2013). *Multi-label learning with millions of labels: Recommending advertiser bid phrases for web pages*. In Proceedings of the 22nd international conference on world wide web (pp. 13–24).
121. Bahrawi, N. (2019). Sentiment analysis using random forest algorithm-online social media based. *Journal of Information Technology and Its Utilization, 2*(2), 29–33.
122. Sutton, C., & McCallum, A. (2012). An introduction to conditional random fields. *Foundations and Trends® in Machine Learning, 4*(4), 267–373.
123. Pathak, P., Goswami, R., Joshi, G., Patel, P., & Patel, A. (2013). *CRF-based clinical named entity recognition using clinical NLP*. In Proceedings of international conference on natural language processing.
124. Warjri, S., Pakray, P., Lyngdoh, S. A., & Maji, A. K. (2021). Part-of-speech (POS) tagging using conditional random field (CRF) model for khasi corpora. *International Journal of Speech Technology, 24*(4), 853–864.
125. Huang, Z., Xu, W., & Yu, K.(2015). Bidirectional LSTM-CRF models for sequence tagging. arXiv preprint arXiv:1508.01991.
126. Dai, Z., Wang, X., Ni, P., Li, Y., Li, G., & Bai, X. (2019). *Named entity recognition using BERT BiLSTM CRF for chinese electronic health records*. In 12th international congress on image and signal processing, biomedical engineering and informatics (pp. 1–5).

127. Pandit, R., & Naskar, S. K. (2015). *A memory based approach to word sense disambiguation in bengali using k-nn method.* In IEEE 2nd international conference on recent trends in information systems (pp. 383–386).
128. Yong, Z., Youwen, L., & Shixiong, X. (2009). An improved KNN text classification algorithm based on clustering. *Journal of Computers, 4*(3), 230–237.
129. Hota, S., & Pathak, S. (2018). KNN classifier based approach for multi-class sentiment analysis of twitter data. *International Journal of Engineering & Technology, 7*(3), 1372–1375.
130. Venkata Subbarao, M., Terlapu, S. K., Geethika, N., & Harika, K. D. (2021). *Speech emotion recognition using k-nearest neighbor classifiers.* In Recent advances in artificial intelligence and data engineering: Select proceedings of AIDE 2020 (pp. 123–131).
131. Steinley, D. (2006). K-means clustering: A half-century synthesis. *British Journal of Mathematical and Statistical Psychology, 59*(1), 1–34.
132. Singh, V. K., Tiwari, N., & Garg, S. (2011). *Document clustering using k-means, heuristic k-means and fuzzy c-means.* In: 2011 international conference on computational intelligence and communication networks (pp. 297–301).
133. Alhawarat, M., & Hegazi, M. (2018). Revisiting k-means and topic modeling, a comparison study to cluster arabic documents. *IEEE Access, 6*, 42740–42749.
134. Al-Azzawy, D. S., & Al-Rufaye, F. M. L. (2017). *Arabic words clustering by using k-means algorithm.* In 2017 annual conference on new trends in information & communications technology applications (pp. 263–267).
135. D'Silva, J., & Sharma, U. (2020). Unsupervised automatic text summarization of konkani texts using k-means with elbow method. *International Journal of Engineering Research and Technology, 13*(9), 2380–2384.
136. Nick, T. G., & Campbell, K. M. (2007). Logistic regression. In *Topics in biostatistics* (pp. 273–301).
137. Pranckeviˇcius, T., & Marcinkeviˇcius, V. (2017). Comparison of naive bayes, random forest, decision tree, support vector machines, and logistic regression classifiers for text reviews classification. *Baltic Journal of Modern Computing, 5*(2), 221.
138. Hasan, M. R., Maliha, M., & Arifuzzaman, M. (2019). *Sentiment analysis with nlp on twitter data.* In International conference on computer, communication, chemical, materials and electronic engineering (pp. 1–4).
139. Hidayat, T. H. J., Ruldeviyani, Y., Aditama, A. R., Madya, G. R., Nugraha, A. W., & Adisaputra, M. W. (2022). Sentiment analysis of twitter data related to Rinca Island development using doc2vec and SVM and logistic regression as classifier. *Procedia Computer Science, 197*, 660–667.
140. Saikia, P., Baruah, R. D., Singh, S. K., & Chaudhuri, P. K. (2020). Artificial neural networks in the domain of reservoir characterization: A review from shallow to deep models. *Computers & Geosciences, 135*, 104357.
141. Strubytskyi, R., & Shakhovska, N. (2023). Method and models for sentiment analysis and hidden propaganda finding. *Computers in Human Behavior Reports, 12*, 100328.
142. Harrag, F., & El-Qawasmah, E. (2009). *Neural network for arabic text classification.* In The second international conference on the applications of digital information and web technologies (pp. 778–783).
143. Farkas, R., Szarvas, G., & Kocsor, A. (2006). Named entity recognition for hungarian using various machine learning algorithms. *Acta Cybernetica, 17*(3), 633–646.
144. Sutton, R. S., McAllester, D., Singh, S., & Mansour, Y. (1999). Policy gradient methods for reinforcement learning with function approximation. *Advances in Neural Information Processing Systems, 12.*
145. Rieser, V., & Lemon, O. (2011). *Reinforcement learning for adaptive dialogue systems: A data-driven methodology for dialogue management and natural language generation.* Springer.
146. Wu, L., Tian, F., Qin, T., Lai, J., & Liu, T.-Y. (2018). A study of reinforcement learning for neural machine translation. arXiv preprint arXiv:1808.08866.

147. Ryang, S., & Abekawa, T. (2012). *Framework of automatic text summarization using reinforcement learning*. In: The joint conference on empirical methods in natural language processing and computational natural language learning (pp. 256–265).
148. Shein, K. P. P., & Nyunt, T. T. S. (2010). *Sentiment classification based on ontology and SVM classifier*. In 2010 second international conference on communication software and networks (pp. 169–172).
149. Lauriola, I., Lavelli, A., & Aiolli, F. (2022). An introduction to deep learning in natural language processing: Models, techniques, and tools. *Neurocomputing, 470*, 443–456.
150. Ortiz-Garces, I., Govea, J., Andrade, R. O., & Villegas-Ch, W. (2024). Optimizing chatbot effectiveness through advanced syntactic analysis: A comprehensive study in natural language processing. *Applied Sciences, 14*(5), 1737.
151. Pan, S., Zheng, Y., & Liu, Y. (2024). Integrating graphs with large language models: Methods and prospects. *IEEE Intelligent Systems, 39*(1), 64–68.
152. Rani, S., & Jain, A. (2024). Aspect-based sentiment analysis of drug reviews using multi- task learning based dual BiLSTM model. *Multimedia Tools and Applications, 83*(8), 22473–22501.
153. Malik, S., & Jain, S. (2024). Deep convolutional neural network for knowledge-infused text classification. *New Generation Computing, 42*, 157.
154. Bakhit, D. M. A., Nderu, L., & Ngunyi, A. (2024). A hybrid neural network model based on transfer learning for arabic sentiment analysis of customer satisfaction. *Engineering Reports*, e12874.
155. Wajid, M. A., Zafar, A., & Wajid, M. S. (2024). A deep learning approach for image and text classification using neutrosophy. *International Journal of Information Technology, 16*(2), 853–859.
156. Alijani, S., Fayyad, J., & Najjaran, H. (2024). Vision transformers in domain adaptation and generalization: A study of robustness. arXiv preprint arXiv:2404.04452.
157. Raiaan, M. A. K., Mukta, M. S. H., Fatema, K., Fahad, N. M., Sakib, S., Mim, M. M. J., & Azam, S. (2024). A review on large language models: Architectures, applications, taxonomies, open issues and challenges. *IEEE Access*.
158. Hamidi Rad, R., Cucerzan, S., Chandrasekaran, N., & Gamon, M. (2024). Interactive topic tagging in community question answering platforms. In *European conference on information retrieval* (pp. 195–209). Springer.
159. Gasparetto, A., Marcuzzo, M., Zangari, A., & Albarelli, A. (2022). A survey on text classification algorithms: From text to predictions. *Information, 13*(2), 83.
160. Fanni, S. C., Febi, M., Aghakhanyan, G., & Neri, E. (2023). Natural language processing. In *Introduction to artificial intelligence* (pp. 87–99). Springer.
161. Alshemali, J. B. (2020). Kalita: Improving the reliability of deep neural networks in nlp: A review. *Knowledge-Based Systems, 191*, 105210.
162. Khan, W., Daud, A., Khan, K., Muhammad, S., & Haq, R. (2023). Exploring the frontiers of deep learning and natural language processing: A comprehensive overview of key challenges and emerging trends. *Natural Language Processing Journal*, 100026.
163. Dhar, T., Dey, N., Borra, S., & Sherratt, R. S. (2023). Challenges of deep learning in medical image analysis – Improving explainability and trust. *IEEE Transactions on Technology and Society, 4*(1), 68–75.
164. Somani, A., Horsch, A., & Prasad, D. K. (2023). *Interpretability in deep learning* (pp. 1–466). Springer.
165. Hestness, J., Narang, S., Ardalani, N., Diamos, G., Jun, H., Kianinejad, H., Patwary, M. M. A., Yang, Y., & Zhou, Y. (2017). Deep learning scaling is predictable, empirically. arXiv preprint arXiv:1712.00409.
166. Meir, Y., Sardi, S., Hodassman, S., Kisos, K., Ben-Noam, I., Goldental, A., & Kanter, I. (2020). Power-law scaling to assist with key challenges in artificial intelligence. *Scientific Reports, 10*(1), 19628.

167. Pattnaik, S., Nayak, A. K., & Patnaik, S. (2020). A semi-supervised learning of HMM to build a POS tagger for a low resourced language. *Journal of Information and Communication Convergence Engineering, 18*(4), 207–215.
168. Pradhan, A., & Yajnik, A. (2024). Parts-of-speech tagging of nepali texts with bidirectional LSTM, conditional random fields and HMM. *Multimedia Tools and Applications, 83*(4), 9893–9909.
169. Mishra, P., & Sharma, D. M. (2022). Building odia shallow parser. arXiv preprint arXiv:2204.08960.
170. Dalai, T., Mishra, T. K., & Sa, P. K. (2023). Part-of-speech tagging of odia language using statistical and deep learning based approaches. *ACM Transactions on Asian and Low-Resource Language Information Processing, 22*(6), 1–24.
171. Dhumal Deshmukh, R., & Kiwelekar, A. (2020). *Deep learning techniques for part of speech tagging by natural language processing*. In 2020 2nd international conference on innovative mechanisms for industry applications (pp. 76–81).
172. Warjri, S., Pakray, P., Lyngdoh, S. A., & Maji, A. K. (2021). Part-of-speech (POS) tagging using deep learning-based approaches on the designed khasi POS corpus. *Transactions on Asian and Low-Resource Language Information Processing, 21*(3), 1–24.
173. Lison, P., Hubin, A., Barnes, J., & Touileb, S. (2020). Named entity recognition without labelled data: A weak supervision approach. arXiv preprint arXiv:2004.14723.
174. Lv, Y., Qin, X., Du, X., & Qiu, S. (2023). Deep adaptation of cnn in chinese named entity recognition. *Engineering Reports, 5*(6), 12614.
175. Jin, Z., He, X., Wu, X., & Zhao, X. (2022). A hybrid transformer approach for chinese ner with features augmentation. *Expert Systems with Applications, 209*(2), 118385.
176. Odisho, A. Y., Park, B., Altieri, N., DeNero, J., Cooperberg, M. R., Carroll, P. R., & Yu, B. (2020). Natural language processing systems for pathology parsing in limited data environments with uncertainty estimation. *JAMIA Open, 3*(3), 431–438.
177. Li, Z., Wu, Y., Peng, B., Chen, X., Sun, Z., Liu, Y., & Yu, D. (2021). Secnn: A semantic cnn parser for code comment generation. *Journal of Systems and Software, 181*, 111036.
178. Tham, M. J. (2020). Bidirectional gated recurrent unit for shallow parsing. *Indian Journal of Computer Science and Engineering (IJCSE), 11*(5), 517–521.
179. Li, Z., & Tu, K. (2020). *Unsupervised cross-lingual adaptation of dependency parsers using crf autoencoders*. In Findings of the association for computational linguistics: EMNLP 2020 (pp. 2127–2133).
180. Li, P., Wang, Y., Si, T., Ullah, K., Han, W., & Wang, L. (2024). Mffsp: Multi-scale feature fusion scene parsing network for landslides detection based on high-resolution satellite images. *Engineering Applications of Artificial Intelligence, 127*, 107337.
181. Zhang, H., Du, Q., Chen, Z., & Zhang, C. (2022). A chinese address parsing method using RoBERTa-BiLSTM-CRF. *Geomatics and Information Science of Wuhan University, 47*(5), 665–672.
182. Wang, C. A corpus approach to discourse comparison of chinese and foreign english media coverage on belt and road. *Applied Mathematics and Nonlinear Sciences, 9*(1).
183. Sewunetie, W. T., & Kovacs, L. (2024). A comparative study of chatgpt-based and hybrid parser-based sentence parsing methods for semantic graph-based induction. *IEEE Access, 209*(2), 118385.
184. Chen, Z., Zhou, L. J., Da Li, X., Zhang, J. N., & Huo, W. J. (2020). The lao text classification method based on KNN. *Procedia Computer Science, 166*, 523–528.
185. Luo, X. (2021). Efficient english text classification using selected machine learning techniques. *Alexandria Engineering Journal, 60*(3), 3401–3409.
186. Yu, S., Liu, D., Zhang, Y., Zhao, S., & Wang, W. (2021). Dptcn: A novel deep cnn model for short text classification. *Journal of Intelligent & Fuzzy Systems, 41*(6), 7093–7100.
187. Soni, S., Chouhan, S. S., & Rathore, S. S. (2023). Textconvonet: A convolutional neural network based architecture for text classification. *Applied Intelligence, 53*(11), 14249–14268.

188. Jang, B., Kim, M., Harerimana, G., Kang, S.-U., & Kim, J. W. (2020). Bi-LSTM model to increase accuracy in text classification: Combining word2vec cnn and attention mechanism. *Applied Sciences, 10*(17), 5841.
189. Bodapati, S., Bandarupally, H., Shaw, R. N., & Ghosh, A. (2021). Comparison and analysis of RNN-LSTMs and CNNs for social reviews classification. *Advances in Applications of Data-Driven Computing*, 49–59.
190. Li, X., Cui, M., Li, J., Bai, R., Lu, Z., & Aickelin, U. (2021). A hybrid medical text classification framework: Integrating attentive rule construction and neural network. *Neurocomputing, 443*, 345–355.
191. Nguyen, C., Nguyen, P., Tran, T., Nguyen, D., Trieu, A., Pham, T., Dang, A., & Nguyen, L.-M. (2024). CAPTAIN at COLIEE 2023: Efficient methods for legal information retrieval and entailment tasks. arXiv preprint arXiv:2401.03551.
192. VanGessel, F. G., Perry, E., Mohan, S., Barham, O. M., & Cavolowsky, M. (2024). NLP for knowledge discovery and information extraction from energetics corpora. arXiv preprint arXiv:2402.06964.
193. Zhao, H., Pan, Y., & Yang, F. (2020). Research on information extraction of technical documents and construction of domain knowledge graph. *IEEE Access, 8*, 168087–168098.
194. Munir, K., Zhao, H., & Li, Z. (2021). Adaptive convolution for semantic role labeling. *IEEE/ACM Transactions on Audio, Speech, and Language Processing, 29*, 782–791.
195. Ma, N., Wang, J., & Zhu, A. (2023). *Multi-feature chinese semantic role labeling combined with global information*. In International conference on electronic information engineering and data processing (vol. 12700, pp. 66–74).
196. Wan, F., Yang, Y., Zhu, D., Yu, H., Zhu, A., Che, G., & Ma, N. (2022). Semantic role labeling integrated with multilevel linguistic cues and Bi-LSTM-CRF. *Mathematical Problems in Engineering, 2022*, 1–8.
197. Shamrat, F., Chakraborty, S., Imran, M., Muna, J. N., Billah, M. M., Das, P., & Rahman, M. O. (2021). Sentiment analysis on twitter tweets about covid-19 vaccines using NLP and supervised KNN classification algorithm. *Indonesian Journal of Electrical Engineering and Computer Science, 23*(1), 463–470.
198. Hartmann, J., Heitmann, M., Siebert, C., & Schamp, C. (2023). More than a feeling: Accuracy and application of sentiment analysis. *International Journal of Research in Marketing, 40*(1), 75–87.
199. Dang, N. C., Moreno-García, M. N., & De la Prieta, F. (2020). Sentiment analysis based on deep learning: A comparative study. *Electronics, 9*(3), 483.
200. Nassif, A. B., Darya, A. M., & Elnagar, A. (2021). Empirical evaluation of shallow and deep learning classifiers for arabic sentiment analysis. *Transactions on Asian and Low-Resource Language Information Processing, 1*, 1–25.
201. Gomez-Adorno, H., Bel-Enguix, G., Sierra, G., Barajas, J. C., & Álvarez, W. (2024). Machine learning and deep learning sentiment analysis models: Case study on the sent-covid corpus of tweets in mexican spanish. *Informatics, 11*, 24.
202. Dang, C. N., Moreno-Garcia, M. N., & Prieta, F. (2021). Hybrid deep learning models for sentiment analysis. *Complexity, 2021*, 1–16.
203. Jain, P. K., Saravanan, V., & Pamula, R. (2021). A hybrid CNN-LSTM: A deep learning approach for consumer sentiment analysis using qualitative user-generated contents. *Transactions on Asian and Low-Resource Language Information Processing, 20*(5), 1–15.
204. Islam, M. S., & Alam, K. M. (2024). Sentiment analysis of Bangla language using a new comprehensive dataset bangdsa and the novel feature metric skipBangla-BERT. *Natural Language Processing Journal, 7*(100069), 1–25.
205. Ramirez-Alcocer, U. M., Tello-Leal, E., Hernandez-Resendiz, J. D., & Romero, G. (2024). A hybrid CNN-LSTM approach for sentiment analysis. In *Congress on intelligent systems* (pp. 425–437).
206. Arora, R., & Bhatia, R. (2024). NLP-based speech analysis using k-neighbor classifier. In *Optimized predictive models in health care using machine learning* (p. 215).

207. Singh, A., Srivastava, K. K., & Murugan, H. (2020). Speech emotion recognition using convolutional neural network (CNN). *International Journal of Psychosocial Rehabilitation, 24*(8), 1–20.
208. Al-Makhadmeh, Z., & Tolba, A. (2020). Automatic hate speech detection using killer natural language processing optimizing ensemble deep learning approach. *Computing, 102*(2), 501–522.
209. Deshmukh, A. M. (2020). Comparison of hidden markov model and recurrent neural network in automatic speech recognition. *European Journal of Engineering and Technology Research, 5*(8), 958–965.
210. Ejigu, Y. A., & Asfaw, T. T. (2024). Enhancing amharic speech recognition in noisy conditions through end-to-end deep learning. *Preprint*. https://doi.org/10.20944/preprints202402.0754.v1
211. Qiu, X., Sun, T., Xu, Y., Shao, Y., Dai, N., & Huang, X. (2020). Pre-trained models for natural language processing: A survey. *Science China Technological Sciences, 63*(10), 1872–1897.
212. Wang, H., Li, J., Wu, H., Hovy, E., & Sun, Y. (2023). Pre-trained language models and their applications. *Engineering, 25*, 51–65.
213. Ekramipooya, A., Boroushaki, M., & Rashtchian, D. (2023). Application of natural language processing and machine learning in prediction of deviations in the hazop study worksheet: A comparison of classifiers. *Process Safety and Environmental Protection, 176*, 65–73.
214. Gao, B., Dong, Q., Zhu, C., Huang, J., Chen, H., Yang, Q., & Liu, H. (2023). Enhancing medical text representation for lung diagnosis prediction via knowledge infusion. Available at SSRN 4625860.
215. Murtaza, H., Ahmed, M., Khan, N. F., Murtaza, G., Zafar, S., & Bano, A. (2023). Synthetic data generation: State of the art in health care domain. *Computer Science Review, 48*, 100546.
216. Saravanan, K. S., & Bhagavathiappan, V. (2024). Innovative agricultural ontology construction using nlp methodologies and graph neural network. *Engineering Science and Technology, an International Journal, 52*, 101675.
217. Chakraborty, T., KS, U. R., Naik, S. M., Panja, M., & Manvitha, B. (2024). Ten years of generative adversarial nets (gans): A survey of the state-of-the-art. Machine learning. *Science and Technology, 5*(1), 011001.
218. Knisely, B. M., & Pavliscsak, H. H. (2023). Research proposal content extraction using natural language processing and semi-supervised clustering: A demonstration and comparative analysis. *Scientometrics, 128*(5), 3197–3224.
219. Mansur, Z., Omar, N., & Tiun, S. (2023). Twitter hate speech detection: A systematic review of methods, taxonomy analysis, challenges, and opportunities. *IEEE Access, 11*, 16226–16249.
220. Danilevsky, M., Qian, K., Aharonov, R., Katsis, Y., Kawas, B., & Sen, P. (2020). A survey of the state of explainable AI for natural language processing. arXiv preprint arXiv:2010.00711.
221. Gurrapu, S., Kulkarni, A., Huang, L., Lourentzou, I., & Batarseh, F. A. (2023). Rationalization for explainable NLP: A survey. *Frontiers in Artificial Intelligence, 6*(1225093).
222. Wikle, C. K., Datta, A., Hari, B. V., Boone, E. L., Sahoo, I., Kavila, I., Castruccio, S., Simmons, S. J., Burr, W. S., & Chang, W. (2023). An illustration of model agnostic explainability methods applied to environmental data. *Environmetrics, 34*(1), 2772.
223. Band, S. S., Yarahmadi, A., Hsu, C. C., Biyari, M., Sookhak, M., Ameri, R., Dehzangi, I., Chronopoulos, A. T., & Liang, H. W. (2023). Application of explainable artificial intelligence in medical health: A systematic review of interpretability methods. *Informatics in Medicine Unlocked, 40*(1), 101286.
224. Silva, P., Goncalves, C., Godinho, C., Antunes, N., & Curado, M. (2020). *Using NLP and machine learning to detect data privacy violations*. In IEEE conference on computer communications workshops (pp. 972–977).
225. Sousa, S., & Kern, R. (2023). How to keep text private? A systematic review of deep learning methods for privacy-preserving natural language processing. *Artificial Intelligence Review, 56*(2), 1427–1492.

226. The IEEE. (2024). *Global initiative on ethics of autonomous and intelligent systems*. https://standards.ieee.org/industry-connections/ec/autonomous-systems/. Last accessed Apr 2024.
227. AI Ethics Guidelines Global Inventory. (2024). https://algorithmwatch.org/en/ai-ethics-guidelines-global-inventory/. Last accessed Apr 2024.
228. Corrêa, N. K., Galvão, C., Santos, J. W., Del Pino, C., Pinto, E. P., Barbosa, C., Massmann, D., Mambrini, R., Galvão, L., Terem, E., & de Oliveira, N. (2023). Worldwide AI ethics: A review of 200 guidelines and recommendations for AI governance. *Patterns, 4*(10), 1–14.
229. Shafik, W. (2024). Artificial intelligence and machine learning with cyber ethics for the future world. In *Future communication systems using artificial intelligence, internet of things and data science* (pp. 110–130). CRC Press.
230. Blodgett, S. L., Barocas, S., Daume III, H., & Wallach, H. (2020). Language (technology) is power: A critical survey of "bias" in NLP. arXiv preprint arXiv:2005.14050.

Shallow Learning Versus Deep Learning in Speech Recognition Applications

Nasmin Jiwani and Ketan Gupta

1 Introduction

Speech recognition systems are becoming more and more common in our daily lives, ranging from digital assistants in our smartphones to smart home devices. These packages have significantly enhanced our productivity and convenience, and their total performance continues to evolve due to breakthroughs in machine learning. Specifically, the research and development in speech recognition have focused on processes for both shallow learning and deep learning [1]. Although both employ machine learning techniques, these systems vary significantly in terms of their complexity and performance. This paragraph examines the underlying differences between shallow learning and deep learning in voice recognition applications and how these processes affect the accuracy and usefulness of speech recognition systems. Speech recognition is the process of converting spoken speech into digital text. There are two primary methods of speech recognition: Shallow learning and deep learning. Shallow learning uses a predetermined set of rules and patterns to comprehend speech, but deep learning utilises intricate artificial neural networks to analyse and comprehend speech patterns. Superficial learning takes extensive manual engineering and expertise to establish and sustain such rules. In contrast, deep learning can automatically extract features and learn from vast quantities of data, rendering it more efficient in speech recognition [2]. Furthermore, profound comprehension can develop and grow through the acquisition of further knowledge, whereas predetermined guidelines constrain superficial learning. Shallow learning, also known as conventional machine learning, is a set of techniques that have been employed for numerous years to develop models and provide predictions based solely on data. However, deep studying is a relatively recent strategy that has

N. Jiwani (✉) · K. Gupta
Department of Information Technology, University of The Cumberlands, Williamsburg, KY, USA

Ö. F. Ertuğrul et al. (eds.), *Shallow Learning vs. Deep Learning*, The Springer Series in Applied Machine Learning, https://doi.org/10.1007/978-3-031-69499-8_9

become popular in recent years because of its improved performance in several applications. Shallow learning refers to educational models that utilise pre-defined algorithms to extract features from input data and create predictions based on those features. These algorithms include decision trees, logistic regression, support vector machines, and many others. These fashion styles are deemed superficial due to their typical inclusion of only one or two concealed layers and a restricted set of variables [3]. Superficial learning algorithms necessitate a substantial quantity of preliminary processing and feature engineering before they can be trained and applied to a specific problem. This technique involves selecting relevant functions from the input data, transforming them into a suitable representation, and subsequently inputting them into the model. This technique can be time-consuming and labour-intensive, and the effectiveness of the model depends greatly on the quality of the selected features [4]. The significance of speech recognition in contemporary society cannot be overstated since it has fundamentally transformed the way people engage with technology. Speech recognition has become an integral aspect of our daily lives with the rise of automated voice assistants and speech-to-text transcriptions. Due to technological progress, voice recognition has developed into two distinct learning methods: shallow learning and deep learning. Both of these strategies hold significant relevance in modern society, and understanding their differences is vital in comprehending their impact on speech recognition systems. Superficial acquisition of knowledge sometimes referred to as traditional or rule-based learning, relies on preestablished rules and statistical models to comprehend and interpret speech signals. This method involves dissecting the speech signal into distinct elements, such as phonemes, and comparing them to language models in order to determine the most likely words or phrases being uttered. This method has been in use for several decades and has been employed in various applications, including speech-to-text transcriptions, voice-activated control systems, and virtual assistants. One of the primary benefits of shallow learning in speech recognition is its ability to process and comprehend straightforward and standardised speech patterns rapidly. This renders it a rational aspiration for applications in which the language is very stable and foreseeable, such as in voice commands for smart home devices or call centre customer service. In addition, shallow learning models are highly trainable and implementable, making them cost-effective and accessible for small and medium-scale businesses. Deep learning, also known as machine learning, operates by supplying vast amounts of data to artificially intelligent systems, enabling them to acquire information autonomously and better their performance without explicit programming. This approach has demonstrated remarkable success in various domains, including speech recognition. Deep learning models employ artificial neural networks to process voice signals, which possess a structure resembling that of the human brain. The models are developed using extensive datasets and possess the ability to identify intricate patterns and differences in human speech accurately.

One of the greatest advantages of deep learning in speech recognition is its ability to adapt and enhance itself over time. As further information is inputted into the model, it can continuously learn and update its understanding and analysis of speech cues. This has led to significant advancements in speech recognition technology, resulting in a more enhanced and precise comprehension of spoken words. Deep learning models have been proven to be effective in handling loud and intricate speech patterns, which can pose challenges for shallow learning models. Both shallow learning and deep learning are crucial in voice recognition systems in modern society. Shallow learning is commonly employed for rudimentary tasks and uncomplicated language structures, but deep learning is better suited for intricate and ever-changing language patterns. For instance, shallow learning is applicable in digital assistants to handle basic inquiries or perform basic tasks. In contrast, deep learning is employed in voice-activated search engines such as Google or customer service chatbots that aim to comprehend and address various and constantly evolving speech patterns. The importance of the prevalence of speech in contemporary culture cannot be exaggerated. It has revolutionised the way we engage with technology and has enhanced the efficiency and convenience of our daily tasks. Both superficial learning and profound understanding methods possess distinct advantages and limitations in speech recognition systems, and comprehending their disparities is crucial in developing and implementing efficient and precise voice recognition systems for the future.

2 Shallow Learning

Shallow learning is a form of artificial intelligence that utilises a single layer of processing units to analyse and derive insights from data. The term "shallow" is used to describe a limited number of layers and processing devices in comparison to deep learning, which utilises several layers to extract features and patterns from data. In the field of speech recognition applications, shallow learning has been employed for an extended period and has demonstrated its efficacy in identifying basic patterns and features from speech data. These structures employ a single layer of processing units, each of which is trained to identify a specific aspect of speech, such as phonemes, words, or sentences [5]. The output of these processing units is subsequently combined to get the ultimate recognition outcome. One of the key benefits of shallow mastery in voice recognition is its straightforwardness. The network architecture is remarkably straightforward, facilitating ease of instruction and configuration. Furthermore, it requires less processing power in comparison to deep learning, rendering it more feasible for applications lacking access to high-performance hardware [6]. Shallow learning is also more easily understandable when compared to deep learning. This implies that it is far simpler to identify and provide a rationale

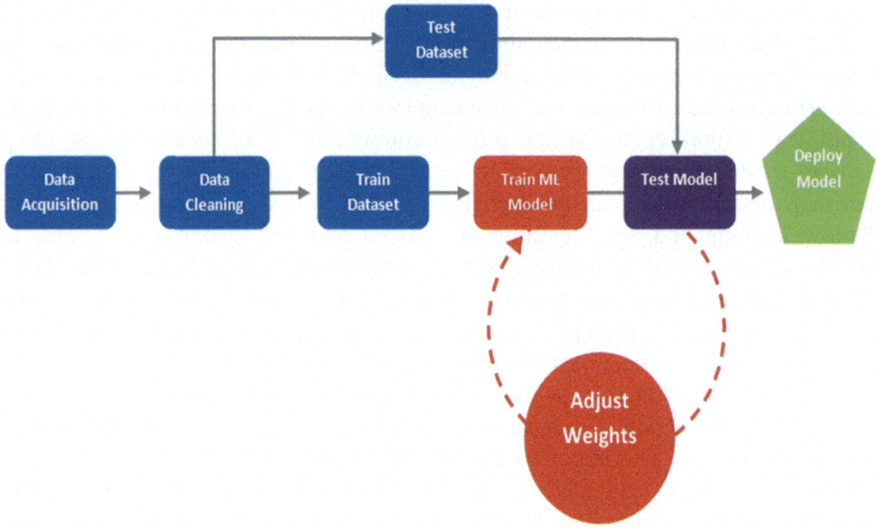

Fig. 1 Outline of shallow learning

for the decision-making process of the gadget. Consequently, shallow learning has been extensively employed in domains such as finance, healthcare, and customer service, where interpretability is crucial. Nevertheless, shallow learning possesses several drawbacks that render it less suitable for certain voice recognition tasks [7]. For example, it can effectively handle complex patterns and features included in noisy or fluctuating speech data. Figure 1 shows the outline of shallow learning.

This is because a single layer of processing devices may need more capacity to extract and learn these patterns efficiently [8].

$$\frac{de}{df} = \frac{d}{df}(e^{e*} \sin Ef) \tag{1}$$

Let $E = e^e$ and $F = \sin Ef$; then, we have the following,

$$G = E^* F \tag{2}$$

$$\frac{df}{de} = \left(E^* \frac{dF}{de}\right) + \left(F^* \frac{dE}{de}\right) \tag{3}$$

Deep learning shines in this aspect since it utilises numerous layers of processing devices to analyse intricate patterns and features in data. Deep learning has been demonstrated to be more effective than shallow learning in voice recognition tasks, especially in challenging conditions such as loud surroundings or identifying distinct accents [9].

2.1 *Techniques Used in Shallow Learning for Speech Recognition*

Shallow learning approaches have been widely employed in speech recognition systems, involving the extraction of features and the utilisation of statistical models for the classification and identification of speech patterns. This technology, also known as machine learning, has been proven to be successful and efficient in various tasks, including speech recognition [10]. However, because of the rise of deep learning, there has been a transition towards employing more intricate techniques that incorporate several layers of neural networks to extract features and categorise voice data. Despite the achievement of deep gaining knowledge of in reaching higher accuracy prices in speech popularity, shallow studying strategies nonetheless hold their floor owing to their simplicity and speed, rendering them perfect for actual-time programs.

$$\frac{df}{de} = \left(e^{e*}\frac{d}{de}\sin Ef\right) + \left(\sin Ef^{*}\frac{d}{de}(e^{e})\right) \tag{4}$$

$$\frac{df}{de} = (E^{*}e^{e}\cos Ef) + (e^{e}\sin Ef) \tag{5}$$

In addition, shallow learning techniques have the benefit of being easier to read and debug, making them more transparent compared to the opaque nature of deep learning models [11]. The choice between shallow and deep learning for voice recognition systems ultimately depends on the specific requirements and limitations of the given task.

2.2 *Advantages and Disadvantages of Shallow Learning for Speech Recognition*

Speech recognition, the ability of a computer to listen and comprehend human speech, has significantly advanced in recent years through the utilisation of several machine learning methodologies. Shallow learning refers to a collection of techniques where a model is trained on a limited number of predetermined functions in order to comprehend speech patterns. One of the primary advantages of superficial examination in the context of speech popularity is its straightforwardness and efficiency in execution. It is significantly easier to understand and educate, rendering it accessible even to individuals without advanced technical proficiency. However, shallow learning has its challenges, as it mainly focuses on surface-level patterns and can miss out on more complex skills that are vital in effectively interpreting human speech. This is where deep learning, a more advanced technique, comes into play. Deep learning can effectively capture intricate patterns, leading to improved accuracy in speech recognition [12]. However, this comes at the expense of an extended duration of education and the need for additional information. The choice between shallow and deep learning in voice recognition programmes ultimately relies on the specific requirements and resources of the task at hand.

3 Deep Learning

Deep learning, a subfield of machine learning, has become a potent tool in various applications, including speech recognition. Deep learning algorithms, unlike traditional shallow learning methods, can analyse and extract intricate and abstract aspects from vast quantities of data. Figure 2 shows the outline of deep learning.

This makes them well-suited for very intricate tasks such as speech recognition. This is particularly beneficial for packages that require natural language processing, where there is a high degree of diversity in spoken language [13].

$$G = e(f) = g^e \tag{6}$$

$$f'' = \lim_{e \to 0} \left(\frac{g(e+f) - g(e)}{f} \right) \tag{7}$$

$$f'' = \lim_{e \to 0} \left(\frac{g^{e+f} - g^e}{f} \right) \tag{8}$$

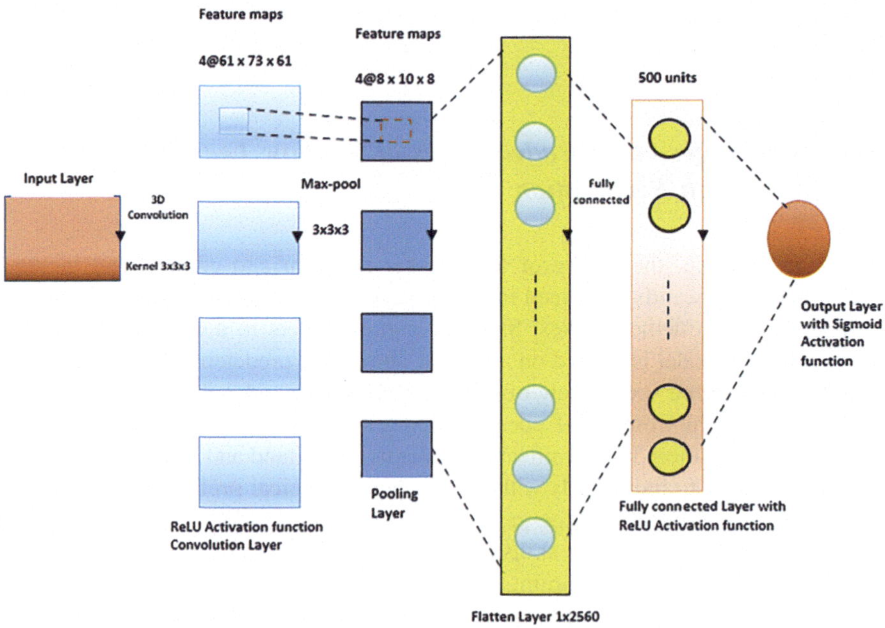

Fig. 2 Outline of deep learning

Deep learning models, such as deep neural networks, have demonstrated superior performance compared to traditional shallow learning methods in speech recognition tasks. They can encompass several layers of representations, enabling them to examine hierarchical functions and catch more intricate patterns in voice information. Consequently, profound comprehension has transformed the field of speech recognition, enabling more precise and efficient systems for numerous applications such as virtual assistants, speech-to-text technologies, and voice-controlled devices [14].

3.1 Techniques Used in Deep Learning for Speech Recognition

The field of speech recognition has been revolutionised by deep learning. Deep learning techniques, in contrast to shallow mastering, employ neural networks to analyse and extract relevant features from raw voice data automatically rather than rely on manually crafted features and rule-based algorithms. This enables a more accurate and robust representation of speech, allowing deep learning models to reach higher precision and handle various accents, background noise, and other challenging circumstances. In addition, advanced learning algorithms can adjust and improve using more training data, hence increasing their scalability for large datasets [15].

$$f'' = \lim_{e \to 0} \left(\frac{\left(g^{e*} g^{f}\right) - g^{e}}{f} \right) \tag{9}$$

$$f'' = \lim_{e \to 0} \left(\frac{g^{e*} \left(g^{f} - 1\right)}{f} \right) \tag{10}$$

Common techniques employed in deep learning for speech recognition include convolutional neural networks, recurrent neural networks, and deep neural networks. These techniques have significantly enhanced the overall efficiency of voice recognition systems, rendering them suitable for a diverse range of practical scenarios, including virtual assistants and speech-to-text transcription services.

3.2 Advantages and Disadvantages of Deep Learning for Speech Recognition

Deep learning has demonstrated promising outcomes in various speech recognition applications, owing to its ability to analyse intricate patterns and functionalities from data. One of the primary advantages of utilising deep learning for speech recognition is its superior precision in comparison to shallow learning techniques. Deep learning enhances the accuracy of speech recognition systems by automatically extracting advanced features from extensive datasets. This allows for more precise identification of intricate speech patterns, rendering it suited for challenging tasks such as speaker diarisation and voice recognition. However, in-depth studying techniques necessitate substantial quantities of educational data and might be computationally costly, hence presenting difficulties for certain applications. Conversely, shallow studying procedures necessitate a smaller amount of information and impose significantly less computing burden, rendering them particularly suitable for real-time or low-resource environments [16]. When choosing between shallow learning and deep learning algorithms for voice recognition applications, it is important to analyse the trade-off between accuracy and resource intensity carefully.

4 Challenges in Shallow and Deep Learning

One of the key obstacles in shallow learning for speech recognition systems is the limited capacity to handle intricate and diverse speech patterns. Speech recognition commonly relied on shallow learning procedures such as Hidden Markov models and Gaussian mixture models. However, these methods need help in capturing the complex relationship between speech features and phonemes due to their inability to handle non-linear patterns. Consequently, their performance may be impeded when dealing with speech that is either boisterous or accented. However, deep learning algorithms can overcome these challenging conditions by capturing higher-level representations of voice data over multiple layers [17]. Figure 3 shows the comparison of shallow learning vs deep learning.

However, this has its own set of challenges, including the requirement for a substantial amount of data to train the deep neural networks, as well as the difficulty in evaluating the acquired skills [18]. In addition, deep learning models require more processing resources, which makes them less suitable for real-time applications.

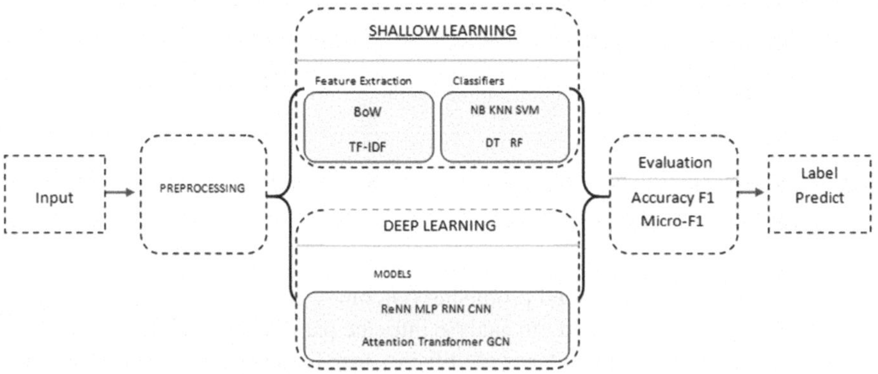

Fig. 3 comparison of shallow learning vs deep learning

Therefore, finding a compromise between the simplicity and efficiency of shallow learning and the power and intricacy of deep learning remains a difficulty in the field of speech recognition.

4.1 Limitations and Issues in Shallow Learning

Several limitations and issues plague shallow learning in speech recognition programmes. An important challenge is the limited ability to handle variations in speech styles and dialects. Superficial learning models heavily rely on manually designed features, which are typically insufficient to capture the diverse range of linguistic and auditory variations found in genuine speech [19]. Furthermore, superficial learning models have limitations in their ability to manage intricate and interwoven linguistic relationships, leading to erroneous recognition outcomes [20]. Figure 4 shows the performance analysis of shallow learning and deep learning.

In addition, superficial learning algorithms necessitate a substantial amount of training data to achieve accurate performance, rendering them less suitable for tasks that demand data with specific characteristics. Conversely, employing in-depth learning techniques has demonstrated more favourable outcomes in overcoming these limitations and achieving superior performance in speech recognition tests [21].

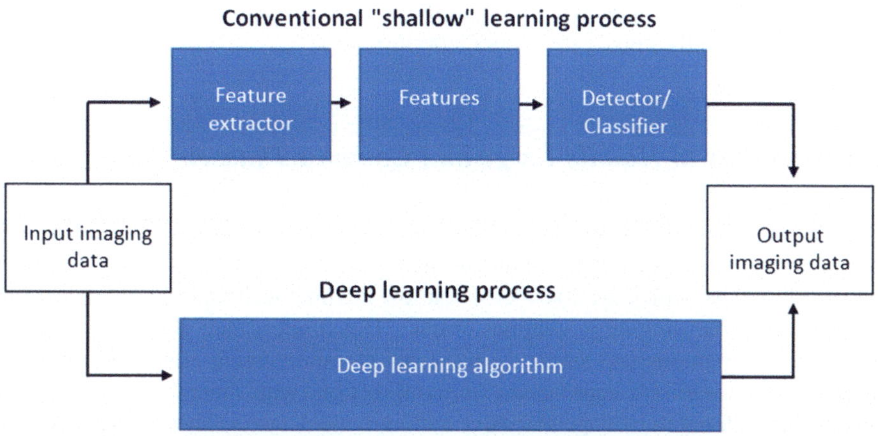

Fig. 4 Performance analysis of shallow learning and deep learning

4.2 Limitations and Issues in Deep Learning

One of the primary challenges in deep learning for voice recognition systems is its escalating intricacy in contrast to shallow learning methods. Although shallow learning systems have proven effective in handling simpler tasks, such as basic voice recognition, they need help in dealing with more intricate speech patterns and variations, including diverse accents and dialects. On the other hand, advanced machine learning models have demonstrated encouraging outcomes in addressing such challenging scenarios. However, their implementation and training are more complex due to the need for extensive data and processing power [22]. Furthermore, comprehending and enhancing the overall performance of deep learning models can be challenging because of the complexity involved in deciphering their choices and results. The challenging circumstances emphasise the trade-off between simplicity and performance when choosing between shallow and deep learning approaches for speech recognition applications.

4.3 Dependency on Hand-Crafted Features

Speech recognition programmes have seen a remarkable transition from traditional shallow learning techniques to advanced deep learning methods. One of the key differences between these strategies is the dependence on commercially available or homemade features. In shallow learning, a significant amount of human feature engineering is necessary to extract pertinent information from speech indications, whereas in deep learning, this process is mostly avoided. Alternatively, deep learning algorithms can extract meaningful representations straight from the raw data. This greatly diminishes the dependence on human expertise and enables a more automated and efficient approach to voice recognition. Nevertheless, function engineering is still crucial in deep learning, as the selection of appropriate input features can still significantly impact the model's performance.

4.4 Inability to Handle Large and Complex Datasets

One crucial distinction between shallow and deep learning in speech recognition applications is their capacity to handle extensive and intricate datasets. Conventional statistical models, which are considered shallow learning techniques, need help with handling and analysing large volumes of data. This hinders their ability to comprehend intricate patterns and subtleties in verbal communication effectively. Nevertheless, deep learning approaches, in conjunction with neural networks, are specifically tailored to manage extensive and intricate datasets. They possess exceptional proficiency in identifying and acquiring knowledge from intricate and diverse characteristics within the data, hence enhancing their accuracy and efficiency in tasks related to speech recognition. The limited effectiveness of superficial learning methods in handling extensive data and intricate patterns is a significant drawback, especially in the domain of speech recognition applications where the quality of outcomes heavily relies on the ability to analyse and comprehend substantial volumes of speech data.

4.5 Challenges in Deep Learning

One of the primary challenges in deep learning for voice recognition systems is its escalating intricacy in contrast to shallow learning methods. Although shallow learning systems have proven effective in handling simpler tasks, such as basic voice recognition, they need help in dealing with more intricate speech patterns and variations, including diverse accents and dialects. On the other hand, advanced machine learning models have demonstrated encouraging outcomes in addressing such challenging scenarios. However, their implementation and training are more complex due to the need for extensive data and processing power. Furthermore, comprehending and enhancing the overall performance of deep learning models can be challenging because of the complexity involved in deciphering their choices and results. The challenging circumstances emphasise the trade-off between simplicity and performance when choosing between shallow and deep learning approaches for speech recognition applications.

4.6 Need for a Large Amount of Training Data

Both shallow learning and deep learning algorithms have been extensively employed in speech popularity programmes. Nevertheless, deep learning approaches, which include the use of numerous layers of neural networks, have demonstrated superior performance when compared to shallow learning techniques. One motive for this is the desire to acquire a substantial amount of educational information. In order to study the intricate patterns and functions of spoken language, deep learning techniques necessitate an extensive and diverse dataset. Nevertheless, superficial learning techniques can perform adequately with a notably limited sample. This is because shallow learning algorithms rely on extracting specified functions from the input data, whereas deep learning algorithms can learn these functions directly from the data. Therefore, the necessity for a substantial dataset is crucial for the efficacy of deep learning models in voice recognition programmes.

4.7 Difficulty in Interpreting and Explaining Results

Deciphering and elucidating the impacts of shallow learning and deep learning in voice recognition might provide a challenging task. Superficial studying methods, which include traditional machine learning algorithms, are more easily comprehensible as they rely on explicit rules and feature engineering. Alternatively, deep learning techniques, such as neural networks, are more intricate and rely on abstract representations for making predictions. This is a challenge in providing a comprehensive explanation for the internal mechanisms of these models and comprehending the rationale behind their choices. Moreover, the expanding complexity of layers and parameters in deep learning models might require more clarification on the identification of the specific features that influence the final output, thus rendering the interpretation of these models' results challenging. Furthermore, the overall efficacy of deep learning models is highly contingent upon the quality and quantity of data utilised for training, thus posing a challenge in deriving generalisable findings. The increasing prevalence of deep learning in the field of speech recognition poses a significant challenge in deciphering and elucidating the results.

5 Conclusion

The continuous debate between shallow learning and deep learning in voice recognition applications has persisted for a considerable duration. While shallow learning, also known as standard machine learning, mainly depends on manually designed functions and approaches, deep learning utilises sophisticated algorithms and neural networks to learn from data automatically. Both strategies possess their own merits and drawbacks, yet it is evident that deep learning has emerged as the superior method in speech recognition tasks. The device's ability to process large volumes of data and its potential for continuous development make it a valuable tool for enhancing voice recognition technology. However, it is crucial to engage in research and development in both superficial and profound learning in order to ensure continuous advancement in this field. Shallow learning and deep learning are widely employed techniques in speech recognition software. Although both technologies aim to identify and capture human speech, there are significant differences between them. Shallow mastery, also known as traditional machine learning, relies on a predetermined collection of features to process and analyse speech data. On the other hand, deep learning employs a hierarchical structure and sophisticated algorithms to automatically extract features from raw data, enabling more precise and efficient processing. Furthermore, deep learning enables increased scalability and adaptability to novel and dynamic contexts. Nevertheless, superficial mastery still offers certain benefits, such as being easier to apply and demanding less educational information. The decision between shallow and deep learning in speech recognition applications ultimately depends on the specific requirements and limitations of the task. The field of speech recognition has witnessed significant progress in recent years, thanks to the breakthroughs in deep learning techniques. This has allowed speech recognition systems to achieve a greater level of precision and effectiveness in processing human speech. Unlike shallow learning, which depends on predetermined norms and functions, deep learning algorithms utilise neural networks that emulate the cognitive process of the human brain in processing data. This has ushered in a whole new arena of possibilities in voice recognition, as systems are now able to adapt and learn from vast amounts of data. The future trajectory of speech recognition is projected to focus on expanding the precision and speed of recognition, as well as strengthening natural language processing capabilities. Due to the ongoing advancements in deep learning, we may anticipate the development of increasingly advanced and intuitive speech recognition systems in the future.

References

1. Ramos, A., Carrasco, A., Fontanet, J., Herranz, L. E., Ramos, D., Díaz, M., et al. (2024). Artificial intelligence and machine learning applications in the Spanish nuclear field. *Nuclear Engineering and Design, 417*, 112842.
2. Wang, X., Mi, J., Li, B., Zhao, Y., & Meng, J. (2024). CATNet: Cross-modal fusion for audio-visual speech recognition. *Pattern Recognition Letters, 178*, 216–222.
3. Bhatti, U. A., Mengxing, H., Li, J., Bazai, S. U., & Aamir, M. (Eds.). (2024). *Deep learning for multimedia processing applications: Volume two: Signal processing and pattern recognition*. CRC Press.
4. Suganeshwari, G., & Divya, D. (2024). Deep learning in big data: Challenges and perspectives. *Big Data Computing*, 132–144.
5. Karthikeyan, V. (2024). Modified layer deep convolution neural network for text-independent speaker recognition. *Journal of Experimental & Theoretical Artificial Intelligence, 36*(2), 273–285.
6. Shen, Z., Chen, D., Zhao, L., & Wei, Y. (2024). Toward the balance between computational cost and model performance for the void detection of concrete-filled steel tubular structure using one-dimensional Mel-frequency cepstral coefficients and ensemble machine learning. *Construction and Building Materials, 411*, 134366.
7. Kirthiga, R., & Elavenil, S. (2024). A survey on crack detection in concrete surface using image processing and machine learning. *Journal of Building Pathology and Rehabilitation, 9*(1), 15.
8. Lee, J. P., Jang, H., Jang, Y., Song, H., Lee, S., Lee, P. S., & Kim, J. (2024). Encoding of multi-modal emotional information via personalized skin-integrated wireless facial interface. *Nature Communications, 15*(1), 530.
9. Gong, X., Wu, Y., Li, J., Liu, S., Zhao, R., Chen, X., & Qian, Y. (2024). Advanced long-content speech recognition with factorized neural transducer. In *IEEE/ACM transactions on audio, speech, and language processing*.
10. Eren, Y., & Küçükdemiral, İ. (2024). A comprehensive review on deep learning approaches for short-term load forecasting. *Renewable and Sustainable Energy Reviews, 189*, 114031.
11. Daidone, M., Ferrantelli, S., & Tuttolomondo, A. (2024). Machine learning applications in stroke medicine: Advancements, challenges, and future prospectives. *Neural Regeneration Research, 19*(4), 769–773.
12. Xia, Y., Che, T., Meng, J., Hu, J., Qiao, G., Liu, W., et al. (2024). Detection of surface defects for maize seeds based on YOLOv5. *Journal of Stored Products Research, 105*, 102242.

13. Zhang, S., & Gao, Y. (2024). Hybrid multi-objective evolutionary model compression with convolutional neural networks. *Results in Engineering, 21*, 101751.
14. Hou, X., Lian, P., Zhao, J., Zai, Y., Zhu, W., & Wang, F. (2024). Identification of carbonate sedimentary facies from well logs with machine learning. *Petroleum Research, 9*, 165.
15. Qureshi, S. A., Hussain, L., Rafique, M., Sohail, H., Aman, H., Abbas, S. R., et al. (2024). EML-PSP: A novel ensemble machine learning-based physical security paradigm using cross-domain ultra-fused feature extraction with hybrid data augmentation scheme. *Expert Systems with Applications, 243*, 122863.
16. Ullah, U., & Garcia-Zapirain, B. (2024). Quantum machine learning revolution in healthcare: A systematic review of emerging perspectives and applications. *IEEE Access, 12*, 11423.
17. Anitha, G., Priyadarshini, R., Titus, A., Sahoo, S., Muppala, C., Ramkumar, G., et al. (2024). Deep learning for the encounter of inorganic nanomaterial for efficient photochemical hydrogen production. *International Journal of Hydrogen Energy, 52*, 664–673.
18. Shamshiri, A., Ryu, K. R., & Park, J. Y. (2024). Text mining and natural language processing in construction. *Automation in Construction, 158*, 105200.
19. Ranjith, R., & Chandrasekar, A. (2024). GTSO: Gradient tangent search optimization enabled voice transformer with speech intelligibility for aphasia. *Computer Speech & Language, 84*, 101568.
20. Wang, W., Zhou, H., Li, W., & Goi, E. (2024). Neuromorphic computing. In *Neuromorphic photonic devices and applications* (pp. 27–45). Elsevier.
21. Dwivedi, D., Babu, K. V. S. M., Yemula, P. K., Chakraborty, P., & Pal, M. (2024). Identification of surface defects on solar pv panels and wind turbine blades using attention based deep learning model. *Engineering Applications of Artificial Intelligence, 131*, 107836.
22. Zulqarnain, M., Sheikh, R., Hussain, S., Sajid, M., Abbas, S. N., Majid, M., & Ullah, U. (2024). Text classification using deep learning models: A comparative review. *Cloud Computing and Data Science*, 80–96.

Shallow Learning vs Deep Learning in Recommendation Systems

Enes Bakış and Emrullah Acar

1 Introduction

The wealth of information available on the Internet may be daunting for users, especially when seeking a specific item. Recommender systems (RS) offer practical software tools that propose items that could be advantageous to the user. These systems utilize past behavioral data and other relevant information to predict the most fitting products or items for users. By using that rating, a recommender system produces a rating model, and then by using that rating, the model recommendation has been made to the users for further items according to the user preferences [1]. Shallow and deep learning are frequently utilized in these recommendation systems, with an escalating number of systems employing these methods. Remote sensing applications are commonly recommended for this purpose in the present day.

Remote sensing, in simple terms, means getting data regarding an object devoid of coming into contact with the object itself. It consists of two aspects; the analysis of information to interpret the physical characteristics of the object and the technology for receiving data from the object via a remote device, both of which are closely interconnected [2].

Remote sensing images are used as the main information source in numerous areas such as disaster risk management, city planning, agriculture, and irrigation. Moreover, they have been indicated as a critical and effective tool in ensuring data. Therefore, remote sensing image processing is very important for such applications

E. Bakış
Department of Electrical and Electronics Engineering, Piri Reis University, Batman, Turkey
e-mail: ebakis@pirireis.edu.tr

E. Acar (✉)
Department of Electrical and Electronics Engineering, Batman University, Batman, Turkey
e-mail: emrullah.acar@batman.edu.tr

Ö. F. Ertuğrul et al. (eds.), *Shallow Learning vs. Deep Learning*, The Springer Series in Applied Machine Learning, https://doi.org/10.1007/978-3-031-69499-8_10

to get beneficial information from them, and its tasks consist of object detection, image classification, and image fusion change detection. [3]. For adequate detection and mapping of objects, information about their spatial extent, location, and evolution is needed. Therefore, it is possible to provide them with the help of high and very high-resolution remote sensing data [4]. Accurate and timely classification of high-resolution remote sensing images is essential for many application areas such as disaster management, agriculture, and urban planning. But, as spatial resolution improves in high-resolution remote sensing images, it becomes progressively hard in order to identify complex patterns in these images [5]. Some of the studies on object recognition and detection using remote sensing images in the literature are as follows.

Sayedain et al. (2020) evaluated the improvement of land cover type classification employing interorbital Sentinel-1 images, and as a result, the proposed mapping approach verified that utilizing data from both descending and ascending bipolar images could provide a more precise classification map than employing a single orbital image separately [6]. Orlikova et al. (2019) explored the potential of using Sentinel-1 SAR images for land cover mapping in the Moravian-Silesian region. They used images recorded in 2018 to classify farmland on a per-pixel and per-object basis and observed that significantly better results were obtained through object-oriented classification as a result They achieved the highest processing and user sensitivity for the arable land class in both areas of interest [7]. Yakup et al. (2020) proposed the suitability of Sentinel-1 interferometric consistency for vegetation and land cover mapping using different classification algorithms [8]. Abdikan et al. (2018) used SAR images to research the performance of the sensor backscatter image in Sentinel-1 synthetic clipping tracking and acquired multi-time C-band VV and VH polarized SAR images simultaneously with in situ measurements. They carried out the study in the Konya basin, in the Central Anatolia region of Turkey. They collected the plant height of the maize plant during the measurements and analyzed the relationship between backscatter values and plant height, concluding that the Sentinel-1 data provided useful information about plant growth [9]. Sandro Martinis et al. (2018) presented an approach in order to develop the flood classification results of the German Aerospace Center's automated Sentinel-1 Flood Service. They did this by calculating a sand exclusion layer from Sentinel-1-time series data. They tested the method in Somalia/Ethiopia and examined the impact of parameters (time series length and frequency interval) on the result. They have proven in this study that the approach can significantly improve the flood classification accuracy in this study area [10]. Himan Shahabi et al. (2020) proposed a new flood susceptibility mapping approach to estimate floods in the Haraz Basin, Iran. In this technique, they used new ensemble models on the basis of bagging as a meta-classifier and K-nearest neighbor (KNN) cubic, cosine, rough, and weighted base classifiers. After detecting flood-prone areas employing data from Sentinel-1, they chose 10 conditioning factors to estimate floods and evaluate their predictive power employing the relief quality assessment method. As a result, they confirmed that this model they presented outperformed other ensemble models [11]. Hassan Bazzi et al. (2020) presented an effective approach for mapping rice crops using Sentinel-1 SAR time series over the Camargue region, France. They mapped rice plots with two different approaches: decision tree and random forest (RF). As a result, high accuracy was achieved with both approaches [12]. Zhang et al. (2021) stated that Sentinel-1 helps us to understand temporal and spatial dynamics with object-based random methods in order to improve the data that SAR images convert into input in classification by using the backscattering and coherence variation properties of the time series [13]. Lu et al. (2018) pointed out that the multispectral instruments of Sentinel satellites facilitate access to satellite data for researchers to use comfortably and that these satellite transmitters enable them to be used in matters such as land overlap [14]. Brinkhoff et al. (2020) defined that further analysis can be made with remote sensing images using preprocessing methods [15]. Kaplan et al. (2018) stated that Sentinel-1 gives valuable data on subjects such as land cover, thanks to the average combinations of VV and VH pretreatments and the reflected values of radar images [16]. Ndikumana et al. (2018) obtained detailed information about agricultural products with the data provided from SAR, which covers a wide area with spatial resolutions of approximately 20 m at 6-day intervals [17]. Tewkesbury et al. (2015) provided a synoptic view of the field of optical remote sensing change detection in the case of multimodal data sources, for example, VHR images acquired by different sensors [18]. Mohammadi et al. (2018) found that the Sentinel-1 radar data was useful for mapping flood coverage. However, the task of choosing the best-performing shallow learning algorithm in terms of flood susceptibility modeling could be difficult due to the data complexity, they noted, often needing to be done with a trial-and-error approach [19]. Gu et al. (2017) have argued that Geographic Object-Oriented Image Analysis (GEOBIA) techniques are becoming increasingly popular in remote sensing and that GEOBIA represents a paradigm shift in remote sensing interpretation. In their study, they introduced an object-based semantic classification method for high-resolution satellite images using an ontology that aims to take full advantage of the ontology's advantages over GEOBIA [20]. Martins et al. (2020) stated that remotely sensed images are being used more and more for land cover mapping via Convolutional Neural Network (CNN). Therefore, they presented a new multi-scale CNN framework for large-scale land cover classification at 1 m resolution over 145,740 km^2 [21]. Acar (2020) employed Sentinel-1 SAR images with shallow learning approaches in order to detect unregistered electricity distribution transformers (EDTs) over agricultural lands. Finally, unregistered EDTs were successfully detected with high accuracy [22]. Prasad et al. (2019) proposed land-based aquaculture at the pond level for the whole coastal region of India employing big Sentinel-1 SAR time series data. Finally, they demonstrated the spatiotemporal transferability of the methodological structure and the great potential for global-scale application with high accuracy [23]. Bakış and Acar (2023) tried to detect mucilage field data with deep learning methods using the Sentinel-1 satellite. The accuracy rate they found is very high. This indicates the detectability of the mucilage [24]. Bağcı et al. (2023) conducted a study in Diyarbakır/Turkey on the determination of cotton and corn cultivation areas using a deep transformer coding approach and different time series and Sentinel-1 satellite images. They achieved a high accuracy rate with the study results [25]. Acar and Ozerdem (2022) have studied the Automatic Detection of Different Soil Types Through Various Shallow Learning Algorithms Using Radarsat-2 SAR Image Polarization Coefficients. They achieved high accuracy rates in the study [26].

In this study, statistical (mean and standard deviation) analyses of signals in different polarizations belonging to objects (plant, water, building, land) in the province of Batman in 2021, each consisting of 10 samples, were made. The VV and VH bands of the Sentinel-1 satellite images were obtained and the reflection values of the objects at different times were compared.

In Sect. 2, the image preprocessing steps are given in detail. In Sect. 3, statistical analyses related to different polarization parameters are given. Moreover, shallow vs deep learning models are analyzed and compared in this section. The summations of the proposed study are given in Sect. 4.

2 Materials and Method

2.1 The Experimental Area

In this study, Sentinel-1 data sets (VV-VH polarizations) belonging to five different times were employed, including GPS data of four different objects (building, land, plant, and water) within the borders of Batman City between 1 May and 1 June 2021. The experimental area is shown in Fig. 1.

Fig. 1 The location of the experimental area

2.2 Sentinel-1 Satellite Data Collection

In this study, Sentinel-1 satellite images with VV and VH polarizations belonging to five different dates (May 1 to June 1, 2021) were employed with the aid of Google Earth engine code editor platform (https://code.earthengine.google.com). The Sentinel-1 image covering the work area is shown in Fig. 2.

2.3 Object Data Collection

In this phase, the geographical coordinates of four different objects with a total of 40 samples were recorded with the help of a GPS device, and the recorded data were then imported to the Sentinel-1 satellite images as indicated in Fig. 3.

2.4 Feature Extraction

The Sentinel-1 mission ensures data from a C-band synthetic aperture radar (SAR) device at 5.405 GHz. This collection contains S1 ground range detected scenes, which are calibrated, ortho-corrected products. Each scene has three resolutions (40, 25, and 10 m), three instrument modes, and four band combinations (corresponding to scene polarization) [27]. Information about Sentinel-1 polarization bands is given in Table 1.

In this study, VV and VH polarization bands were employed for statistical analyses of the different objects.

Fig. 2 Work area on Sentinel-1 satellite image

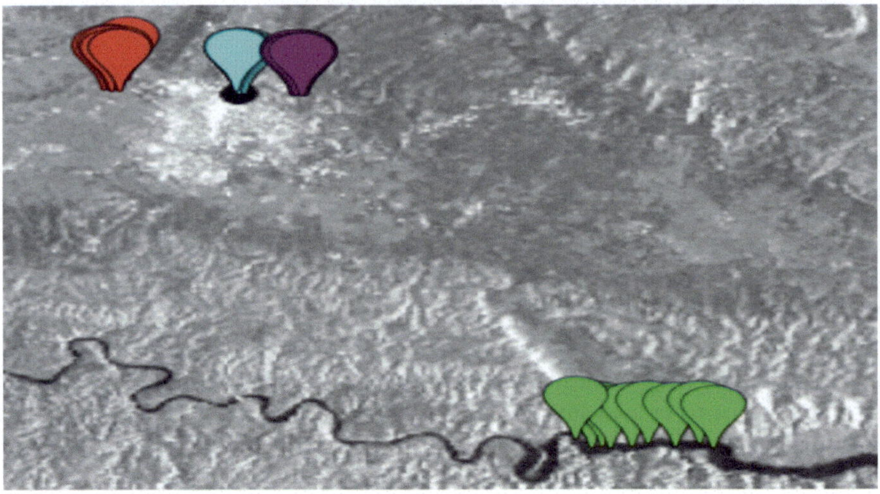

Fig. 3 Geolocation of different objects with 40 samples on Sentinel-1 image

Table 1 Sentinel-1 satellite polarization band parameters

Spectral bands	Pixel resolutions	Definition
VV	10 m	Single co-polarization (vertical transmit and vertical receive)
VH	10 m	Dual-band cross-polarization (vertical transmit and horizontal receive)
HV	10 m	Dual-band cross-polarization (horizontal transmit and vertical receive)
HH	10 m	Single co-polarization (horizontal transmit and horizontal receive)

2.5 *Applied Methods*

The models applied in this section are presented. CNN and RNN were used as deep learning models. Briefly speaking, CNN has a convolutional neural network structure. These neural networks try to reach the result by multiplying certain weight values and using the activation function at the output after passing through the created layers. It has a convolutional layer, a pooling layer, and fully connected layers. RNN is similar to CNN in terms of deep learning structure, that is, it has weight values, layers, and activation functions. But the biggest difference from CNN is that it has a memory that can go back and use old values. Thanks to this memory, it tries to reach the optimum result by using the old values.

Decision tree, random forest, support vector machine, K-nearest neighbor, and Naive Bayes models were used as shallow learning models. Some of these models used show a low accuracy rate, which means that the data set we are using does not work correctly with this model. However, in some methods, the accuracy rate is at the desired level. These shallow learning methods try to detect by classification.

2.6 *Statistical Metrics*

In this study, two statistical measurements (mean and standard deviation) were used to observe the relationship between signal parameters obtained from VV-VH polarizations of different objects. These statistical measurement equations are given below, respectively:

$$\text{Mean } (\mu) = \frac{1}{n}\sum_{i=1}^{n} X_i \quad (1)$$

$$\text{Standard deviation } (\sigma) = \sqrt{\frac{\Sigma(X-\mu)^2}{n-1}} \quad (2)$$

where Xi is the value in the data distribution and n is the total number of observations.

3 Results and Discussion

3.1 *Statistical Results*

In this section, Sentinel-1 images (VV and VH polarizations) of different objects (building, water, land, plant) on May 1–June 1, 2021, are presented as shown in Fig. 4 and the statistical analyses of the parameters obtained from these objects are given in Tables 2, 3, 4, 5, 6, 7, 8, and 9, respectively.

3.1.1 Statistical Analysis of Sentinel-1 VV Polarization Band Parameters

The parameters of the building, water, plant, and land objects belonging to the Sentinel-1 satellite VH band of May 1–June 1, 2021, obtained are given in Tables 2, 3, 4, and 5, respectively.

Considering Table 2, the mean (***μ***) values of the building samples in the VV polarization were computed as 0.707, −4.121, −0.453, −3.260, and −0.755 on May 5-11-17-23 and 29, respectively. Moreover, the standard deviation (***σ***) values of the building samples were calculated as 3.505, 2.261, 3.293, 2.504, and 3.421 on May 5-11-17-23 and 29, respectively.

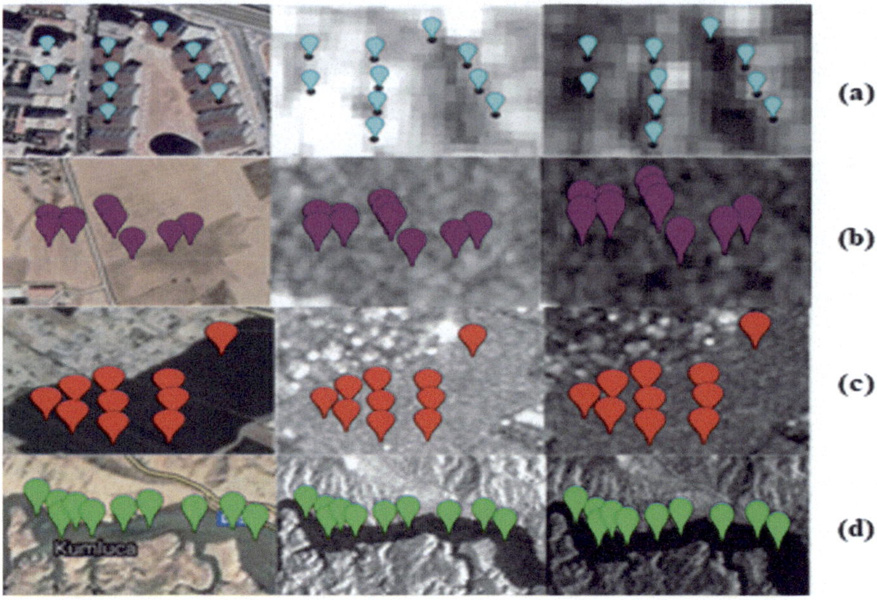

Fig. 4 Sentinel-1 images of different objects: (**a**) building areas with VV and VH polarizations (**b**) land areas with VV and VH polarizations, (**c**) plant areas with VV and VH polarizations, and (**d**) water areas with VV and VH polarizations

Table 2 Statistical analyses of the VV polarization band parameters obtained for the building samples

Building number	May 5	May 11	May 17	May 23	May 29
B0	2.37	−6.427	1.109	−5.318	0.969
B1	2.37	−6.427	1.109	−5.318	0.969
B2	2.37	−6.427	1.109	−7.853	−6.788
B3	2.37	−3.553	1.109	−0.942	0.969
B4	−5.944	−3.553	−6.701	−2.51	−6.788
B5	2.37	−3.553	1.109	−0.942	0.969
B6	−5.944	−3.553	−6.701	−2.51	−6.788
B7	2.37	−3.553	1.109	−0.942	0.969
B8	2.37	1.09	1.109	−0.942	0.969
B9	2.37	−5.257	1.109	−5.318	0.969
μ	0.707	−4.121	−0.453	−3.260	−0.755
σ	3.505	2.261	3.293	2.504	3.421

Considering Table 3, the mean (μ) values of the water samples in the VV polarization were calculated as −19.056, −20.908, −20.68, −22.362, and −20.715 on May 5-11-17-23 and 29, respectively. Moreover, the standard deviation (σ) values of the water samples were computed as 3.184, 2.92, 4.053, 2.178, and 3.589 on May 5-11-17-23 and 29, respectively.

Table 3 Statistical analyses of the VV polarization band parameters obtained for the water samples

Water number	May 5	May 11	May 17	May 23	May 29
W0	−12.731	−21.898	−13.347	−18.444	−13.804
W1	−20.681	−22.504	−23.764	−24.618	−24.146
W2	−20.64	−17.544	−23.792	−21.42	−22.613
W3	−21.853	−14.399	−22.903	−23.952	−21.076
W4	−19.588	−19.809	−21.279	−21.725	−21.337
W5	−20.057	−23.25	−23.071	−23.931	−22.191
W6	−13.692	−22.113	−13.207	−22.583	−14.59
W7	−20.502	−24.054	−23.399	−24.105	−20.993
W8	−21.519	−21.643	−21.466	−19.064	−23.433
W9	−19.299	−21.87	−20.573	−23.773	−22.971
μ	−19.056	−20.908	−20.68	−22.362	−20.715
σ	3.184	2.92	4.053	2.178	3.589

Table 4 Statistical analyses of the VV polarization band parameters obtained for the plant samples

Plant number	May 5	May 11	May 17	May 23	May 29
P0	−9.813	−11.595	−12.523	−10.199	−11.773
P1	−8.93	−10.914	−11.545	−11.878	−11.891
P2	−9.532	−11.646	−10.613	−11.725	−9.898
P3	−9.588	−12.016	−10.12	−11.348	−12.048
P4	−9.435	−12.016	−9.155	−11.583	−11.525
P5	−8.93	−10.922	−11.545	−11.264	−11.891
P6	−9.588	−12.144	−10.12	−11.348	−12.048
P7	−9.04	−12.016	−11.219	−11.348	−11.795
P8	−9.435	−11.842	−9.155	−10.595	−11.525
P9	−9.021	−10.818	−9.655	−11.403	−11.054
μ	−9.331	−11.593	−10.565	−11.269	−11.545
σ	0.321	0.518	1.127	0.507	0.651

Considering Table 4, the mean (μ) values of the plant samples in the VV polarization were computed as −9.331, −11.593, −10.565, −11.269, and −11.545 on May 5-11-17-23 and 29, respectively. Moreover, the standard deviation (σ) values of the plant samples were calculated as 0.321, 0.518, 1.127, 0.507, and 0.651 on May 5-11-17-23 and 29, respectively.

Considering Table 5, the mean (μ) values of the land samples in the VV polarization were calculated as −9.448, −14.689, −12.155, −12.671, and −12.671 on May 5-11-17-23 and 29, respectively. Moreover, the standard deviation (σ) values of the land samples were calculated as 3.357, 0.059, 5.049, 0.262, and 4.663 on May 5-11-17-23 and 29, respectively.

Table 5 Statistical analyses of the VV polarization band parameters obtained for the land samples

Land number	May 5	May 11	May 17	May 23	May 29
L0	−11.876	−14.735	−13.747	−14.872	−14.358
L1	−11.876	−14.735	−13.747	−14.872	−14.358
L2	−10.899	−14.62	−15.35	−14.872	−15.299
L3	−10.416	−14.62	−13.638	−15.499	−14.544
L4	−10.899	−14.735	−15.35	−14.872	−15.299
L5	−3.15	−14.735	−2.691	−14.901	−3.857
L6	−3.15	−14.735	−2.691	−14.901	−3.857
L7	−10.899	−14.735	−15.35	−14.872	−15.299
L8	−10.416	−14.62	−13.638	−15.499	−14.544
L9	−10.899	−14.62	−15.35	−14.872	−15.299
μ	−9.448	−14.689	−12.155	−15.003	−12.671
σ	3.357	0.059	5.049	0.262	4.663

Table 6 Statistical analyses of the VH polarization band parameters obtained for the building samples

Building number	May 5	May 11	May 17	May 23	May 29
B0	−14.942	−12.804	−14.318	−12.719	−15.218
B1	−14.942	−12.804	−14.318	−12.719	−15.218
B2	−14.942	−12.804	−14.318	−14.627	−13.539
B3	−14.942	−8.534	−14.318	−9.526	−15.218
B4	−13.123	−8.534	−13.22	−8.965	−13.539
B5	−14.942	−8.534	−14.318	−9.526	−15.218
B6	−13.123	−8.534	−13.22	−8.965	−13.539
B7	−14.942	−8.534	−14.318	−9.526	−15.218
B8	−14.942	−14.814	−14.318	−9.526	−15.218
B9	−14.942	−13.099	−14.318	−12.719	−15.218
μ	−14.578	−10.899	−14.098	−10.882	−14.714
σ	0.767	2.561	0.463	2.078	0.811

3.1.2 Statistical Analysis of Sentinel-1 VH Polarization Band Parameters

The parameters of the building, water, plant, and land objects belonging to the Sentinel-1 satellite VH band of May 1–June 1, 2021, obtained are given in Tables 6, 7, 8, and 9, respectively.

Considering Table 6, the mean (μ) values of the building samples in the VH polarization were calculated as −14.578, −10.899, −14.098, −10.882, and −14.714 on May 5-11-17-23 and 29, respectively. Moreover, the standard deviation (σ) values of the building samples were computed as 0.767, 2.561, 0.463, 2.078, and 0.811 on May 5-11-17-23 and 29, respectively.

From Table 7, the mean (μ) values of the water samples in the VH polarization were computed as −26.79, −27.636, −26.773, −28.031, and −26.742 on May

Table 7 Statistical analyses of the VH polarization band parameters obtained for the water samples

Water number	May 5	May 11	May 17	May 23	May 29
W0	−22.405	−28.399	−21.53	−26.419	−21.163
W1	−27.871	−28.772	−28.103	−28.803	−27.415
W2	−28.057	−26.445	−27.839	−26.604	−27.992
W3	−27.672	−21.569	−27.814	−28.06	−27.661
W4	−27.59	−27.92	−26.73	−27.372	−27.533
W5	−27.463	−28.665	−27.733	−28.919	−28.406
W6	−24.708	−29.009	−24.289	−28.817	−23.033
W7	−27.962	−28.748	−27.938	−28.335	−28.47
W8	−27.205	−28.988	−28.428	−28.157	−27.869
W9	−26.968	−27.845	−27.33	−28.827	−27.874
μ	−26,79	−27.636	−26.773	−28.031	−26.742
σ	1.819	2.266	2.185	0.932	2.509

Table 8 Statistical analyses of the VH polarization band parameters obtained for the plant samples

Plant number	May 5	May 11	May 17	May 23	May 29
P0	−16.844	−18.748	−19.321	−16.516	−17.042
P1	−16.453	−18.907	−18.818	−18.237	−17.097
P2	−16.973	−19.432	−17.281	−17.565	−15.544
P3	−17.415	−19.425	−16.841	−16.867	−17.073
P4	−17.124	−19.425	−16.247	−17.136	−16.695
P5	−16.453	−16.728	−18.818	−17.391	−17.097
P6	−17.415	−19.562	−16.841	−16.867	−17.073
P7	−16.905	−19.425	−17.672	−16.867	−17.387
P8	−17.124	−18.236	−16.247	−15.949	−16.695
P9	−16.767	−18.903	−15.516	−17.405	−16.948
μ	−16.947	−18.879	−17.36	−17.08	−16.865
σ	0.339	0.864	1.274	0.625	0.506

5-11-17-23 and 29, respectively. Moreover, the standard deviation (**σ**) values of the water samples were calculated as 1.819, 2.266, 2.185, 0.932, and 2.509 on May 5-11-17-23 and 29, respectively.

Evaluating Table 8, the mean (**μ**) values of the plant samples in the VH polarization were calculated as −6.947, −18.879, −17.36, −17.08, and −16,865 on May 5-11-17-23 and 29, respectively. Moreover, the standard deviation (**σ**) values of the plant samples were computed as 0.339, 0.864, 1.274, 0.625, and 0.506 on May 5-11-17-23 and 29, respectively.

Taken into account Table 9, the mean (**μ**) values of the land samples in the VH polarization were computed as −14,807, −22.764, −19.342, −24.469, and −20.597 on May 5-11-17-23 and 29, respectively. Moreover, the standard deviation (**σ**) values of the land samples were calculated as 4.465, 0.372, 6.951, 0.935, and 7.154 on May 5-11-17-23 and 29, respectively.

Table 9 Statistical analyses of the VH polarization band parameters obtained for the land samples

Land number	May 5	May 11	May 17	May 23	May 29
L0	−17.132	−22.476	−21.998	−24.57	−24.495
L1	−17.132	−22.476	−21.998	−24.57	−24.495
L2	−17.247	−23.196	−23.626	−24.57	−23.879
L3	−16.024	−23.196	−21.17	−25.705	−23.587
L4	−17.247	−22.476	−23.626	−24.57	−23.879
L5	−6.386	−22.476	−6.291	−22.929	−7.008
L6	−6.386	−22.476	−6.291	−22.929	−7.008
L7	−17.247	−22.476	−23.626	−24.57	−23.879
L8	−16.024	−23.196	−21.17	−25.705	−23.587
L9	−17.247	−23.196	−23.626	−24.57	−23.879
μ	−14.807	−22.764	−19.342	−24.469	−20.597
σ	4.465	0.372	6.951	0.935	7.154

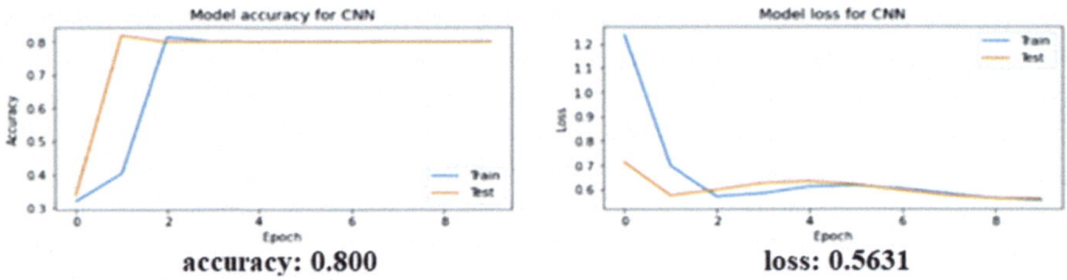

Fig. 5 Accuracy and loss graphics for CNN model

When the statistical results of the reflection values in the VV-VH polarization at different dates are considered, it is observed that the objects are in different spectral ranges. This will greatly facilitate the automatic detection of objects using shallow learning or deep learning techniques in the later stages.

3.2 Deep Learning and Shallow Learning Results

In this section, the results of the models applied to the data set are presented. Accuracy and loss graph and accuracy rate are given for the deep learning model. For shallow learning models, ROC curve, confusion matrix, and accuracy rates of each object are given. The results of the CNN model in Fig. 5 and the results of the RNN model in Fig. 6 are given, respectively.

When Figs. 5 and 6. are examined, the accuracy rate for both deep learning models is 80%. The results of the decision tree model in Fig. 7, the Naive Bayes model in Fig. 8, the random forest model in Fig. 9, the support vector machine model in Fig. 10, and the K-nearest neighbor model in Fig. 11 are given, respectively.

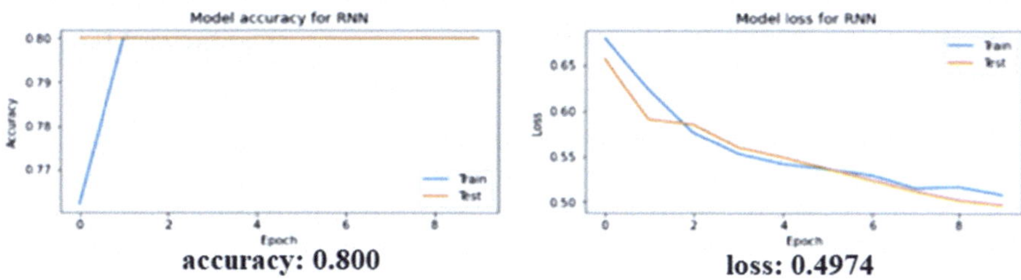

Fig. 6 Accuracy and loss graphics for the RNN model

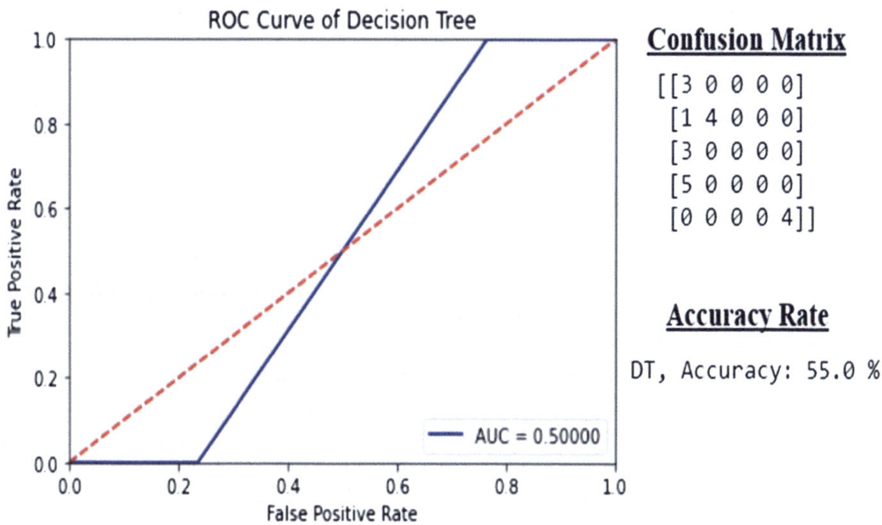

Fig. 7 The results of the decision tree model

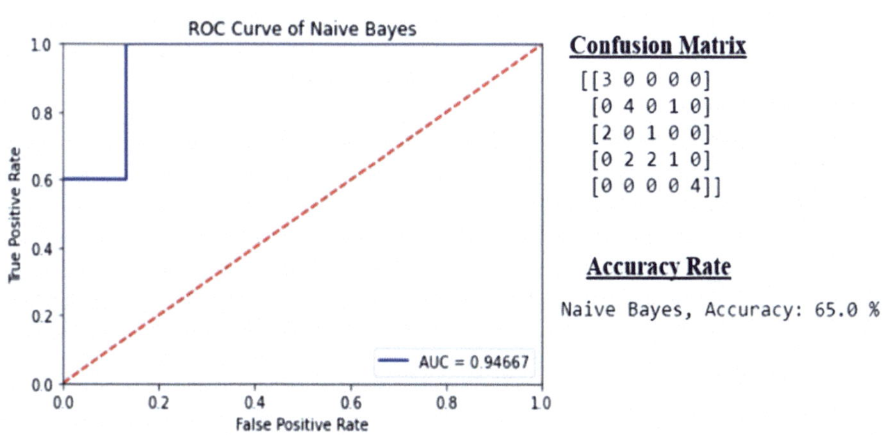

Fig. 8 The results of the Naive Bayes model

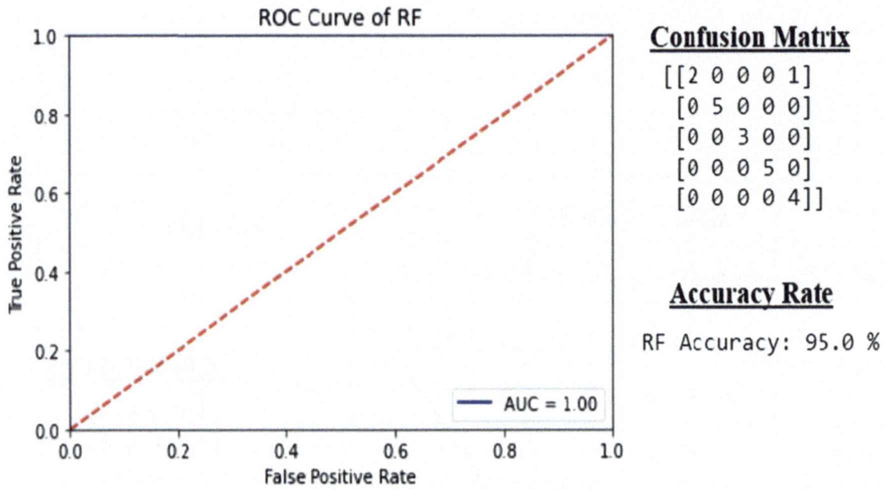

Fig. 9 The results of the RF model

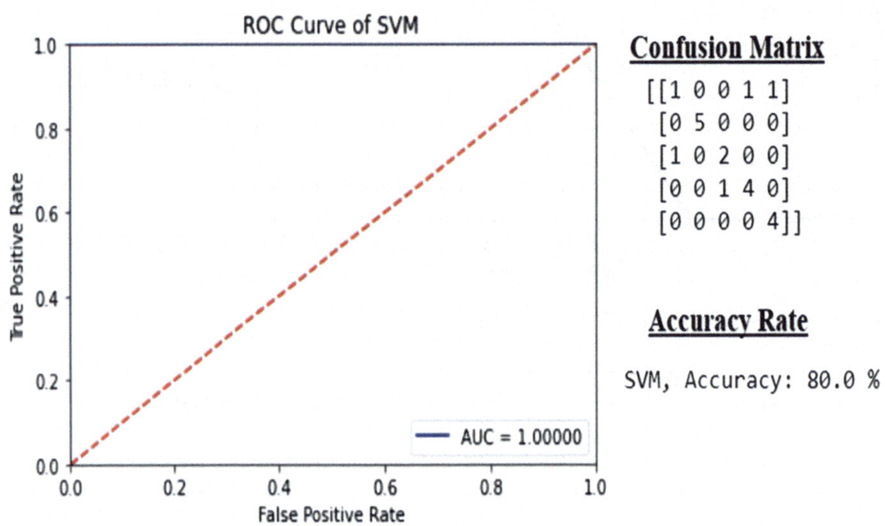

Fig. 10 The results of the SVM model

When the figures of shallow learning models are examined, the highest success rate was obtained in the RF model and KNN model by 95%. The RF model has multiple decision tree mechanism, but it has a very high success rate compared to the decision tree model.

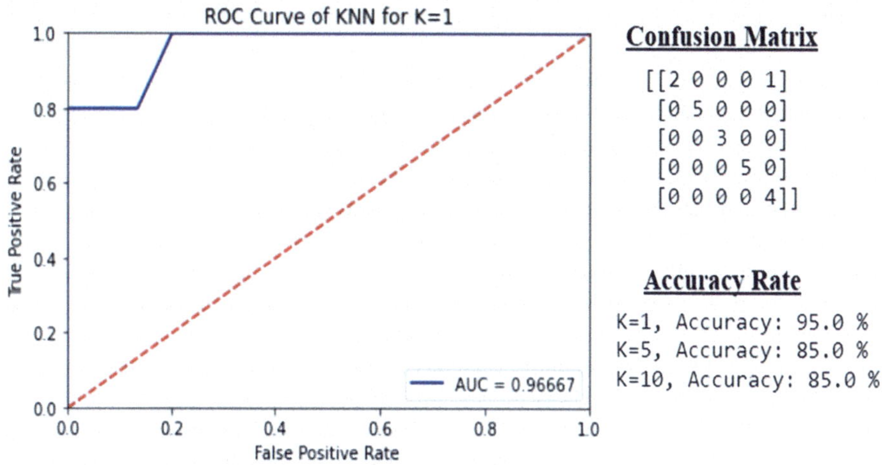

Fig. 11 The results of the Decision KNN model

3.3 Comparison of Deep Learning and Shallow Learning in This Application

When evaluating the results of the recommended system, both shallow and deep learning achieved good levels of success. However, it was observed that there were significant differences in performance between models in shallow learning, while the stability of the system varied depending on the model in deep learning. For instance, the accuracy rate of the decision tree model was 55% in shallow learning, while the accuracy rate of the Naive Bayes model was 65%, and the accuracy rate of the SVM was 80%. In shallow learning, SVM, and in deep learning, CNN and RNN methods have shown an 80% success rate. It can be stated that the most efficient systems recommended for this study are random forests and KNN.

The success levels of the shallow learning methods used in this recommended system vary due to the distinct techniques employed by each shallow learning model. Some methods are not applicable to all data sets. For instance, the decision trees model is most effective in nonlinear and large-sized data sets, achieving high accuracy. However, it can cause overlearning and be less effective when faced with low-dimensional data sets. The random forest (RF) method is a type of shallow learning model that utilizes multiple decision trees. Compared to the decision tree model, the RF model is less vulnerable to overfitting and can achieve high accuracy with low-dimensional data sets. Nevertheless, the Naive Bayes method suffers from a significant drawback in that it presumes all attributes to be independent, which is seldom the case in actual scenarios. As a result, predictions might be erroneous in some situations. This method is seldom employed, except for categorical data sets. The main advantage of this approach is its short application and training times. The KNN technique is a shallow learning model that performs poorly with large data sets, where the parameter K is an influential factor. Its primary benefits are its ability to make predictions without the need for training and its simplicity of use. The SVM model is a shallow learning model that excels with extensive data sets, but its efficiency is compromised with larger data sets. The major benefit of this method is its effectiveness with small data sets and its ability to resolve non-linear problems. Therefore, one can conclude that our data set is not compatible with the decision trees and Naive Bayes techniques because of its limited size.

The success levels of both deep learning methods employed in the recommended system are seen to be at the same level. However, it cannot be concluded that these models will show the same success level when applied to different data sets. The greatest advantage of deep learning systems is that they perform feature extraction by themselves. Therefore, the application of these systems requires less time compared to shallow learning, and their usage has become much more widespread today. One advantage of using GPUs is their ability to perform large-scale parallel computations and scale for high-volume data. The greatest disadvantage is that they require a large amount of data to achieve higher performance than other techniques. The key difference between CNN and RNN models is that RNN has a memory that remembers each input, while CNN is mostly used with images. When these data are examined, it can be seen that the deep learning methods applied in the proposed system did not show very high success due to the small amount of data in the data set. However, they have shown an acceptable level of success at 80%.

4 Conclusion

Recommended systems are imperative for accessing information in the most precise and competent manner. Nowadays, shallow learning and deep learning are prevalent in most recommended systems, which comprise social media, search engines, artificial intelligence tools, advertising, remote sensing, and object detection, among others. This research acquired data using remote sensing techniques from multiple objects with the employment of the Sentinel-1 satellite. The data set was evaluated using shallow and deep learning models, and comparisons were drawn based on their respective success rates.

The use of remote sensing techniques is extremely important in determining the polarization band parameters of various objects. In this study, different polarization band parameters of Sentinel-1 satellite data were determined respectively.

The proposed work consists of several stages. Firstly, the city of Batman in Turkey was chosen as the study area. In the second stage, after selecting the building, water, plant, and land objects in Batman City, ten samples were selected from each object. The positions of these samples were recorded by the GPS equipment and these GPS data were transferred to the Sentinel-1 satellite data. In the final step, a preliminary study was conducted on the detectability of these objects by applying deep learning and shallow learning models alongside image analysis to calculate polarization band parameters from Sentinel-1 data. In addition, the relations between the polarization band parameters obtained by using statistical measurements and the objects were established. When the success levels of deep learning and shallow learning models were examined, it was observed that the results showed an accuracy value above the average. In the next stages, these accuracy rates can be maximized by increasing the amount of data. It should be noted that 10 of each object were selected in this study.

It can be said that the system that works most efficiently with the data set used in this proposed study is the shallow learning models KNN and RF. However, it is not correct to say this for all shallow learning models. This is because the Naive Bayes and decision tree models have low accuracy rates. In contrast, it can be said that the success rate of all deep learning models applied to the data set in the proposed study is the same.

References

1. Ouhbi, B., Frikh, B., Zemmouri, E., & Abbad, A. (2018, October). Deep learning based recommender systems. In *2018 IEEE 5th international congress on information science and technology (cist)* (pp. 161–166). IEEE.
2. Gupta, R. P. (2017). *Remote sensing geology*. Springer.
3. Ghaffarian, S., Valente, J., Van Der Voort, M., & Tekinerdogan, B. (2021). Effect of attention mechanism in deep learning-based remote sensing image processing: A systematic literature review. *Remote Sensing, 13*(15), 2965.
4. Mahabir, R., Croitoru, A., Crooks, A. T., Agouris, P., & Stefanidis, A. (2018). A critical review of high and very high-resolution remote sensing approaches for detecting and mapping slums: Trends, challenges and emerging opportunities. *Urban Science, 2*(1), 8.
5. Zhao, W., Du, S., & Emery, W. J. (2017). Object-based convolutional neural network for high-resolution imagery classification. *IEEE Journal of Selected Topics in Applied Earth Observations and Remote Sensing, 10*(7), 3386–3396.
6. Sayedain, S. A., Maghsoudi, Y., & Eini-Zinab, S. (2020). Assessing the use of cross-orbit Sentinel-1 images in land cover classification. *International Journal of Remote Sensing, 41*(20), 7801–7819.
7. Orlikova, L., & Horak, J. (2019, Mayıs). Sentinel-1 SAR Verilerini Kullanarak Arazi Örtüsü Sınıflandırması. *Gelen Askeri Teknolojiler üzerine 2019 Uluslararası Konferansı (ICMT)* (s. 1–5). IEEE.
8. Jacob, A. W., Vicente-Guijalba, F., Lopez-Martinez, C., Lopez-Sanchez, J. M., Litzinger, M., Kristen, H., et al. (2020). Sentinel-1 InSAR coherence for land cover mapping: A comparison of multiple feature-based classifiers. *IEEE Journal of Selected Topics in Applied Earth Observations and Remote Sensing, 13*, 535–552.
9. Abdikan, S., Şekertekin, A., Üstünern, M., Şanlı, F. B., & Nasirzadehdizaji, R. (2018). Konya Havzası, Türkiye'de mısırın mahsul büyümesi için çok zamanlı Sentinel-1 SAR verileri kullanılarak geri saçılım analizi. *Int. Kemer Fotogramm. Uzaktan Algılama Tükürüğü. Enf. Bilim, 42*, 9–13.
10. Martinis, S., Plank, S., & Ćwik, K. (2018). The use of Sentinel-1 time-series data to improve flood monitoring in arid areas. *Remote Sensing, 10*(4), 583.
11. Shahabi, H., Shirzadi, A., Ghaderi, K., Omidvar, E., Al-Ansari, N., Clague, J. J., et al. (2020). Flood detection and susceptibility mapping using sentinel-1 remote sensing data and a shallow learning approach: Hybrid intelligence of bagging ensemble based on k-nearest neighbor classifier. *Remote Sensing, 12*(2), 266.

12. Bazzi, H., Baghdadi, N., El Hajj, M., Zribi, M., Minh, D. H. T., Ndikumana, E., et al. (2019). Mapping paddy rice using Sentinel-1 SAR time series in Camargue, France. *Remote Sensing, 11*(7), 887.
13. Zhang, X., Chan, N. W., Pan, B., Ge, X., & Yang, H. (2021). Mapping flood by the object-based method using backscattering coefficient and interference coherence of Sentinel-1 time series. *Science of the Total Environment, 794*, 148388. ISSN 0048-9697.
14. Lu, L., Tao, Y., & Di, L. (2018). Object-based plastic-mulched landcover extraction using integrated Sentinel-1 and Sentinel-2 data. *Remote Sensing, 10*(11), 1820.
15. Brinkhoff, J., Vardanega, J., & Robson, A. J. (2020). Land cover classification of nine perennial crops using Sentinel-1 and -2 data. *Remote Sensing, 12*(1), 96.
16. Kaplan, G., & Avdan, U. (2018). The international archives of the photogrammetry, remote sensing and spatial information sciences, Volume XLII-3. *2018 ISPRS TC III mid-term symposium "developments, technologies and applications in remote sensing"*, 7–10 May, Beijing, China.
17. Ndikumana, E., Ho Tong Minh, D., Baghdadi, N., Courault, D., & Hossard, L. (2018). Deep recurrent neural network for agricultural classification using multitemporal SAR Sentinel-1 for Camargue, France. *Remote Sensing, 10*(8), 1217.
18. Tewkesbury, A. P., Comber, A. J., Tate, N. J., Lamb, A., & Fisher, P. F. (2015). A critical synthesis of remotely sensed optical image change detection techniques. *Remote Sensing of Environment, 160*, 1–14.
19. Mohammadi, A., Shahabi, H., & Bin Ahmad, B. (2018). Integration of Insar technique, google earth images and extensive field survey for landslide inventory in a part of Cameron highlands, Pahang, Malaysia. *Applied Ecology and Environmental Research, 16*, 8075–8091.
20. Gu, H., Li, H., Yan, L., Liu, Z., Blaschke, T., & Soergel, U. (2017). An object-based semantic classification method for high resolution remote sensing imagery using ontology. *Remote Sensing, 9*(4), 329.
21. Martins, V. S., Kaleita, A. L., Gelder, B. K., da Silveira, H. L., & Abe, C. A. (2020). Exploring multiscale object-based convolutional neural network (multi-OCNN) for remote sensing image classification at high spatial resolution. *ISPRS Journal of Photogrammetry and Remote Sensing, 168*, 56–73.
22. Acar, E. (2020). Detection of unregistered electric distribution transformers in agricultural fields with the aid of Sentinel-1 SAR images by shallow learning approaches. *Computers and Electronics in Agriculture, 175*, 105559.
23. Prasad, K. A., et al. (2019). Assessment of coastal aquaculture for India from Sentinel-1 SAR time series. *Remote Sensing, 11*(3), 357.
24. Bakış, E., & Acar, E. (2023). *Aktif uzaktan algılama verileri kullanılarak LSTM (uzun kısa süreli bellek) yöntemi ile müsilaj bölgelerinin otomatik olarak tespit edilmesi*. AY, F. Z. Uluslararası Bilişim Kongresi Bildiriler Kitabı, 20 Mart 2023, Batman.
25. Bağcı, R. Ş., Acar, E., & Türk, Ö. (2023). Identification of cotton and corn plant areas by employing deep transformer encoder approach and different time series satellite images: A case study in Diyarbakir, Turkey. *Computers and Electronics in Agriculture, 209*, 107838.
26. Acar, E., & Ozerdem, M. S. (2022). Automatic determination of different soil types via several shallow learning algorithms employing Radarsat-2 SAR image polarization coefficients. In *Synthetic aperture radar (SAR) data applications* (pp. 219–233). Springer.
27. Conradsen, K., Nielsen, A. A., Schou, J., & Skriver, H. (2003, January). A test statistic in the complex Wishart distribution and its application to change detection in polarimetric SAR data. *IEEE Transactions on Geoscience and Remote Sensing, 41*(1), 4–19. https://doi.org/10.1109/TGRS.2002.808066

Advanced Techniques and Application Areas in Remote Sensing Images: Integration of Deep Learning and YOLOv5 Algorithms

Yasin Sonmez and Fatih Ozyurt

1 Introduction

Artificial learning applications developed today have evolved into systems that can successfully perform several complex tasks and are equipped with human-like abilities. For example, recognizing certain objects in photographs, driving autonomous vehicles in traffic, and voice recognition have become possible with advances in equipping artificial intelligence systems with human-like skills. In this context, within the scope of the study, the process of classifying objects will be discussed using the most up-to-date and advanced models in the field. The classification process will focus on accurately identifying and distinguishing various objects, including humans, vehicles, unspecified flying objects, and autonomous systems. This aims to achieve more reliable and precise results using the latest technological advances and models in object classification. The dataset used in the study was prepared using a drone, based on the conditions of camera angle between 60 and 90 degrees and height between 25 m and 100 m. Figure 1 gives examples of the object classes determined by.

Recently, the difficulties experienced in evaluating big data have been related to the difficulties that arise in creating special data sets for specific purposes. These challenges often involve large costs and complexities. Big Data refers to large and complex data sets, and the processing and analysis of such data can present a greater challenge, especially if a specialized dataset is created that must serve a specific

Y. Sonmez (✉)
Department of Computer Technologies, School of Technical Sciences, University of Batman, Batman, Turkey
e-mail: yasin.sonmez@batman.edu.tr

F. Ozyurt
Software Engineering, Faculty of Engineering, University of Fırat, Elâzığ, Turkey
e-mail: fatihozyurt@firat.edu.tr

Ö. F. Ertuğrul et al. (eds.), *Shallow Learning vs. Deep Learning*, The Springer Series in Applied Machine Learning, https://doi.org/10.1007/978-3-031-69499-8_11

Fig. 1 Various examples for object classes

purpose. In this context, the cost and difficulty of creating a purpose-built dataset may depend on various factors. Creating a custom dataset often requires access to original data sources. It can require significant effort and resources to extract data from those sources, clean it, and organize it into the appropriate format. It may also require expertise and attention to ensure reliable and quality data sets. These difficulties in making sense of big data encourage businesses, researchers, and other stakeholders to constantly seek solutions to access and make sense of this data more effectively. This process may require the adoption of various strategies, both technological and managerial. Some studies in the literature can be briefly discussed as follows.

Object recognition studies using deep learning techniques on drone images are divided into two categories. These categories are single-stage object detection algorithms and two-stage object detection algorithms. While YOLO, RetinaNet, EfficientNet, and SSD models come to the fore in single-stage object detection studies, RCNN, Fast R-CNN, and Faster R-CNN models provide the most successful and fastest results in two-stage object detection studies.

Sarıbaş, Çevikalp, and Kahveci conducted a study to compare the object location finding algorithm that uses a polyhedral conical classifier and the YOLOv2 algorithm, which gives the fastest and best results in the literature for the problem of detecting and locating cars in images taken from unmanned aerial vehicles. The study used the dataset they created, which included approximately 10,000 images obtained from unmanned aerial vehicles. The results obtained determined that the EPCC algorithm with eight root detectors gave the best average accuracy rate. The YOLO-Tiny algorithm gave the best result in terms of speed [1]. In his study in 2018, Toraman detected pedestrians from images taken from the drone by comparing AlexNet, which won the ImageNet competition of 2012, and VGG16 deep learning models, which came second in the same competition in 2014. In addition to the algorithm used for object recognition, the effect of transfer learning and data augmentation methods was also examined in the study. As a result of the study using the Stanford University drone dataset, approximately 99% pedestrian classification success rate was achieved in both models [2]. In their study, Wang et al. compared

Faster-RCNN and SSD architectures on the RetinaNet algorithm using the Stanford University drone dataset. As a result of the study, they found that the single-stage SSD architecture achieved 80.42%, and the two-stage Faster-RCNN architecture achieved 83.63% as average accuracy values. The study results concluded that single-stage architectures are close to two-stage architectures in terms of success [3]. Benjdira et al. compared the single-stage YOLOv3 and the two-stage Faster R-CNN model to realize the vehicle surveillance and traffic monitoring scenario. In their study using a dataset consisting of images taken from unmanned aerial vehicles, they discussed performance evaluation in terms of accuracy and processing time. According to the study results, both algorithms achieved object prediction accuracy of at least 99% and above in the data set. However, it was concluded that the YOLOv3 algorithm is more robust and has higher recall than the Faster-RCNN model [4]. Nassim and his colleagues implemented four important steps in their study to compare car detection from images taken from unmanned aerial vehicles with other algorithms in the field. As a first step, the image was divided into different regions using the mean shift algorithm to determine whether a car was in the image. In the second step, the reserved region was given as input to the pretrained convolutional neural network feature extraction was performed. In the third step, classification was made using the SVM algorithm. In the last step, fine adjustments were made to the detected objects with morphological operations. It was observed that the results obtained in the applied study reached a higher accuracy rate compared to other successful algorithms [5]. Lan et al. have worked on detecting faults in electrical networks with the Faster R-CNN architecture, using the dataset they created from approximately 6000 images taken from unmanned aerial vehicles. In their study, they changed the Faster R-CNN architecture, experimented on VGG16 and ResNet101 networks, and compared the results. According to the results obtained, the average accuracy rate of the test values was 81.11% in the VGG16 network and 67.92% in the ResNet101 network. In other words, the Faster R-CNN architecture they modified successfully detected errors and malfunctions in electrical networks [6]. Xu et al., in their study investigating the effect of using a horizontal bounding box on multidirectional object detection, especially in images taken from unmanned aerial vehicles, and suggested a sliding hill, stated the disadvantages of using a horizontal bounding box for objects in different directions. According to the study, it is stated that using a specific horizontal bounding box for objects in different directions in aerial images will not provide accurate orientation and scale information due to angle in detecting object change and will create a low IoU value due to angle deviation, which will reduce the prediction accuracy. In the proposed method, in addition to features such as coordinate information, width, and height (x, y, w, h) in the horizontal bounding box, rotated box values determined by the skew coefficient were also used during the network training. As a result of the study, it was determined that using a rotated horizontal bounding box in the ResNet101 architecture, which is a single-stage object detection model, is both faster than other studies and produces better results with an average accuracy rate of 75.02% to 88.2% compared to the data sets used [7]. In real-time and cloud-based object recognition studies for unmanned aerial vehicles, Lee et al. have suggested that the unmanned aerial vehicle performs real-time object recognition from the images it receives, and the computational cost is lower. They emphasized that unmanned aerial vehicles have some disadvantages regarding real-time object recognition. These are as follows: The images taken by the unmanned aerial vehicle are noisy and blurry due to its mobility, have low resolution due to the built-in cameras, and the target objects are generally small. In addition, detecting even a single object in object recognition studies requires quite high processing power, and recent studies cover this computational cost by using graphics cards. Although object recognition of images taken from unmanned aerial vehicles by cloud services is seen as an advantage, possible communication delays may occur. Cloud computing systems provide on-demand access to virtually unlimited computing resources, which is useful for workloads that periodically require large computations. In the study, it was not preferred that all work be done on cloud platforms, but a lightweight object detector was run, and images and information likely to contain the object of interest were sent to the cloud platform. The drone camera used for real-time object detection has a resolution of 320 × 240 pixels. Object detection was made using only the Faster R-CNN algorithm in cloud services to reduce the processing cost. The algorithm was trained with the Pascal VOC2012 dataset to detect objects. According to the results obtained in the study, the Faster R-CNN architecture running on the cloud service reached an average accuracy rate of 83.9%. Among the SSD architectures used, SSD300 and SSD500 achieved an average accuracy rate of 81.6% and 82.6%. Among the other architectures used, the YOLO model and the Fast YOLO model reached an average accuracy rate of 78.3% and 79.4%, respectively. As a result, a more accurate and efficient architecture was proposed in the study, although the processing load was covered by cloud services [8]. The YOLO model, a high-performance single-shot detector, was used in the studies conducted by Ophoff et al. examining the task of vehicle and ship detection and the classification of different object detection networks from satellite images. The experimental study of vehicle and ship detection and classification in the dataset consisting of satellite images was carried out with ImageNet and DOTA pretrained weight data. Among the YOLO models (YOLOv2, YOLOv3, D-YOLO, and YOLT) compared according to the results obtained, the most suitable one was found in terms of both running time and accuracy. The model was determined to be DYOLO [9]. In Mandal et al.'s dataset and moving object detection studies created from video images taken from unmanned aerial vehicles, which they saw as missing in the field, they created a new dataset called MOR-UAV, consisting of 30 video images and a total of 10,948 frames for detecting moving objects. Using this data set, they achieved an average accuracy rate of 58.56% over at least a 50% IoU threshold in their experimental studies using the deep learning framework called MOR-UAVNet to localize and classify moving objects simultaneously [10]. Algancı et al., in their study on aircraft recognition from photographs taken from satellite images, applied the validation success of YOLOv3, SSD, and Faster R-CNN models on multi-perspective datasets such as Pascal VOC and COCO on low perspective and limited datasets. DOTA dataset was used as the dataset in the study. The hyperparameters used in the three convolutional neural network models were optimized, and the models were trained several times. At the end of the training, the models were tested with both DOTA and another satellite image dataset. According to the results, Faster R-CNN stands out compared to other models regarding average accuracy rate and YOLOv3 processing time. Although the SSD model showed the worst detection performance, it produced better results in object localization [11].

Hinzmann et al. have studied human detection with deep learning methods using optical and infrared images of unmanned aerial vehicles. The study aims to combine infrared and optical images in convolutional neural network architectures and monitor the results. The created dataset was run with different RetinaNet and YOLO models, and the results were compared. In addition, in comparisons, the focus loss optimization used by the RetinaNet model versus the cross-entropy loss used by the YOLOv3 model was considered. The results obtained determined that the RetinaNet model was more successful than the YOLOv3 model. It has been stated that the RetinaNet architecture is more successful because of the focus loss function it uses to cope with the large imbalance between foreground and background [12]. Carlet and Abayowa conducted a study comparing YOLOv2 and Faster R-CNN models using datasets consisting of VEDAI and similar aerial photographs. It was emphasized that the Faster RCNN model, which was their first choice in their study, has been used extensively since its inception in 2015 due to its speed and accuracy. As a result of the study using aerial photographs, it was concluded that the YOLOv2 model is approximately four times superior to the Faster R-CNN model in terms of speed and accuracy [13]. A character recognition application was developed using the YOLOv4 model to detect the license plates of vehicles in traffic in Bangladesh. Data augmentation and transfer learning were used during the model's training, and MobileNetv2 architecture was chosen as the backbone. A photo dataset was created using video footage obtained from Bangladesh traffic, and the model achieved an average accuracy rate of approximately 90% after 6000 epochs of training. The results confirmed that the model is fast and successful [14].

2 Analyzes Data Sets Enriched with TensorFlow Object Detection API

TensorFlow Object Detection API is a tool developed by Google and built on the TensorFlow library. This API is designed specifically for training and implementing models used in object detection. Object detection is a task that involves identifying and classifying objects in an image or video. Remote sensing is a method used to monitor and analyze changes over large areas, usually using satellite data or remote sensors. One of the key benefits of the TensorFlow Object Detection API is pretrained models: the API offers pretrained models trained on popular and large datasets. These models have achieved high success in general object detection tasks and have a wide range of object classification. Pretrained models can be quickly used in a variety of application areas. The other is customizability: At the same time, users can train these models on their own custom data sets. This is a great advantage for researchers and application developers who want to develop customized solutions to identify specific objects on remote sensing images.

Remote sensing is often used to analyze data covering large areas. Remote sensing images can often be high resolution and large. A powerful object detection tool is needed to manage and analyze such large data sets effectively. TensorFlow Object Detection API is an effective tool for object detection on remote sensing images because its powerful computing infrastructure, in terms of performance and speed, provides the ability to run pretrained models quickly and effectively. This is important to obtain fast and reliable results on large data sets. In addition, having customizable training, they can train models for specific object detection tasks on remote sensing images. This provides flexibility to develop solutions tailored to specific industry or application requirements.

In summary, TensorFlow Object Detection API is a powerful tool to quickly and effectively perform object detection tasks on remote sensing images. This API offers e-developers the flexibility to create customizable object detection solutions for a wide range of applications. In the remote sensing discipline, the use of such technologies represents an important step towards obtaining sharper and more meaningful results.

Within the scope of the study, 142 drone footage videos were extracted, adapted, and added to the team's virtual repository. ImageNet [15], COCO [16], Kaggle [17], VisDrone [18], Vedai [19], Dota [20], Bdd100k [21], Voc [22], Dlr [23], and Detrac [24] Data were collected from Uavdt [25], Arxiv [26], Pesmod [27], and Tugraz [28] websites (200,000 drone data). Additionally, 400 drone-shot videos were downloaded from video platforms, such as YouTube, iStockphoto, and Pexels, and 50,000 data were collected. A total of 40,000 images suitable for the format were obtained. In the extracted dataset are pictures of sunny, snowy, foggy, rainy, and night weather conditions. Synthetic data was produced for train, bus, and motorcycle objects missing in the images using the GANs [29] algorithm. Missing object classes were added to the original dataset by making drone flights at various altitudes. Drone flights were made in vehicular traffic and on the Firat University Campus. Due to UAP and UAI data insufficiency, drone flights were made in empty parking areas. Then, the dataset was diversified by pasting UAP and UAI images into the empty areas with Photoshop. Virtual drone videos were created in Davinci Resolve 17, and GTA V programs and the need for a dataset in UAP and UAI areas that do not exist in real life was met. As a result, a more efficient artificial intelligence model was trained with a balanced number of objects and data diversity (Fig. 2).

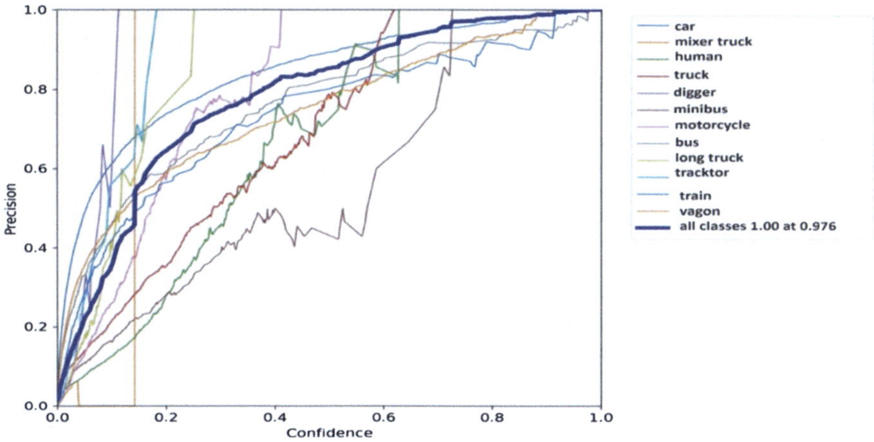

Fig. 2 Dataset scatter plot

2.1 Model Architectures

2.1.1 YOLOv5

YOLO (You Only Look Once) is an algorithm for real-time object detection. The YOLO algorithm can detect multiple objects simultaneously when analyzing an image and is, therefore, faster than other object detection methods. YOLOv5 refers to its fifth version that runs on PyTorch framework. YOLO uses forward propagation when analyzing an image. This allows the detection of box-shaped objects in the image. It is an algorithm fast enough to be used in real-time applications. It can work faster than other object detection methods. It considers the entire image during training and testing of the dataset. This helps implicitly encode contextual information about classes and views. YOLOv5 is the first YOLO model to run on the PyTorch framework. PyTorch is compatible with Python and can be easily integrated with various libraries. It is also an advantage that PyTorch offers a faster experience using graphics processing units (GPUs). In summary, the YOLO algorithm is considered an effective option for real-time object detection, and YOLOv5 is one of the latest and fastest versions of this algorithm running on PyTorch.

2.1.2 YOLOR

YOLOR (You Only Learn One Representation) is a model that can be trained to solve multiple problems simultaneously, usually while fulfilling a specific purpose. Convolutional Neural Networks (ESA) [30] are usually created with a single goal. While they learn how to analyze inputs to get outputs from ESAs, YOLOR tries to get ESAs to learn how to get outputs and what all the different outputs might be. Instead of a single output, it can have many outputs. People can answer different questions when given a single input. Given a single piece of data, people can analyze the data from different perspectives. For example, a photo of something, action shown, location, etc. It may cause different reactions regarding the YOLOR aims to give this capability to machine learning models. The fact that the YOLOR method significantly outperforms state-of-the-art methods in terms of detection efficiency while maintaining high accuracy caused it to be chosen as the priority model [31].

2.1.3 SAHI

The "SAHI" model may be a special model for detecting small objects. This model is hybrid integrated with pretrained base models. Hybrid integration includes a specially designed structure to achieve more successful results, especially on small objects, while preserving the basic models' features and general object detection capabilities. The SAHI model is capable of providing successful results on small objects. This feature generally makes it preferred in situations where other object detection models have difficulty with small objects. Since the SAHI model is specifically designed to detect small objects, it can usually detect such objects without the need to retrain the model. This allows users to apply the model more flexibly. The SAHI model does not need larger GPU memory allocation when working on small objects. This makes the model lighter and more usable on systems with lower hardware requirements. Observations have documented the successes and performance of the SAHI model on small objects. This can help users understand how effective the model is. This model type can be preferred, especially in industrial applications, medical fields, or similar fields that require detecting small objects.

2.1.4 YOLOv7

YOLOv7 provides a significant performance increase compared to previous versions. According to studies, this model, which can operate between 5 FPS and 160 FPS, leaves other object detection models behind in speed and accuracy. Capable of operating in a wide speed range between 5 FPS and 160 FPS, YOLOv7 can be used effectively in real-time applications. With an AP (average precision) value of 56.8%, YOLOv7 has high accuracy. The model's ability to successfully locate small and distant objects makes it have a wide range of applications. It is especially important in security systems or remote sensing applications.

3 Implementation Steps of the Model

The steps taken within the scope of the study are given in eight stages. The steps of the methods used are given in Fig. 3.

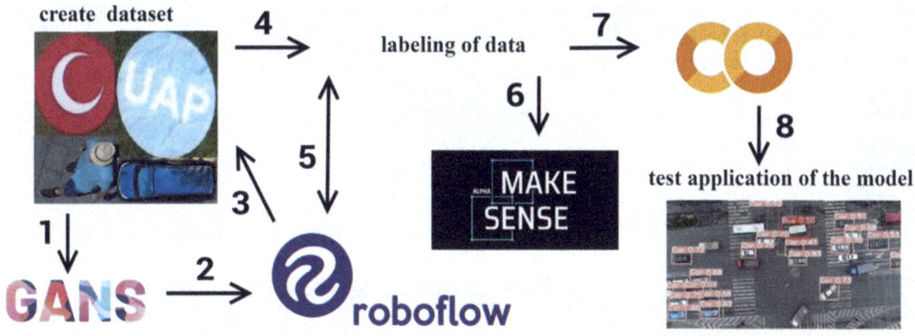

Fig. 3 Steps forming the beginning and ending of the artificial intelligence model

1. Data collection is essential when training artificial intelligence models via images. As a first stage, data was collected from many open source-based platforms.
2. Environments were created for drone shooting, and appropriate data was collected.
3. Since the data was not in the desired amount and variety, data was increased with Roboflow [32], one of the data augmentation methods.
4. When the data preparation process was completed, a controlled dataset was prepared by examining the data health on Roboflow.
5. The data tagged with Make Sense [33] was labeled as a wizard from Roboflow.
6. Label printouts in txt format were taken from Make Sense for artificial intelligence model training.
7. The training was done using Google Colab Pro from the resulting text format.
8. As a result, the testing phase of the trained artificial intelligence model has been completed.

4 Results of the Study and Evaluations

The algorithms intended to be used were trained using Google Colab Pro on labeled data. First, 100-step (epoch) YOLOv5 was trained, and then 100-step YOLOR, YOLOv5, YOLOv7, and SAHI models were trained, respectively (Fig. 4).

To evaluate the performance of the trained models, various IoU metrics were considered. The confusion matrix is another metric used to measure model success [34]. A complexity matrix is a table used to measure the performance of a classification model on a test dataset where the true values are known. True positive (TP), false positive (FP), and false negative (FN) cases were examined with the help of the confusion matrix. False positive cases are detections with an IoU ratio of 0.5 and marking an unlabeled object. A false negative situation is when a labeled object cannot be detected. TP, FP, and FN states in the complexity matrix are used to calculate metrics, such as accuracy, precision, recall, and F1 score for object classes.

Fig. 4 Example representation of trained models

The aforementioned metrics were calculated for the created classification model. Object detection accuracy was determined according to the Intersection Over Union (IoU) value. This ratio shows the match between the predicted area and the actual area of the object. mAP (mean precision) is a popular metric used to measure the accuracy of models used in object detection. It was generally decided to calculate the mAP metric according to the IoU threshold value (0.5). To calculate the mAP metric value, precision and recall values must be calculated. Precision is the metric value that measures how many of the predictions made by the classification model are correct.

Precision value, *Precision:* $\frac{TP}{TP+FP}$ It was calculated with the formula [35]. Recall (sensitivity) is a metric that shows how many situations that should be predicted as positive are predicted as positive.

Recall value, *Recall:* $\frac{TP}{TP+FN}$. It was calculated with the formula [36]. AP (average precision) is calculated by finding the area under the precision and sensitivity curve. The mAP metric value is the average of the AP value.

The performance results of your model on data sets suitable for your study in the literature and the comparison of these results with similar studies in the literature are included here (Figs. 5 and 6).

In the algorithm written for detecting UAP and UAI areas, first, the UAP and UAI areas were determined, and then their coordinates were kept. Half of UAPs and UAIs are labeled "y-uap" and "y-uai," so they are unsuitable for landing. If different objects are detected above the fields, "not suitable" information for landing or parking is returned.

This study proposes an effective detection algorithm based on YOLOv5 to detect objects captured by remote sensing. When the study is examined, the YOLOR model performs less than other models. The YOLOv5 model performs better than YOLOR

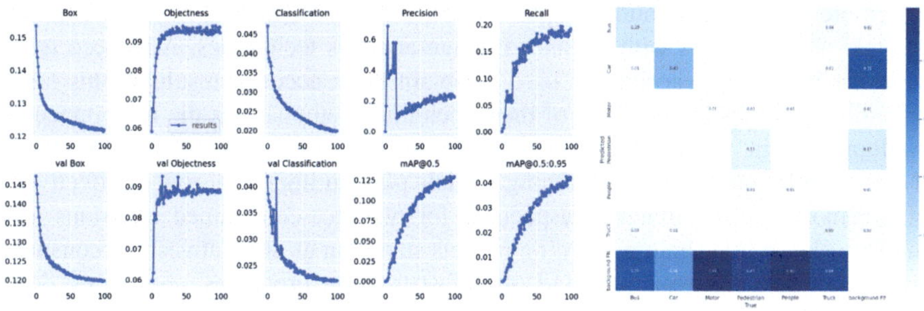

Fig. 5 YOLOV5 model trial training result graphs

Train Model	Accuracy (%)	Precision	Recall(%)	F1-Score (%)	Processing Time(ms)
YOLOR	0,57	0,42	0,38	0,49	82
YOLOv5	0,63	0,58	0,54	0,52	220
YOLOv7	0,69	0,67	0,64	0,8	430
YOLOv5+SAHI	0,79	0,88	0,75	0,76	1500

Fig. 6 Comparison of YOLOv5 models according to their results

but was integrated with the SAHI algorithm because it was thought to still have room for improvement. Increasing processing time indicates that it is a more complex model and requires longer processes. Although YOLOv7 performs better than previous models, the YOLOv5+SAHI model has the highest accuracy and precision. However, it has a lower recall value and a much longer processing time than other models, and its high processing time may limit its usability in real-time applications. It generally achieved better results with the YOLOv5+SAHI model to fully extract and learn the target features. In the experimental studies, basic model algorithms were tested, and according to their efficiency, the YOLOv5+SAHI model was found to be more efficient. YOLOv5 sees the entire image during the dataset training and testing period. Thus, it implicitly encodes contextual information about classes and their views. The reason why we use the fifth version of the YOLO algorithm is that YOLOv5 is the first YOLO model that works on the PyTorch framework. YOLOv5 runs faster with PyTorch, which is more compatible with various libraries in Python due to its Pythonic structure, which is easier and simpler to understand and offers a faster experience using the graphics processing unit. YoloR and YoloX, object recognition models, were created, training was performed on the created dataset, and possible weaknesses were identified. Tests were performed with some sample videos, and the results were analyzed. It was decided to integrate the pre- and post-processing algorithms into the system and test them. YOLOX is a model that performs training using data sets in json format. Work continued with YOLOv5, the best model after YOLOv7. In particular, the hybrid use of YOLOv5 with the SAHI algorithm provided almost as much accuracy and speed as YOLOv7. That's why the YOLOv5+SAHI algorithm was used.

Remote sensing is a multidisciplinary research field that focuses on evaluating remotely acquired data with advanced image analysis techniques. Advanced techniques such as deep learning are used to obtain more accurate results in this field. Expanding the application areas of these techniques emphasizes the importance of developments in the remote sensing discipline. However, in addition to these advanced techniques, shallow learning models play an important role in effectively using remote sensing imagery, especially for resource-constrained situations and simpler tasks. While shallow learning models stand out as alternatives that consume fewer computational resources and offer fast training processes, combining deep learning-based models with the ability to recognize complex visual patterns enables the development of innovative systems in remote sensing.

Our study focuses on remote sensing imagery, providing a broad perspective of this field's developments and potential use. In the process of creating our original dataset and identifying missing data classes, synthetic data was generated with Generative Adversarial Networks (GAN) to diversify the object classes found to be insufficient, which contributed to the enrichment of our dataset. Super-resolution methods were used to improve image quality, and the latest object recognition algorithms (YOLOv5, YOLOR, YOLOX) were tested for target object detection. In addition, the SAHI algorithm was tested as a hybrid model for small object detection, and experimental results showed the superior performance of YOLOv5.

This study demonstrates the potential of shallow learning models and advanced image analysis techniques in remote sensing image processing and analysis. Hybrid approaches have the potential to combine the advantages of both deep and shallow learning techniques, reducing computational cost while maintaining accuracy. Advances in remote sensing can contribute to developing innovative systems with potential for application in various sectors. Research in this multidisciplinary field aims to increase the value of information derived from remote sensing images through the integration of advanced image analysis techniques and models.

In conclusion, the results of this study provide important insights into the applicability and performance of deep learning methods in object detection. The comparison of YOLOv5, YOLOR, YOLOv7, and YOLOv5+SAHI models reveals how various model configurations perform on object detection tasks. According to the observed data, the YOLOv5+SAHI model outperforms in accuracy and precision metrics; however, the processing time may limit its usability in real-time applications.

To achieve better results in real-time object detection, models need to find a balance between speed and accuracy. Optimizations can be made in the model architecture and training process to increase the model's speed. For example, using lightweight versions of the model or freezing certain network parts (stopping learning) can significantly reduce the processing time. Furthermore, applying regularization techniques against overfitting in the model training phase, especially in multilayer structures, can increase the model's generalization ability and thus perform better on a larger data set. On the other hand, increasing the variety of data in the training process of deep learning models through methods such as data enrichment and synthetic data generation allows the model to become more robust and make more accurate detections. Synthetic data generation can increase the model's accuracy, especially for rare or hard-to-detect objects. In addition, reviewing the loss functions and optimization algorithms used in the training process and selecting more efficient ones will also improve the model's success.

Ultimately, an ideal model for real-time object detection should be able to balance high accuracy and speed. The high accuracy offered by the YOLOv5+SAHI model, combined with fast processing time, will greatly benefit critical applications such as remote sensing. This balance will be achieved by optimizing the model's architecture, increasing data diversity, and managing the training process more effectively. While this study provides a foundation for exploring the steps and methods to achieve this balance, future work should move in this direction.

5 Comparison of Deep and Shallow Learning in the Context of Advanced Techniques and Application Areas in Remote Sensing Images

The general purpose of this study is to compare deep and shallow learning in the context of advanced techniques and application areas in remote sensing images. Within the scope of the study, our original dataset was created and our dataset was enriched as missing data classes were identified. The use of advanced techniques such as deep learning and spectral analysis indicates significant potential to obtain sharper and more accurate results in remote sensing. Therefore, the results of these methods were shared in our study. While target objects were detected using the most up-to-date object recognition algorithms (YOLOv5, YOLOR, YOLOX), the SAHI algorithm was tested as a hybrid model to detect small objects. Similar studies have been conducted in the literature with the aim of comparing deep learning and shallow learning approaches in the context of advanced techniques and application areas in remote sensing images. The studies may be important for researchers and experts who want to comparatively evaluate machine learning applications, especially in the field of remote sensing. In particular, whether the researchers provide a detailed analysis of what advanced techniques and application areas deep and shallow learning models can be used in remote sensing images; In particular, it focuses on how these techniques perform in various remote sensing tasks. For this, they use performance metrics such as accuracy, precision, and recall to compare the performance of the models. Of course, one of the topics being researched is what kind of data sets and what kind of sampling strategies these performance metrics focus on. After all these details, the main differences between deep learning and shallow learning techniques and the advantages and disadvantages of each should be explained in detail, so that it can reveal in which areas there are opportunities and challenges for future research.

This type of study can provide researchers and practitioners with valuable insight into which machine learning methods are more effective in remote sensing

Method	Image content	Accuracy
Decision tree	in earth images; forest, steppe, soil-stone and settlement	%92,35 [41]
Naive Bayes Decision tree Random Forest	Ship	%95.03 %96.60 %97.59 [42]
Support Vector Machine Spatial Logistic Regression	in earth images	%98.00 %91.00 [43]
Support Vector Machine	Vehicles	%90.00 [44]
Multiclass classifier	Vehicles	%98.20 [45]

Fig. 7 Comparison of shallow learning models results

applications. The results of the five articles examined in this context are shared below (Fig. 7).

6 Conclusion

This study evaluates the applicability of deep learning-based approaches for object detection in remote sensing images by comprehensively comparing the performance of YOLOv5, YOLOR, YOLOv7, and YOLOv5+SAHI models. The results show that the YOLOv5+SAHI model is accurate and precise, superior to the other models. However, the increased processing time to achieve high accuracy is a factor that limits the potential use of the model for real-time applications. This represents a significant challenge in achieving the speed-accuracy trade-off required by real-time applications.

In the future, increasing the speed of models while not sacrificing accuracy should remain a central goal in the study of real-time object detection. This work has shed light on identifying steps toward optimizing this trade-off. Strategies such as model optimization, data enrichment, and effective management of training processes can lead to faster and more accurate real-time object detection. The results of this study provide important contributions to the development and application of remote sensing technology, especially for applications that require strong performance on complex and diverse data sets. While the high accuracy offered by the YOLOv5 +SAHI model is an attractive option for application developers and researchers, it emphasizes that future work should focus on improving the model's speed. To achieve this balance, rethinking model architectures and further examining data management strategies will shape future research and applications in this field.

As for the difference between deep learning and surface learning, surface learning aims to reach the minimum level and includes habitual learning strategies that will ensure the reproduction of the information received. Deep learning aims to improve skills in certain academic subjects. Deep learning involves models such as multilayer artificial neural networks, and these models may have many hidden layers. Artificial Neural Networks (ANN) are also a subset of deep learning and essentially involve multiple iterated logistic regressions. While there are two layers (input and output) in logistic regression, there is also a hidden layer in ANN. These hidden layers are called "hidden" because they do not see the input. Deep learning is considered "deep" depending on the number of hidden layers. That is, as the number of hidden layers increases, deep learning occurs more. While activation functions such as Tanh are used in deep learning, the Sigmoid function is used in superficial learning. Deep learning is important for applications that require powerful performance on more complex datasets.

References

1. Sarıbaş, H., Çevikalp, H., & Kahvecioğlu, S. (2018) İnsansız Hava Araçlarından Alınan Görüntülerdeki Araçların Konumlarının Bulunması, %1 içinde 26. *IEEE Sinyal İşleme ve İletişim Uygulamaları Kurultayı*, İzmir.
2. Toraman, S. (2018). Derin Öğrenme ile İnsansız Hava Aracı Görüntülerinden Yaya Tespiti. *Journal of Aviation, 2*(2), 64–69.
3. Wang, X., Cheng, P., Liu, X., & Uzochukwu, B. (2018). *Fast and accurate, convolutional neural network based approach for object detection from UAV*. arXiv.org
4. Benjdira, B., Khursheed, T., Koubaa, A., Ammar, A., & Ouni, K. (2018). Car detection using unmanned aerial vehicles: Comparison between faster R-CNN and YOLOv3. arXiv.
5. Ammour, N., Alhichri, H., Bazi, Y., Benjdira, B., Alajlan, N., & Zuair, M. (2017). Deep learning approach for car detection in UAV imagery. *Remote Sensing, 9*, 312.
6. Lan, M., Zhang, Y., Zhang, L., & Du, B. (2018). *Defect detection from UAV images based on region-based CNNs* (p. 87). School of Computer Science.
7. Xu, Y., Fu, M., Wang, Q., Wang, Y., Chen, K., & Xia, G.-S. (2020). Gliding vertex on the horizontal bounding box for multi-oriented object detection. *IEEE Transactions on Pattern Analysis and Machine Intelligence, 43*, 1452–1459.
8. Lee, J., Wang, J., Crandall, D., Sabanovic, S., & Fox, G. (2017). *Real-time, cloud-based object detection for unmanned aerial vehicles, %1 içinde*. 2017 first IEEE international conference on robotic computing, Indiana.
9. Ophoff, T., Puttemans, S., Kalogirou, V., Robin, J. P., & Goedemé, T. (2020). Vehicle and vessel detection on satellite imagery: A comparative study on single-shot detectors. *Remote Sensing, 12*, 1217.
10. Mandal, M., Kumar, L. K., & Vipparthi, S. K. (2020) *MOR-UAV: A benchmark dataset and baselines for moving object recognition in UAV videos*. %1 içinde Computer vision and pattern recognition, Seattle.
11. Algancı, U., Soydaş, M., & Sertel, E. (2020). Comparative research on deep learning approaches for airplane detection from very high-resolution satellite images. *Remote Sensing, 12*(3), 458.
12. Hinzmann, T., Stegemann, T., Cadena, C., & Siegwart, R. (2020). Deep Learning-based Human Detection for UAVs with Optical and Infrared Cameras: System and Experiments. arXiv preprint arXiv:2008.04197.

13. Carlet, J., & Abayowa, B. (2017). Fast Vehicle Detection in Aerial Imagery. arXiv preprint arXiv:1709.08666.
14. Onim, S. H., Akash, M. I., Haque, M., & Hafiz, R. I. (2020). *Traffic surveillance using vehicle license plate detection and recognition in Bangladesh*. %1 içinde 11th International Conference on Electrical and Computer Engineering (ICECE), Dhaka.
15. Russakovsky, O., Deng, J., Su, H., Krause, J., Satheesh, S., Ma, S., et al. (2015). Imagenet large scale visual recognition challenge. *International Journal of Computer Vision, 115*(3), 211–252.
16. Common objects in context (COCO). (n.d.). Retrieved 2021, from https://cocodataset.org/
17. Your machine learning and Data Science Community. Kaggle. (n.d.). Retrieved April 13, 2022, from https://www.kaggle.com/
18. VisDrone. (n.d.). *VisDrone/Visdrone-Dataset: The dataset for drone based detection and tracking is released, including both image/video, and annotations*. GitHub. Retrieved October 1, 2019, from https://github.com/VisDrone/VisDrone-Dataset
19. Sakla, W., Konjevod, G., & Mundhenk, T. N. (2017, March). Deep multi-modal vehicle detection in aerial ISR imagery. In *2017 IEEE winter conference on applications of computer vision (WACV)* (pp. 916–923). IEEE.
20. Xia, G. S., Bai, X., Ding, J., Zhu, Z., Belongie, S., Luo, J., et al. (2018). DOTA: A large-scale dataset for object detection in aerial images. In *Proceedings of the IEEE conference on computer vision and pattern recognition* (pp. 3974–3983). IEEE.
21. Yu, F., Xian, W., Chen, Y., Liu, F., Liao, M., Madhavan, V., & Darrell, T. (2018). Bdd100k: A diverse driving video database with scalable annotation tooling. *arXiv* preprint arXiv:1805.04687, *2*(5), 6.
22. Everingham, M., Van Gool, L., Williams, C. K., Winn, J., & Zisserman, A. (2010). The Pascal visual object classes (VOC) challenge. *International Journal of Computer Vision, 88*(2), 303–338.
23. Lin, T. Y., Maire, M., Belongie, S., Hays, J., Perona, P., Ramanan, D., et al. (2014, September). Microsoft coco: Common objects in context. In *European conference on computer vision* (pp. 740–755). Springer.
24. Wen, L., Du, D., Cai, Z., Lei, Z., Chang, M. C., Qi, H., et al. (2020). UA-DETRAC: A new benchmark and protocol for multi-object detection and tracking. *Computer Vision and Image Understanding, 193*, 102907.
25. Gibril, M. B. A., Kalantar, B., Al-Ruzouq, R., Ueda, N., Saeidi, V., Shanableh, A., et al. (2020). Mapping heterogeneous urban landscapes from the fusion of digital surface model and unmanned aerial vehicle-based images using adaptive multiscale image segmentation and classification. *Remote Sensing, 12*(7), 1081.
26. Lai, G., Xie, Q., Liu, H., Yang, Y., & Hovy, E. (2017). Race: Large-scale reading comprehension dataset from examinations. *arXiv* preprint arXiv:1704.04683.
27. Mribrahim. (n.d.). *MRIBRAHIM/PESMOD: UAV images dataset for Moving object detection*. GitHub. Retrieved June 15, 2022, from https://github.com/mribrahim/PESMOD
28. https://www.tugraz.at/index.php?id=22387
29. Yu, Y., Li, X., & Liu, F. (2019). Attention GANs: Unsupervised deep feature learning for aerial scene classification. *IEEE Transactions on Geoscience and Remote Sensing, 58*(1), 519–531.
30. Özyurt, F. (2020). A fused CNN model for WBC detection with MRMR feature selection and extreme learning machine. *Soft Computing, 24*(11), 8163–8172.
31. Meel, V. (n.d.). *Yolor – You only learn one representation (what's new, 2022)*. viso.ai. Retrieved 2018, from https://viso.ai/deep-learning/yolor/
32. Give your software the power to see objects in images and video. Roboflow. (n.d.). Retrieved April 3, 2020, from https://roboflow.com/
33. Wittmann, M. C., Steinberg, R. N., & Redish, E. F. (1999). Making sense of how students make sense of mechanical waves. *The Physics Teacher, 37*(1), 15–21.
34. Veribilimcisi. (2018, November 28). *Karışıklık matrisi (confusion matrix)*. Veri Bilimcisi-YENİLENİYOR ... Retrieved June 15, 2022, from https://veribilimcisi.com/2017/07/18/karisiklik-matrisi-nedir/

35. Yohanandan, S. (2020, June 9). *Map (mean average precision) might confuse you!* Medium. Retrieved June 15, 2022, from https://towardsdatascience.com/map-mean-average-precision-might-confuse-you-5956f1bfa9e2
36. Abdalla, G., & Özyurt, F. (2020). Sentiment Analysis of Fast Food Companies With Deep Learning Models. The Computer Journal, 00(00), 1-8. doi:10.1093/comjnl/bxaa131
37. Kavzoğlu, T., & Çölkesen, İ. (2010). Karar Ağaçları İle Uydu Görüntülerinin Sınıflandırılması. *Harita Teknolojileri Elektronik Dergisi, 2*(1), 36–45.
38. Dolapci, B., & Özcan, C. (2021). Automatic ship detection and classification using machine learning from remote sensing images on apache spark. *Journal of Intelligent Systems: Theory and Applications, 4*(2), 94–102. https://doi.org/10.38016/jista.772145
39. Huang, B., Xie, C., & Tay, R. (2010). Support vector machines for urban growth modeling. *GeoInformatica, 14*, 83–99.
40. Cao, X., Wu, C., Yan, P., & Li, X. (2011, September). Linear SVM classification using boosting HOG features for vehicle detection in low-altitude airborne videos. In *Proceedings of the 18th IEEE international conference on image processing* (pp. 2421–2424). IEEE.
41. Liu, K., & Mattyus, G. (2015, September). Fast multiclass vehicle detection on aerial images. *IEEE Geoscience and Remote Sensing Letters, 12*(9), 1938–1942.

Shallow Learning vs Deep Learning in Smart Grid Applications

Musa Yilmaz and Josep M. Guerrero

1 Introduction

Smart grids are more efficient and reliable than traditional grids. As with all technological systems, the development and integration of computing and communication technologies is supported by and leads to the generation of large amounts of data. Effective analysis of this data is crucial for various applications in HVs, such as anomaly detection, fraud detection, energy consumption monitoring, and more. In this context, the role of machine learning (ML) and deep learning becomes crucial [1–3].

Shallow learning techniques, which were the cornerstone of early ML applications due to their simplicity and efficiency in processing structured data, are now shifting toward more complex deep learning models with the advent of big data in MLs. Deep learning is increasingly recognized as a game changer in the smart grid space with its ability to process large volumes of unstructured data and learn complex representations [4–7].

Smart grids and technologies have also been highly influenced by the advancing artificial intelligence technology. In particular, deep learning methods have been successfully applied to load forecasting and power fault detection in electric power systems.

Alrasheedi and Almalaq (2022) developed an algorithm for hybrid deep learning methods for load forecasting in Saudi smart grid systems and discussed the

M. Yilmaz (✉)
Bourns College of Engineering, Center for Environmental Research and Technology, University of California at Riverside, Riverside, CA, USA

Department of Electrical and Electronics Engineering, Batman University, Batman, Turkey

J. M. Guerrero
Center for Research on Microgrids (CROM), AAU Energy, Aalborg University, Aalborg, Denmark

Ö. F. Ertuğrul et al. (eds.), *Shallow Learning vs. Deep Learning*, The Springer Series in Applied Machine Learning, https://doi.org/10.1007/978-3-031-69499-8_12

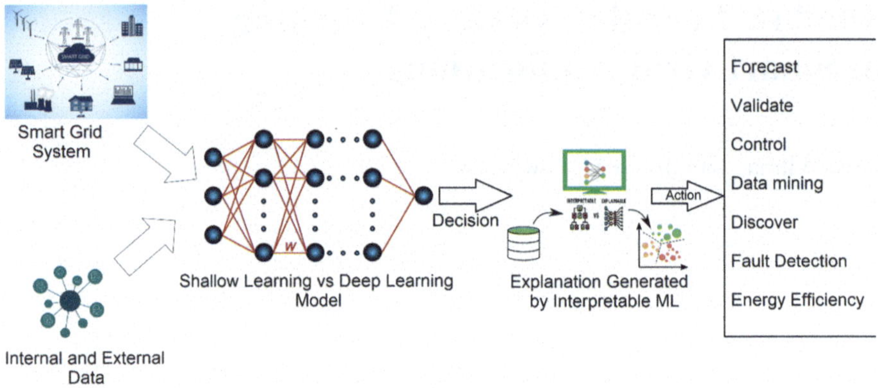

Fig. 1 Shallow learning vs deep learning in smart grid

superiority of these algorithms over traditional models using time series forecasts [8]. Similarly, Yang et al. (2021), addressing the challenges posed by complex environments and limited data sets, combined shallow and deep learning techniques in a novel defect detection algorithm for power insulators [9].

In addition, Ghoneim et al. (2017), who also addressed the environmental aspects of smart cities with deep learning applications and worked on algorithms that develop accurate, shallow pollution prediction models for these applications, stated that the method is highly applicable and successful [10]. Zhang et al. (2020) introduced an innovative approach to probabilistic load forecasting by incorporating jump connections into deep neural networks. This method not only improves the accuracy of predictions but also overcomes technical challenges in deep learning architectures such as gradient fading [11].

Smart grids have made significant progress in the energy sector, showing significant potential in critical areas such as energy distribution, data analysis, and energy efficiency. In this section, we explore the roles of deep learning and sig learning in smart grids [12].

When these AI techniques are compared to traditional energy distribution systems, smart grids are characterized by being more intelligent, adaptive, and data-centric. These systems can efficiently collect, analyze, and manage diverse datasets, including energy consumption, distribution network status, and user behavior. This big data creates a new opportunity to analyze and integrate artificial intelligence and machine learning techniques into smart grid applications. In the future, where the analysis results obtained from big data results will increase smart grid efficiency, it will be possible to achieve cheaper, less loss, and more efficient energy [13].

In this context, shallow learning, which uses traditional machine learning algorithms for low-dimensional data analysis, and deep learning, which uses multilayer neural networks to process large and complex datasets with automatic representation learning capabilities.

Figure 1 shows a detailed comparison of how SL and DL analyze data in smart grids. The figure indicates that both SL and DL systems get data from SGs, including both internal and external sources; however, data in the SL framework is more linear and allows for simpler decision-making. On the other hand, DL's layered and complicated structure allows it to generate good outcomes from machine learning algorithms in addition to decision-making. This allows it to be employed in a variety of applications, including sophisticated control mechanisms, comprehensive verification, complicated error detection, precise prediction, and significant data mining and discovery operations.

2 Smart Grids and Data Collection

Artificial intelligence techniques are under development that combine sophisticated technologies to enable more efficient and reliable energy management in HVs. With these algorithms, many applications such as data collection and analysis, demand-demand balancing, pricing policies, energy distribution anomaly detection, and consumer data protection can be handled as a whole.

Wang et al. (2023) used a VC-DCPS technique to improve data collection and privacy in smart grids with blockchain technology. This method uses lattice-based homomorphic encryption and spatiotemporal fusion technology to efficiently and securely aggregate cross-domain data. It maximizes privacy and data security while maintaining high efficiency and low storage costs [14]. In the research by Donnelly et al. (2023), smart meter systems for data collection were investigated.

Yen et al. (2019) investigated the impact of smart meter data collection frequency on the detection of short-term electrical voltage disturbances. They built an anomaly verification algorithm that analyzes voltage data from smart meters and proved that more frequent data collection can successfully detect shorter-duration anomalies, highlighting the potential of smart meters for earlier anomaly detection [15]. Simsek et al. (2018) propose TPS3, a privacy-preserving protocol for smart grids. This technique combines Temporal Perturbation and Shamir's Secret Sharing strategies to protect the confidentiality and privacy of customer information. The protocol provides an excellent compromise between communication costs and security, making it a viable solution to protect consumer data in smart grids [16]. The basic components and operation of Smart Grid systems aim to optimize the entire energy distribution process from generation to consumption. Figure 2 shows a representative graph of the basic components and functions of smart grids.

Data and processing, the most important function of smart grids, is the first step toward more efficient energy management and data analysis in the energy sector. These processes transform the energy distribution and transmission system into a data-driven and intelligent system that contributes to the development of the energy sector. The basic steps in these processes are shown in Fig. 3.

Smart grids play an important role in the energy sector's future, serving as a potent instrument for enhancing energy efficiency, optimizing energy consumption,

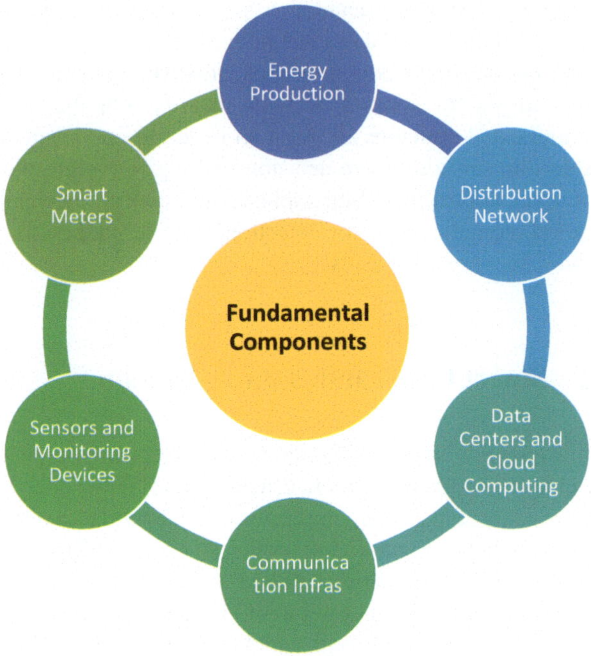

Fig. 2 Fundamental components

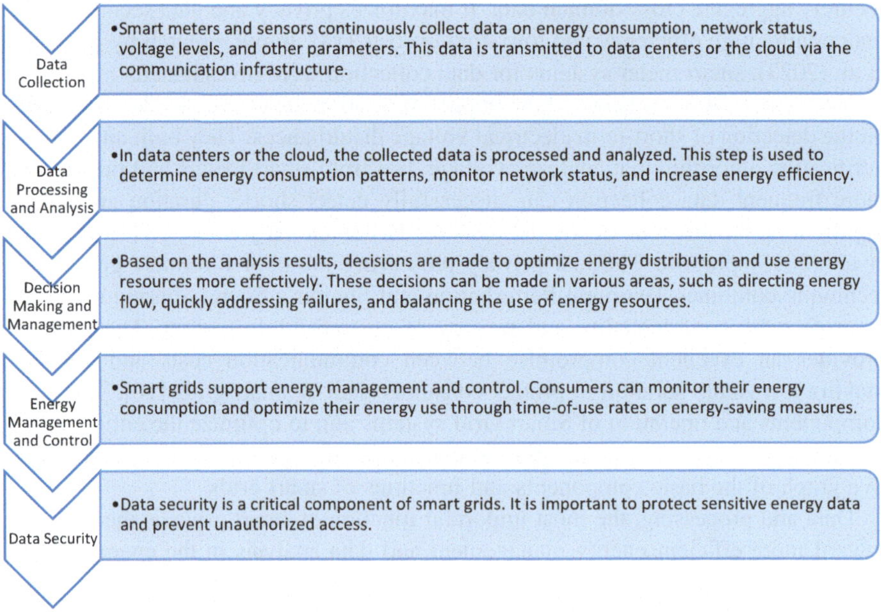

Fig. 3 Data operations

and managing energy resources in a more sustainable manner (Fig. 3). These components and activities help to make energy distribution networks smarter and more data-driven, which benefits the energy sector [14–17].

Energy distribution data is a valuable resource in the energy sector and is critical to the efficient operation of smart grids. This information is required for controlling, enhancing, and optimizing energy distribution systems. Here are the significance and types of energy distribution data:

1. Security Data: Security data helps protect energy networks against cyber threats and security breaches. Unauthorized access attempts and security incidents fall into this category.
2. Energy Consumption Data: This data is used to track customer energy consumption. It is acquired via smart meters and energy measurements. Energy consumption patterns are crucial for energy businesses to forecast and control demand.
3. Network Status Data: This data is utilized to monitor the energy transmission and distribution networks. This category includes network metrics such as voltage, current, and frequency.
4. Customer Information: This includes data on accounts, subscriptions, payment history, and requests. It's utilized to manage client interactions and offer customer assistance.
5. Fault Data: Manages power outages and faults. This information is critical for the rapid detection of grid failures.

This type of data supports the smooth operation of energy grids and helps the energy sector become smarter, more data-driven, and more secure. Proper management and analysis of data are crucial for the success of smart grids and have a significant impact on the future development of the energy sector [4–8].

3 Shallow Learning

Shallow learning is a foundational approach to machine learning that uses more traditional approaches. Shallow learning strategies describe data with specified features or parameters and use these representations to carry out the learning process (Fig. 4). Here are some common shallow learning approaches [2–13, 18–20]:

Linear Regression: Linear regression models the relationship between dependent and independent variables. This method aims to represent data points using a line.

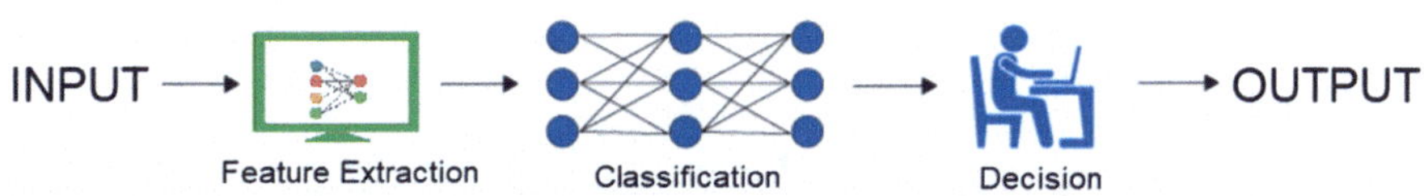

Fig. 4 Shallow learning

Logistic regression is a technique for classifying problems that models the relationship between two or more classes. It classifies based on probability values.

Decision trees classify data based on tree topology or perform regression analyses. Each node makes a decision, which yields a prediction or categorization.

Support vector machines (SVM) are used for classification and regression problems, attempting to discover the optimal separation between two classes.

K-nearest neighbors (K-NN): K-NN classifies or conducts regression using the k nearest neighbors of a data item. It combines comparable data points.

The Application of Traditional Machine Learning Algorithms in Smart Grids Traditional machine learning methods can be employed in a variety of applications in smart grids.

Load Forecasting: Regression methods can be used to predict energy consumption. This allows energy corporations to better plan energy production and distribution.

Fault Detection: Decision trees and logistic regression algorithms can be used to rapidly detect and respond to defects in energy networks.

Energy Efficiency Analysis: Energy consumption data can be analyzed using linear regression and support vector machines to determine energy efficiency measures.

Customer Segmentation: By categorizing clients into distinct groups, K-NN or clustering algorithms can help us better analyze energy demand and provide personalized services.

Feature selection and feature engineering can have a substantial impact on the effectiveness of machine learning models.

These phases are important in data analysis and model training procedures, and they can help the model perform better. Correct feature selection and engineering in smart grids are critical for improved energy management and data analysis.

4 Deep Learning

Deep learning is a subbranch of machine learning that uses modeling methods called artificial neural networks to make sense of complex data, as shown in Fig. 5. Its basic principles are as follows [1, 2]:

- Deep learning relies on neural networks based on mathematical models. These networks process data in a multilayered structure akin to the human brain.
- Deep learning excels at turning data into high-level representations.
- Deep learning models utilize several layers to extract complicated information from previous outputs, leading to improved data comprehension and learning.

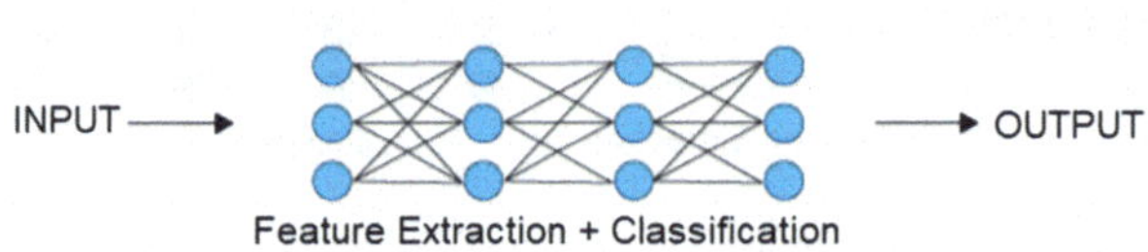

Fig. 5 Deep learning

This multilayered structure allows for more accurate data representation, as well as the system's ability to learn and forecast like a person.

4.1 Deep Neural Networks (DNNs) and Their Use in Smart Grids

Deep neural networks play an important role in data analysis and management in smart grids. Some applications of DNNs in smart grids are listed below:

- Energy Demand Forecasting: DNNs can be used for energy demand forecasting. Processing large datasets allows for better planning of energy distribution by predicting future energy demands.
- Fault Detection and Maintenance: DNNs can be used to detect faults in energy grids and optimize maintenance processes. Data analytics can speed up the fault detection process.
- Smart Meters and Data Analytics: DNNs can process large data sets from smart meters and analyze energy consumption. This helps energy companies better understand customer demands and improve energy efficiency.

4.2 Deep Learning Models and Algorithms for Processing Large Datasets

- Deep learning is a powerful technique for analyzing and understanding huge datasets. Smart grids frequently use large datasets, which are processed using deep learning algorithms. Here are a few examples:
- Deep learning models perform better with more datasets. These models can become more complicated as the data size increases.
- Deep learning is used to accurately describe big datasets and identify patterns. This can reveal hidden relationships within the data.
- Deep learning excels at analyzing massive datasets and providing valuable insights.

Deep learning is an effective technique for a wide range of smart grid applications, including big data analysis, energy forecasting, and problem identification. When working with large datasets, DL models often perform better and extract more information [1–13, 18, 19].

5 Applications of Deep and Shallow Learning in Smart Grids

Shallow and deep learning have different impacts on smart grid applications and each has a specific application area. Below are different applications that illustrate how these two techniques affect different smart grid application domains [7–11]:

5.1 Shallow Learning and Smart Grid Application Areas

- SL approaches are useful for forecasting energy demand. Simple models, such as classic regression analysis or logistic regression, can accurately forecast energy demand.
- Use SL to spot flaws in energy networks. Simple classification techniques can be useful for fault detection.
- SL can help analyze energy usage data and evaluate energy efficiency strategies. Statistical analysis and regression approaches can be used to assess energy efficiency projects.

5.2 Deep Learning and Smart Grid Application Areas

- DL can process massive datasets from smart meters and provide detailed analysis of energy consumption. Deep neural networks learn from complex data and can better forecast energy use patterns.
- DL algorithms excel in forecasting energy demand from vast datasets. They work particularly well for daily and hourly forecasts.
- DL can help segment customers and analyze their behavior. This can assist energy firms provide tailored services to their customers.

5.3 Success Stories

Google DeepMind has made tremendous gains in applying deep learning to optimize data center management, resulting in significant energy savings and increased data center operational efficiency [21]. Similarly, Pacific Gas and Electric (PGE) in California has used deep learning models to improve energy demand estimates, representing a significant step forward in energy production planning and consumption control [22]. Furthermore, IBM has created a sophisticated system for monitoring energy networks and detecting issues using deep learning techniques. This innovation allows energy firms to better manage their networks, resulting in fewer outages [23].

Both learning approaches have been shown to be advantageous in smart grid applications, but deep learning models and algorithms are especially effective when dealing with huge datasets and complexity.

6 Performance Comparison and Results

Shallow and deep learning models might perform differently in various application areas. To compare their performance, several crucial aspects must be addressed. Here's an extensive examination of the performance comparison between shallow and deep learning models, as well as an evaluation of whether strategy is more effective under specific scenarios [7–20, 24].

6.1 Performance Comparison

1. Deep learning models tend to perform better with larger datasets. Shallow learning may be more effective on small to medium-sized datasets.
2. Deep learning is highly effective at learning complex data representations. Shallow learning is limited to low-level representations.
3. Computational Power: Deep learning models require large resources for training and prediction.
4. Feature Engineering: Deep learning models can automatically extract features, but shallow learning may require feature engineering.
5. Shallow learning models require less training time than deep learning models, making them more suitable for time-sensitive circumstances.
6. Model Complexity: Deep learning models have multiple layers and parameters, while shallow learning models are simpler and easy to understand.
7. Deep learning models excel at generalizing from training data, making them ideal for analyzing complex patterns.

6.2 Which Approach Is More Effective Under What Conditions

- Shallow learning is successful for small to medium-sized datasets with sufficient lower-level data representations. It is also preferable when computational resources are limited or model training needs to be completed quickly. Shallow learning can be useful for producing simple and acceptable results.

- Deep learning is more successful for huge datasets and sophisticated data analysis, requiring high-level representations. It relies less on feature engineering and can extract features automatically. However, it necessitates additional computational power and data. Deep learning provides additional skills for sophisticated data processing and prediction.
- Shallow learning is better suited for applications with organized data and particular tasks, whereas deep learning is better for unstructured data like photos and natural language.

Each approach's effectiveness is determined by the dataset's size, complexity, computational resources, and target application area. Both approaches offer benefits and drawbacks, and examining these variables before making a decision is critical. In some circumstances, combining shallow and deep learning may be the most effective technique, using their respective strengths to obtain the greatest results [4–20, 24].

6.3 Future Directions and Research Needs

6.3.1 Future Developments in Smart Grid Technology

Smart grid technology will continue to evolve, and future developments may focus on these areas:

1. Improved Data Analytics: Big data analytics can improve data processing and interpretation in smart grids. Deep learning and artificial intelligence techniques will help with sophisticated data analysis and forecasts.
2. Security and Cyber Protection: Smart grids will need stronger defenses against cyber-attacks. Future advancements will strive to improve network security for energy organizations and provide a stronger protection against cyber threats.
3. Energy Storage and Distribution: Smart grids can interact more effectively with energy storage technology. This will improve energy storage and distribution capacity.
4. Environmental Sustainability: Future smart grids will prioritize sustainability and renewable energy. This provides chances to lower carbon footprints and promote the usage of renewable energy.
5. Integrating smart grids with IoT and smart home devices improves energy management and efficiency for consumers.
6. Real-Time Monitoring and Response: Automated response systems will enhance smart grid efficiency and dependability.

6.3.2 Potential Research Topics and Trends in Deep and Shallow Learning

Future research issues in deep and shallow learning could include:

1. Big Data and Deep Learning: Future research will focus on improving the effectiveness of deep learning models for training and analyzing massive datasets.
2. Deep Learning in Energy Management: Research aims to improve energy demand forecasts, efficiency analysis, and network management using deep learning models.
3. Optimizing Deep Learning Models: We will investigate new ways for faster and more efficient training.
4. Deep Learning Security Research: We will explore ways to improve the security of deep learning models against cyber threats.
5. Research will focus on improving shallow learning algorithms for energy and smart grid applications.
6. Interdisciplinary Approaches: Integrating deep and shallow learning with economics, environmental science, and engineering to develop comprehensive energy management solutions.

Future research predicts that the energy sector will become more data-centric, stabilizing energy security and sustainability (supply/demand), and smart grid management will be fully AI-based.

7 Applications

7.1 Shallow Learning Applications

1. Energy Demand Forecasting: PJM Interconnection, a leading electric transmission organization in the United States, employs a variety of statistical and machine learning models for the purpose of forecasting short-term electrical loads. These models incorporate a range of factors, including weather conditions, dates, and times, to accurately predict the demand for electricity. Techniques from shallow learning significantly contribute to the accuracy of these forecasts. In a notable development in July 2011, the Federal Energy Regulatory Commission (FERC) introduced Order 1000. This order established a competitive framework for the development of transmission infrastructure across the United States, marking a departure from the traditional approach where transmission projects were designated and constructed by incumbent owners based on regional or local planning processes, with the associated costs being distributed through existing transmission tariffs [25].
2. Anomaly Detection and Fraud Analysis: France's electricity distribution company, Enedis, uses machine learning models to detect anomalies in electricity

consumption. This is important for reducing technical losses and preventing fraud. Shallow learning algorithms can be used to identify irregularities in consumption data [26].

3. Energy Efficiency and Building Management: IBM and Siemens have collaborated to optimize energy consumption in buildings. In this project, machine learning models are used to analyze data collected from sensors inside buildings to adjust HVAC (heating, ventilation, and air conditioning) systems in a way that reduces energy use. Shallow learning techniques can be effective for such optimization tasks [27].
4. Renewable Energy Forecasting: The NREL, affiliated with the US Department of Energy, uses various machine learning techniques to forecast wind and solar energy production. These forecasts help improve the integration of renewable energy sources into the grid. Shallow learning methods, such as simple regression models and time series analysis, are commonly used in this field [28].
5. Distributed Energy Resource Management: AutoGrid offers solutions such as demand response management and distributed energy resource management for energy companies. The company's platform uses machine learning algorithms to optimize energy consumption and production in real time. Shallow learning can be used to manage energy flows and create demand responses in such systems [29].

7.2 Deep Learning Applications

1. Google DeepMind and UK National Grid Collaboration: Google's artificial intelligence company, DeepMind, has collaborated with the United Kingdom's National Grid to use deep learning algorithms for predicting energy demand. This project aims to predict energy demand more accurately, thereby making energy production more efficient. Deep learning models analyze information from various data sources, including weather conditions, seasons, holidays, and other factors, to predict when energy demand will increase or decrease [30].
2. Siemens and LO3 Energy: Siemens has collaborated with the blockchain-based energy trading platform LO3 Energy to develop the Brooklyn Microgrid project. This project aims to create a microgrid at the neighborhood level that produces solar energy and shares excess energy with neighbors. Deep learning is used to analyze patterns of energy production and consumption, making decisions about when and where energy should be stored or traded [31].
3. Envision Energy: Envision Energy uses deep learning to analyze data obtained from wind turbines. This analysis predicts the maintenance needs of the turbines and anticipates failures in advance, thereby reducing maintenance costs and increasing the continuity of energy production [32].
4. NextEra Energy: NextEra Energy uses deep learning algorithms for the management of energy storage systems. These systems store excess energy produced when energy demand is low and return this energy to the grid when demand is

high. Deep learning models make predictions about energy demand and prices, determining the most efficient times for energy to be stored and used [33].
5. MIT's Laboratory for Information and Decision Systems (LIDS) has been honored with an award by the Appalachian Regional Commission (ARC) in recognition of its role in the forward-thinking project known as "Forming the Smart Grid Deployment Consortium (SGDC) and Expanding the HILLTOP+ Platform." This funding comes from the ARC's initiative focused on Appalachian Regional Initiative for Stronger Economies, which is dedicated to promoting economic development across states through collaborative efforts. The project has led to the creation of generative models with a broad spectrum of uses, from simulating grid behaviors to developing algorithms for startups in the energy sector. These models, once trained on available data, are capable of producing new, realistic datasets that can supplement existing ones or serve as alternatives to data that are sensitive or scarce. This capability allows project stakeholders to delve into and strategize for various hypothetical scenarios that extend well beyond the limitations of current data. For example, such models can predict the impact on the electrical grid of an additional 1000 households adopting solar technologies, including variations in demand over the course of a day, among other vital considerations for strategic planning [34].

8 Conclusion

Both deep learning (DL) and shallow learning (SL) play important roles in smart grid applications, demonstrating the potential to accomplish even more significant tasks. The advantages and limitations of both approaches have been discussed, and their applicability has been examined, especially in terms of data size, data complexity, computational power, and application scope. While DL produces successful results in tasks such as processing large datasets from smart meters, making energy demand forecasts, or detecting faults in complex scenarios, SL is more successful in simpler applications and when the data size is limited.

The existing literature comparing the performance of both DL and SL techniques in tasks such as load forecasting or energy efficiency analysis in the context of smart grid applications has been reviewed. The importance of considering the data size and analysis requirements when choosing a learning method for smart grid applications has been discussed by comparing applications. It has been observed that deep learning excels in handling large and complex datasets, while shallow learning performs well on smaller datasets characterized by simpler patterns.

For instance, the statistical and machine learning models used by PJM Interconnection for short-term load forecasting exemplify how shallow learning techniques can be effective on limited datasets. This demonstrates that SL can be successful in simpler applications. On the other hand, Google's DeepMind, in collaboration with the UK National Grid, has used deep learning algorithms to predict energy demand.

This collaboration showcases DL's ability to extract complex patterns from large datasets.

The machine learning models used by France's electricity distribution company, Enedis, to detect anomalies in electricity consumption illustrate the success of shallow learning algorithms in recognizing simpler patterns. This is important for reducing technical losses and preventing fraud. Additionally, the machine learning algorithms used by AutoGrid to optimize energy consumption and production demonstrate DL's capacity to manage real-time data flows and complex optimization tasks.

These examples highlight how DL and SL can be effective in different scenarios within smart grid applications and emphasize the need to consider data size, complexity, and the requirements of the application when making a choice. The applications of DL and SL in smart grid applications, as discussed in this study, can significantly be utilized in areas where data sets are available, contributing greatly to this field. In conclusion, the use of machine learning in smart grid applications, especially in areas such as load forecasting, anomaly detection and energy consumption monitoring, fault detection, energy efficiency, control, and validation, has been discussed and can be extensively applied.

References

1. LeCun, Y., Bengio, Y., & Hinton, G. (2015). Deep learning. *Nature, 521*(7553), 436–444.
2. Paszke, A., et al. (2019). Pytorch: An imperative style, high-performance deep learning library. *Advances in Neural Information Processing Systems, 32*, 8026–8037.
3. Schmidhuber, J. (2015). Deep learning in neural networks: An overview. *Neural Networks, 61*, 85–117.
4. Liu, H., Li, X., Zhou, W., Chen, Y., He, Y., Xue, H., et al. (2021). Spatial-phase shallow learning: Rethinking face forgery detection in frequency domain. In *Proceedings of the IEEE/CVF conference on computer vision and pattern recognition* (pp. 772–781). IEEE.
5. DeLancey, E. R., Simms, J. F., Mahdianpari, M., Brisco, B., Mahoney, C., & Kariyeva, J. (2019). Comparing deep learning and shallow learning for large-scale wetland classification in Alberta, Canada. *Remote Sensing, 12*(1), 2.
6. Ma, S., & Belkin, M. (2017). Diving into the shallows: A computational perspective on large-scale shallow learning. 31st Conference on Neural Information Processing Systems (NIPS 2017), Long Beach, CA, USA.
7. Islam, M. R., Moni, M. A., Islam, M. M., Rashed-Al-Mahfuz, M., Islam, M. S., Hasan, M. K., & Lió, P. (2021). Emotion recognition from EEG signal focusing on deep learning and shallow learning techniques. *IEEE Access, 9*, 94601–94624.
8. Alrasheedi, A., & Almalaq, A. (2022). Hybrid deep learning applied on Saudi smart grids for short-term load forecasting. *Mathematics, 10*(15), 2666.
9. Yang, L., Song, S., Niu, Y., & Liu, Y. (2021, August). A lightweight defect detection algorithm of insulators for power inspection. In *2021 IEEE international conference on mechatronics and automation (ICMA)* (pp. 281–286). IEEE.
10. Ghoneim, O. A., & Manjunatha, B. R. (2017, September). Forecasting of ozone concentration in smart city using deep learning. In *2017 international conference on advances in computing, communications and informatics (ICACCI)* (pp. 1320–1326). IEEE.

11. Zhang, W., Quan, H., Gandhi, O., Rajagopal, R., Tan, C. W., & Srinivasan, D. (2020). Improving probabilistic load forecasting using quantile regression NN with skip connections. *IEEE Transactions on Smart Grid, 11*(6), 5442–5450.
12. Li, Y., Wei, X., Li, Y., Dong, Z., & Shahidehpour, M. (2022). Detection of false data injection attacks in smart grid: A secure federated deep learning approach. *IEEE Transactions on Smart Grid, 13*(6), 4862–4872.
13. Akinci, T. C. (2020). Applications of big data and AI in electric power systems engineering. In *AI and Big Data's potential for disruptive innovation* (pp. 240–260). IGI Global.
14. Wang, P., Li, K., Shi, R., & Shao, B. (2023). VC-DCPS: Verifiable cross-domain data collection and privacy-persevering sharing scheme based on lattice in Blockchain-enhanced smart grids IEEE. *Internet of Things Journal, 10*(14), 12449–12461.
15. Yen, S. W., Morris, S., Ezra, M., & Huat, T. J. (2019). Effect of smart meter data collection frequency in an early detection of shorter-duration voltage anomalies in smart grids. *International Journal of Electrical Power & Energy Systems, 109*, 1–8.
16. Simsek, M., Okay, F. Y., Mert, D., & Özdemir, S. (2018). TPS3: A privacy preserving data collection protocol for smart grids. *Information Security Journal A Global Perspective, 27*(2), 1–17.
17. Donnelly, W., Keifer, P., Minor, R., Muthukumaran, U., Parolek, B., Tuck, B., ... & Fouda, M. M. (2023, August). A review of privacy-preserving and efficient data collection and aggregation in smart grids. In 2023 11th International Conference on Information and Communication Technology (ICoICT) (pp. 326–332). IEEE.
18. Majidi, S. H., Hadayeghparast, S., & Karimipour, H. (2022). FDI attack detection using extra trees algorithm and deep learning algorithm-autoencoder in smart grid. *International Journal of Critical Infrastructure Protection, 37*, 100508.
19. Massaoudi, M., Abu-Rub, H., Refaat, S. S., Chihi, I., & Oueslati, F. S. (2021). Deep learning in smart grid technology: A review of recent advancements and future prospects. *IEEE Access, 9*, 54558–54578. https://ieeexplore.ieee.org/document/10262531
20. Ismail, M., Shahin, M., Serpedin, E., & Qaraqe, K. (2018, March). A deep learning approach for detection of electricity theft cyber attacks in smart grids. In *Qatar Foundation annual research conference proceedings* (Vol. 2018, No. 3, p. ICTPP62). HBKU Press.
21. Wang, S., Qin, L., Ma, C., & Wu, W. (2023). Research on overall energy consumption optimization method for data center based on deep reinforcement learning. *Journal of Intelligent & Fuzzy Systems*. https://doi.org/10.3233/jifs-223769
22. Koohfar, S., Woldemariam, W., & Kumar, A. (2023). Prediction of electric vehicles charging demand: A transformer-based deep learning approach. *Sustainability, 15*(3), 2105.
23. Bindi, M., Iturrino-García, C., Piccirilli, M. C., Grasso, F. F., Luchetta, A., & Paolucci, L. (2023). Power grid monitoring based on machine learning and deep learning techniques. *Journal of Applied Informatics, 7*(1). https://doi.org/10.32629/jai.v7i1.1124
24. Cunningham, J. D., Aved, A., Ferris, D., Morrone, P., & Tucker, C. S. (2022). A deep learning game theoretic model for defending against large scale smart grid attacks. *IEEE Transactions on Smart Grid, 14*(2), 1188–1197.
25. Herling, S., Koza, F., & McGlynn, P. (2016). The sponsorship model: Competitive construction of transmission facilities in PJM interconnection. *IEEE Power and Energy Magazine, 14*(4), 65–71.
26. Danieli, A. (2018). The French electricity smart meter: Reconfiguring consumers and providers. In *Infrastructures in practice* (pp. 155–168). Routledge.
27. Lee, D., & Chen, L. (2022). Sustainable air-conditioning systems enabled by artificial intelligence: Research status, enterprise patent analysis, and future prospects. *Sustainability, 14*(12), 7514.
28. https://www.nrel.gov/grid/solar-wind-forecasting.html
29. https://www.auto-grid.com/solution/derms/
30. Murgia, M., & Thomas, N. (2017, March). DeepMind and national grid in AI talks to balance energy supply. *Financial Times*

31. https://www.siemens.com/global/en/company/stories/research-technologies/energytransition/a-microgrid-grows-in-brooklyn.html
32. Lind, C. (2016). *The battle in the wind energy industry: The case of envision energy*. Halmstad University.
33. Majumdarr, S., Jindal, A., Srivastava, S., & Garg, V. (2023). International Experiences: The Cases of Iberdrola, Enel, and NextEra Energy. In The Role of Coal in a Sustainable Energy Mix for India (pp. 264–278). Routledge India.
34. https://lids.mit.edu/news-and-events/news/generative-ai-smart-grid-modeling

Correction to: Shallow Learning Versus Deep Learning in Biomedical Applications

Mithat Önder, Ümit Şentürk, and Kemal Polat

Correction to:
Chapter 6 in: Ö. F. Ertuğrul et al. (eds.), ***Shallow Learning vs. Deep Learning*****, The Springer Series in Applied Machine Learning,**
https://doi.org/10.1007/978-3-031-69499-8_6

The original version of the book was inadvertently published with an incorrect author name—"Mithat Öner". The name has now been corrected to "Mithat Önder" and the correction was approved by the author.

The updated version of this chapter can be found at
https://doi.org/10.1007/978-3-031-69499-8_6

Ö. F. Ertuğrul et al. (eds.), *Shallow Learning vs. Deep Learning*, The Springer Series in Applied Machine Learning, https://doi.org/10.1007/978-3-031-69499-8_13

Index

Öm. F. Ertuğrul et al. (eds.), *Shallow Learning vs. Deep Learning*, The Springer Series in Applied Machine Learning, https://doi.org/10.1007/978-3-031-69499-8

The manufacturer's authorised representative in the EU is Springer Nature Customer Service Centre GmbH, Europaplatz 3, 69115 Heidelberg, Germany. If you have any concerns regarding our products, please contact ProductSafety@springernature.com

Printed and bound by CPI Group (UK) Ltd, Croydon, CR0 4YY
01/10/2026
02225914-0008